THE POLITICS OF COMMAND
Lieutenant-General A.G.L. McNaughton and the
Canadian Army, 1939–1943

In December 1943, Lieutenant-General A.G.L. McNaughton resigned from command of the First Canadian Army amid criticism of his poor generalship and of his abrasive personality. Despite McNaughton's importance to the Canadian Army during the first four years of the Second World War, little has been written about the man himself or the circumstances of his resignation.

In *The Politics of Command*, the first full-length study of the subject since 1969, John Nelson Rickard analyses McNaughton's performance during Exercise SPARTAN in March 1943 and assesses his relationships with key figures such as Sir Alan F. Brooke, Bernard Paget, and Harry Crerar. This detailed re-examination of McNaughton's command argues that the long-accepted reasons for his relief of duty require extensive modification.

Based on a wide range of sources, *The Politics of Command* will redefine how military historians and all Canadians look at not only 'Andy' McNaughton but also the Canadian Army itself.

JOHN NELSON RICKARD is a Canadian Armoured Corps Officer with a PhD in military history from the University of New Brunswick.

The Politics of Command

Lieutenant-General A.G.L. McNaughton and the Canadian Army, 1939–1943

JOHN NELSON RICKARD

UNIVERSITY OF TORONTO PRESS
Toronto Buffalo London

© University of Toronto Press 2010
Toronto Buffalo London
utorontopress.com

Reprinted in paperback 2021

ISBN 978-1-4426-4002-3 (case)
ISBN 978-1-4875-4102-6 (paper)

Library and Archives Canada Cataloguing in Publication

Title: The politics of command : Lieutenant-General A.G.L. McNaughton and
   the Canadian army, 1939–1943 / John Nelson Rickard.
Names: Rickard, John Nelson, 1969– author.
Description: Paperback reprint. Originally published 2010. | Includes
   bibliographical references and index.
Identifiers: Canadiana 20210093765 | ISBN 9781487541026 (softcover)
Subjects: LCSH: McNaughton, A. G. L. (Andrew George Latta), 1887–1966 –
   Military leadership. | LCSH: Canada. Canadian Army. Canadian Infantry
   Division, 1st – History. | LCSH: Canada. Canadian Army –
   History – World War, 1939–1945. | LCSH: Canada. Canadian Army –
   Officers – Biography. | LCSH: Generals – Canada – Biography. | LCSH:
   Command of troops. | LCGFT: Biographies.
Classification: LCC U55.M345 R53 2021 | DDC 940.54/1271092 – dc23

This book has been published with the help of a grant from the Federation for
the Humanities and Social Sciences, through the Awards to Scholarly Publications
Program, using funds provided by the Social Sciences and Humanities Research
Council of Canada.

University of Toronto Press acknowledges the financial assistance to its
publishing program of the Canada Council for the Arts and the Ontario Arts
Council, an agency of the Government of Ontario.

 Canada Council   Conseil des Arts
for the Arts   du Canada

Funded by the   Financé par le
Government   gouvernement
of Canada   du Canada

*Dedicated to my darling wife,
Melissa Leigh*

Contents

Maps	ix
Tables and Figures	x
Foreword by Marc Milner	xiii
Preface by Lieutenant-General Andrew Leslie	xvii
Acknowledgments	xxiii
Abbreviations	xxv

Introduction	3
Part One: The Making of Andy McNaughton	9
1 Early Life and the Crucible of the First World War	11
2 The Road to High Command	23
Part Two: The Problem of Deploying the Army	37
3 A Willingness to Fight, 1940–1941	39
4 From ROUNDUP to TORCH	54
5 Practical Operations of War	69
Part Three: McNaughton as Military Commander and Trainer	81
6 The Difficulty of Training in 1940	83
7 The Politics of Training	96
8 Enter Montgomery	114
9 Exercise SPARTAN	128
10 The Long Shadow of SPARTAN	150

Part Four: The End of an Idea 169

11 The Sicily Incident 171
12 Broken Dagger: A Corps in Italy 187
13 The Final Months of McNaughton's Command 208

Epilogue 217
Conclusion 219

Appendices 231
Notes 247
Bibliography 333
Index 357

Illustrations follow page 68

Maps

Map 1 The Dieppe Operation, 19 August 1942   65
Map 2 Home Forces, May 1941   107
Map 3 Exercise SPARTAN: McNaughton's Initial Plan,
       4 March 1943   134
Map 4 Exercise SPARTAN: Gammell's Initial Plan, 4 March 1943   136
Map 5 Exercise SPARTAN: McNaughton's Manoeuvre,
       6–7 March 1943   142
Map 6 Exercise SPARTAN: Dispositions at End Ex, 0900,
       12 March 1943   148
Map 7 The Landings in Southern Italy, 3–5 September 1943   199

Tables and Figures

Tables

Table	2.1	Advanced training of British and Canadian senior officers, 1919–1939	25
Table	2.2	McNaughton's relative interwar command experience	36
Table	5.1	McNaughton's willingness to employ Canadian troops, 1940–1943	79
Table	6.1	Command and staff element, 1st Canadian Division, January 1940	90
Table	7.1	Principal staff element, VII Corps and 1st Canadian Division, July 1940	99
Table	7.2	Command and staff element, Canadian Corps, January 1941	101
Table	8.1	The expansion of the Canadian Army, 1939–1943	118
Table	8.2	Command and staff element, First Canadian Army/ I Canadian Corps, April 1942	120
Table	8.3	Montgomery's assessment of Canadian brigadiers, January–March 1942	124
Table	8.4	Brigadier Salmon's extra-brigade responsibilities, December 1941–February 1942	125
Table	9.1	Staff element, First Canadian Army, during SPARTAN, March 1943	130
Table	10.1	McNaughton's compressed battle procedure during SPARTAN	156
Table	10.2	Staff element, I and II Canadian Corps, during SPARTAN	159

Table 10.3	Canadian senior officers relieved during training, 1940–1944	164
Table 10.4	Time in command for Canadian Division commanders to the end of 1943	165
Table 12.1	Indications of the return of Canadian formations from the Mediterranean, May–August 1943	201
Table C.1	Time line regarding McNaughton's fitness to command	225

Figures

Figure 9.1	McNaughton's logistical plan for SPARTAN	132

Foreword

Marc Milner

If you ask veterans of First Canadian Army who they served under, chances are they will say Montgomery. Certainly they recall no warmth for the dour technicians who commanded Canada's field army through those grim days from Normandy to the Baltic. Lieutenant-General Guy Granville Simonds may have been brilliant – historians still disagree on that – but he came across as moody and taciturn at best, unquestionably distant from the front-line soldier. General Harry Crerar, who ultimately made his fame as commander of First Canadian Army, was – as Monty once described him – 'a good plain cook.' More important, his pettifogging ways did nothing to endear him to his men. My father's only memory of Crerar was entirely negative. He never forgot the day in the fall of 1944 when Crerar ordered the men of the 3rd Canadian Division to remove the rat symbol that had begun to appear on their vehicles after weeks of fighting below sea level in the flooded polder of Holland. Instead of nurturing the pride that the 3rd Division felt about its accomplishments, Crerar crushed it because his own sense of propriety was offended. Small wonder that Canadian soldiers associated their war effort with Monty. Apparently there was no one else, other than Crerar or Simonds, who could lead the Canadian Army in war by 1944 – except maybe Andy McNaughton.

The standard interpretation of Lieutenant-General Andrew Latta McNaughton is of a brilliant man who was unsuited for high command in war. A distinguished gunner from the Great War and later Chief of the General Staff, McNaughton was also a renowned scientist and the founder of Canada's National Research Council. These distinctions drew him back into active service in 1939 and ultimately to command of First Canadian Army in England between 1941 and 1943. Once the threat of invasion passed, McNaughton's army was, in his own words, 'a dagger pointed at the heart of Germany.' But he was also not everyone's idea of good cheer and efficiency. Norm Hillmer and Ben Greenhous have summed up the conventional wisdom on McNaughton in one sentence: 'By late 1943 the crusty Canadian's uncompromising opposition to fragmentation of the Canadian Army Overseas was causing problems in Ottawa; he was out of favour with his own minister of national defence, J.L. Ralston, and British criticism of his generalship was mounting.'* More important, General Sir Alan F. Brooke, Chief of the Imperial General Staff, did not like him. Nor, it seems, did senior Canadian politicians and brass. For all of these reasons, McNaughton had to go in 1943.

Our view of McNaughton ossified decades ago with the appearance of John Swettenham's three-volume biography in 1969 – just before archives full of the Second World War material and countless collections of personal papers became available to historians. So a fresh interpretation of Andy McNaughton has been long overdue. John Rickard now provides that, and the result of his consummate scholarship is a radically different interpretation of McNaughton.

Rickard argues that historians have been too ready to accept Brooke's assessment of McNaughton as a general, as well as the complaints of Crerar and others back in Ottawa that he obstructed the operational deployment of Canadian formations in order to keep First Canadian Army together. The ostensible reason for McNaughton's dismissal was his failure during Exercise SPARTAN in the spring of 1943, in which he – in command of a two-corps 'Canadian' army – made some errors while attacking a British force. These errors included relying on a fresh and inexperienced Corps Headquarters that performed poorly, massing armour for pursuit, passing one corps immediately behind the other and across its logistical tail, and – now a legendary story – fussing over a bridging operation for several crucial hours while other duties went wanting. All of these errors apparently confirmed Brooke's growing suspicions that McNaughton was not cut out for higher command. Meanwhile, senior Canadian political and military officials were pushing in early 1943 to get Canadians involved in the fighting – somewhere. The electorate had grown restless watching other Allied armies fight while Canadians remained garrisoned in England. By 1943 the popularity of William Lyon Mackenzie King's Liberal government was at its nadir. For the government to survive, the army would have to fight. So both the

---

* *The Canadian Encyclopedia*, 1985 edition, vol. II, 1068.

Canadian Cabinet and the staff in Ottawa wanted the army committed, piecemeal if necessary. McNaughton's dagger would have to be sacrificed for political expediency.

Rickard argues that none of this was quite what it seemed. McNaughton's performance during SPARTAN had been acceptable – indeed, it had been bold and imaginative in some respects, and well received by the exercise's principal referee, General Sir Bernard T. Paget. Brooke's loss of confidence in McNaughton came later. Similarly, the staff in Ottawa were simply wrong in contending that McNaughton was the major impediment to deploying Canadian formations to the Mediterranean. Rickard reveals how McNaughton routinely sought ways to get Canadian troops into battle, including deploying a division to the Mediterranean. In the event, it was the British – anxious that they not be seen as waging the war with 'colonial' troops and few of their own – who did not want more Commonwealth troops in the Mediterranean theatre, where Australians, New Zealanders, South Africans, and Indians already made up about half the combat troops.

So Rickard demonstrates that our accepted view of McNaughton is wrong, though he stops short of peeling the next layer off this onion. Clearly, by 1943 McNaughton's dagger was not going to be the pointy end of the much anticipated landing in France in 1944, alongside an American army, to begin the liberation of Europe. The landing in France was to be the centrepiece of the war in the West, the crowning achievement of years of Allied preparations. It is inconceivable – to this historian, anyway – that the British would ever have allowed the Canadians the leading role. By committing a Canadian division, and then later in 1943 I Canadian Corps, to the Italian campaign, Brooke was clearing the way for a British army to land alongside the Americans on D-Day. In this way, the Canadian army got much needed combat experience, the Canadian government soothed an impatient electorate, and – perhaps most important – Crerar got First Canadian Army when it finally went into action. McNaughton clearly had to go, but not for the facile reasons currently accepted by historians.

If Rickard is right – and he has marshalled impressive evidence to back his case – he is raising the tantalizing question of how McNaughton would have performed as an army commander. It is obvious that he was brilliant, innovative, articulate, and loved by his men. As Rickard points out, many of the things he attempted to do during SPARTAN and elsewhere were later done successfully in battle by the likes of Montgomery. Indeed, McNaughton was probably as innovative as Simonds, and (when he had a good staff) as effective at battle management as Crerar. And he was amiable enough to become the Second World War's equivalent of Sir Arthur Currie. Perhaps that's why the knives came out: it is impossible to say. But it is unlikely that he would have stripped the 3rd Division of its nascent pride in the fall of 1944. 'Andy' probably would have posed for a publicity photo with the 'Rats' of the 3rd Division and their new symbol at the end of the Scheldt campaign, offered warm words of praise, and got on with the war. It is hard to read Rickard's book without concluding that the grim determination displayed by Canadian soldiers to see the war through to victory might have been a little less grim if McNaughton had been in command.

Marc Milner
University of New Brunswick
February 2009

Preface

Lieutenant-General Andrew Leslie, CMM, MSC, MSM, CD

General the Honourable Andrew George Latta McNaughton, PC, CH, CB, CMG, DSO, CD, MSc, DSc, DEng, DCL, LLD, psc, IDC, was a dedicated Canadian patriot, fiercely proud of his country and the countless people who had shaped and guided the evolution of Canada in good times and bad. After a brilliant career as a soldier during which he saw active service in both world wars, culminating in his appointment as Army Commander during the Second World War, he continued to serve his country and the larger international community as a statesman, scientist, scholar, and negotiator. He was gifted with a quick and lively mind, and his studies and knowledge encompassed a wide variety of subjects. His boundless energy and enthusiasm for life made him a charismatic, even legendary, figure on the Canadian landscape. Until the moment of his death in 1966 he vigorously pursued his greatest love – the safeguarding of Canada and all its resources.

Like most great men, Andrew McNaughton was driven. During his childhood in Moosomin, Saskatchewan, he developed a powerful work ethic as well as the self-reliance he was to exhibit in all that he did. While a student at McGill University in Montreal he was exposed to the richness of francophone culture, which he greatly admired. His studies in electrical engineering brought out his passion for the pure sciences and instilled in him an enormous sense of satisfaction in solving complicated technical problems through exhaustive research, collaboration with like-minded colleagues, and hard work. Later in life he would become renowned both as a military ballistician and as the inventor of the cathode ray direction finder, the forerunner of what we now call radar. Later still, as a doting grandfather, he would be equally delighted to design and produce a marvellously effective crossbow to shoot a line over a branch, far overhead, so as to rescue a kite that one of his errant grandsons had misdirected. The moment the kite was returned to earth he immediately lost interest in the crossbow and was off on another project. For a short time the young grandson was left unsupervised with a powerful new toy. But as usual, domestic order was quickly restored by the firm hand of Mabel, the general's devoted and fiercely protective wife and lifelong partner.

Andrew McNaughton joined the Canadian Army while still a student at McGill, where he was completing his Master of Science degree. During the dark days before the First World War, he volunteered for active service with the Canadian Army. A natural leader who was not the least bit intimidated by authority, he rose quickly through the ranks of his beloved artillery and was instrumental in crafting new techniques and tactics that contributed directly to the amazing successes of the 1st Canadian Division. While serving at the front he found the time to conduct technical research on how to locate enemy artillery through sound ranging, ballistic analysis, and a complex web of spotters, and he had the force of personality to ram through innovative new tactics and techniques across the Canadian Corps. Twice he was wounded in combat, and he was decorated for his performance under fire. It was during the war years that his fierce pride in all things Canadian was kindled. He and his peers – decisive young leaders who were confident, used to making tough life-and-death decisions, capable of thinking out of the box, and loyal to the idea of Canada as a nation to a degree largely unprecedented before the horrors of the Great War – returned home with a determination to help craft a model society.

During the interwar years, McNaughton stayed with the army. In a remarkably short time, at the age of forty-two, he was named Chief of the General Staff (CGS). Always active, he never abandoned his true passion, which was scientific research, and somehow he managed to balance two demanding and important careers, achieving excellence in both. After his tenure as CGS was completed, he was appointed President of the National Research Council, a post that gave him tremendous personal satisfaction and a great deal of joy. Nothing gave him greater pleasure than leading and supporting some of the finest minds in Canada as they pushed the boundaries of scientific inquiry in an atmosphere of collegial competition and innovation. This was, arguably, the happiest time of his life, and he and his young family thrived. But storm clouds were brewing on the international horizon, and he soon found himself back in uniform as a Division, Corps, and Army Commander during the Second

World War, responsible for an army of many hundreds of thousands of young Canadians engaged in a life-and-death struggle. He was followed into action by all three of his sons, one in the artillery and two in the air force. His indomitable wife moved to England at the height of the Blitz to better support him, as she had done during the First World War, leaving the two daughters in Canada.

General McNaughton's tenure as a key architect of Canada's war effort was, in retrospect, controversial. His was a larger-than-life presence, and the image of 'Andy' dominated media outlets and gave confidence to his soldiers and to the people of Canada. An energetic and dynamic figure, he was passionate about securing the best equipment and training for his troops. His enthusiasm was infectious, as was his willingness to take on all challenges, and he earned the absolute confidence of the prime minister and the Canadian Cabinet.

Not all was easy sailing, however. The first few years of the war were full of dark tidings as the Axis powers scored victory after victory. Meanwhile, he found himself constantly fending off British suggestions that the Canadian Army be broken into smaller units and embedded in British formations as reinforcements under British tactical command. As a fervent Canadian nationalist, and having encountered the same sorts of suggestions during the First World War, he bluntly rejected all such ideas. He insisted that Canadian soldiers fight under Canadian officers up to and including division, corps, and army level, and in doing so he earned the enmity of powerful voices within the Allied war councils. Disheartened and worn down by the incessant squabbling on this critical issue – and by the death of his youngest son, killed in action as a bomber pilot and Canadian squadron leader – he found himself being doubted regarding his willingness to compromise with his international peers. By this time, the issue had become a political one. A compromise was crafted whereby the Canadians would be kept together to fight as divisions and corps, and eventually as an army; however, the stern voice of Canadian nationalism was to be muted. A few months after his promotion to full general, he was recalled to Canada. He retired from the Canadian Army on 1 November 1944. However, such was the Canadian public's confidence in him that the very next day he was appointed Minister of National Defence.

Andy McNaughton served as Defence Minister until August 1945, and he did a superb job under wartime conditions. A unifying figure of strong charisma, he was a driving force behind the war effort as well as a powerful orator on its behalf. His Cabinet colleagues were enormously respectful of his wisdom and experience, but he was not overly enthusiastic about the nuances of political life, and he found the constant give and take of political compromise frustrating and draining.

After the war he hoped to return to scientific research, but this was not to be. The prime minister named him Chairman of the Canadian Section of the Canada–United States Permanent Joint Board on Defence, and by building on his wide-ranging and strongly influential personal contacts with key American leaders, both in and out of uniform, he was able to ensure that Canada's security interests were not trampled by a powerful and confident neighbour. After a year at that post he was appointed the Canadian representative to the UN Atomic Energy Commission as well as president of the Atomic Energy Control Board of Canada – a natural fit, considering his scientific and military expertise. He found himself drawn to the UN's potential to make the world a better and more peaceful place, and he thoroughly enjoyed his interactions with key diplomatic personalities of the day, many of whom he knew well from his years as a senior military leader. In 1948, to his delight, he was appointed Canada's permanent UN delegate, and in that role he represented his nation on the Security Council – during which tenure he applied his considerable charm and negotiating skills to great effect. A powerful and highly energetic figure of immense charm and determination – once he had made up his mind, it was almost impossible to move him – he was one of a handful of Canadians who helped create the international perception of Canada as a helpful and influential voice on the world stage.

After a lifetime of public service in trying and sometimes unbelievably difficult circumstances, most men would have sought retirement and a quieter life. Andy McNaughton did neither, and in 1950 he began a twelve-year tenure as a member – eventually, the Chairman – of the International Joint Commission. This body was tasked with resolving questions on the use of the waterways between Canada and the United States, and though the title of the organization was deceptively simple, its scope was enormous. For example, frictions between the two nations relating to hydroelectric power generation and to navigation on the St Lawrence River had a direct impact on sovereignty and trade – and this was but one key waterway among dozens if not hundreds. McNaughton relished the challenges in deftly crafting the issues and arguments to ensure that Canadian interests were fully protected, and he held this position until 1962. This was another period of great satisfaction for the Ottawa-based McNaughton household, which became a centre for social and diplomatic discourse, with old friends and new constantly dropping by to discuss the urgent issue of how to safeguard Canadian interests and resources.

After his final retirement in 1962, nothing much changed in terms of his approach to life. He was a much sought public speaker, and he involved himself in so many committees, organizations, and public causes that it has been impossible to keep track of them all. A lifelong student of the sciences and a natural teacher, he was an awe-inspiring figure to his friends and family, an elemental force who was always making things happen and working on a variety of projects, both at home and around the nation. Right up to the end, his energy, his zest for life, and his pride in Canada were the centrepieces of his personality. On 11 July 1966, after a lively dinner with close friends, he died in his sleep with Mabel at his side in his modest cottage in Montebello, Quebec.

At his state funeral, Governor General Georges Vanier concluded his remarks this way: 'General McNaughton was a soldier to the end of his life, fighting for the causes in which he believed, with the integrity, candor and tenacity for which he became famous. He was never a man to be intimidated by authority and strove for what he considered to be right with a complete self-giving of all his powers. He was one of those all too rare human beings who combine a brilliant intellect with a dauntless heart.'

He lived his life to the fullest, a true son of Canada.

Lieutenant-General Andrew Leslie, CMM, MSC, MSM, CD, is the Chief of the Land Staff and the Commander of Canada's Land Forces. He is the paternal grandson of Andrew McNaughton and, like his father and grandfather, a gunner.

Acknowledgments

This work has taken a long time to prepare. I began work on my doctorate in January 1997 at the University of New Brunswick. However, I joined the army in 1999 and thereafter could only devote sporadic effort to the thesis. I need to thank my former Officers Commanding, Recce Squadron, Lord Strathcona's Horse (Royal Canadians), Vince Fagnan and Derek Macauley (currently the Commanding Officer of the Regiment), for their support while I was a troop leader. Captain Tim Andersen and Major Dave McKillop, my immediate superiors while I was posted to Wainwright, Alberta, and my current boss, Lieutenant-Colonel Shaun Tymchuk, supported my efforts to finish the work as well.

I need to thank the archivists at Library and Archives Canada, the Public Records Office in Kew, the Liddell Hart Centre at King's College, London, the United States Army Military History Institute in Carlisle, Pennsylvania, the Massey Library at Royal Military College in Kingston, Ontario, the Queen's University Archives, the University of Victoria Archives, the Imperial War Museum in London, and the Directorate of History and Heritage, Ottawa.

Special mention must be made of Marc Milner, my long-time friend and mentor. We began this project a long time ago and he has stuck with me throughout, offering guidance and encouragement. Others who materially helped me make get this work into its present state include Stephen Harris, Reginald Roy, Lieutenant-Colonel Doug Delaney at RMC, and Lieutenant-Colonel Roger Cirillo (USA Ret.). Colonels Delaney and Cirillo were of great assistance as I separated fact from fiction concerning Exercise SPARTAN. Captain Craig Robertson, a fine soldier from Princess Patricia's Canadian Light Infantry, was a great sounding board during my posting to Wainwright from 2004 to 2006. His basic common sense was useful to me more than he probably knew as I wrestled with command and training issues.

I am also very fortunate that the current Chief of the Land Staff, Lieutenant-General Andrew Leslie, and McNaughton's grandson, agreed to provide the preface to the work. If 'Andy' were still alive I believe he would have profound sympathy for his grandson's efforts to simultaneously fight a war and transform an army. The sympathy, I think, would be reciprocated.

Finally, I have to pay special tribute to my wife, Melissa Leigh. Not only did she have to endure a continual forced separation due to the nature of military service, but she has also had to put up with endless hours of thesis work and manuscript revisions in preparation for publication. She has been a real 'trooper.' Living with a military historian, officer, and football player is not an easy task, and she will always have my gratitude for sticking it out.

Any omissions or errors in the present work are mine alone.

Abbreviations

AA&QMG	Assistant Adjutant and Quartermaster General
ACGS	Assistant Chief of the General Staff
ACIGS	Assistant Chief of the Imperial General Staff
AFHQ	Allied Forces Headquarters
AbP	Alanbrooke Papers
AP	Alexander Papers
ADTB	*Army Doctrine and Training Bulletin*
ATI	Army Training Instruction
ATM	Army Training Memorandum
ANZAC	Australian–New Zealand Army Corps
AQDJ	*Army Quarterly and Defence Journal*
BCATP	British Commonwealth Air Training Plan
BEF	British Expeditionary Force
BGS	Brigadier General Staff
BM	Brigade Major
BP	Burns Papers
CAJ	*Canadian Army Journal*
CAD	Canadian Armoured Division
CASF	Canadian Active Service Force
CATM	Canadian Army Training Memorandum
CBO	Counter-Battery Officer
CBSO	Counter-Battery Staff Officer
CCOS	Combined Chiefs of Staff
CDQ	*Canadian Defence Quarterly*
CE	Chief Engineer
CEF	Canadian Expeditionary Force
CFA	Canadian Forces Artillery

CGS	Chief of the General Staff
CHR	*Canadian Historical Review*
CIB	Canadian Infantry Brigade
CID	Canadian Infantry Division
C-in-C	Commander-in-Chief
CIGS	Chief of the Imperial General Staff
CJSM	Canadian Joint Staff Mission
CJWSC	Canadian Junior War Staff Course
CMH	*Canadian Military History*
CMHQ	Canadian Military Headquarters
CO	commanding officer
COHQ	Combined Operations Headquarters
COS	Chiefs of Staff
CCOS	Combined Chiefs of Staff
CGS	Canadian General Staff
COSSAC	Chief of Staff to the Supreme Allied Commander
CP	Crerar Papers
CRA	Commander Royal Artillery
CRAC	Commander, Royal Armoured Corps
CRO	Current Reports from Overseas
CSO	Chief Signals Officer
C3	Command, Control, and Communications
CWC	Cabinet War Committee
DAFV	Director Armoured Fighting Vehicles
DA&QMG	Deputy Assistant and Quartermaster General
DCGS	Deputy Chief of the General Staff
DDMO	Deputy Director of Military Operations
DHH	Directorate of History and Heritage
DMO	Director of Military Operations
DMO&I	Director of Military Operations and Intelligence
DMT	Director of Military Training
DND	Department of National Defence
DP	Dewing Papers
DSD	Director of Staff Duties
DSO	Distinguished Service Order
FFC	Field Force Component
GHQ	General Headquarters
GOC	General Officer Commanding
GOC-in-C	General Officer Commanding-in-Chief
GS	General Service

GSO	General Staff Officer
HMSO	Her Majesty's Stationary Office
IDC	Imperial Defence College
IWM	Imperial War Museum
JMH	*Journal of Military History*
KP	John Noble Kennedy Papers
LAC	Library and Archives Canada
LCS	London Controlling Station
LHCMA	Liddell Hart Centre for Military Archives
LHP	Liddell Hart Papers
LPP	Lester Pearson Papers
MP	McNaughton Papers
MaP	Malone Papers
MkP	King Papers
MTP	Military Training Pamphlet
NCO	Non-Commissioned Officer
NDHQ	National Defence Headquarters
NPAM	Non-Permanent Active Militia
NRC	National Research Council
NRP	Norman Rogers Papers
NTW	Notes from Theatres of War
OC	Officer Commanding
O Gp	Orders Group
OMFC	Overseas Military Forces of Canada
Op O	Operations Order
OP	Odlum Papers
PA	Personal Assistant
PP	Pearkes Papers
PRO	Public Records Office
RA	Royal Artillery
RAC	Royal Armoured Corps
RAF	Royal Air Force
RCAC	Royal Canadian Armoured Corps
RG	Record Group
RMC	Royal Military College
RP	Ralston Papers
SC	Staff Captain
SECO	South Eastern Command
SDR	Signals Dispatch Rider
SP	Spry Papers

SaP	Sansom Papers
StP	Stuart Papers
TEWT	Tactical Exercise Without Troops
TWTEC	Trench Warfare Training and Experimental Centre
UNB	University of New Brunswick
USAMHI	United States Army Military History Institute
VCGS	Vice Chief of the General Staff
VCIGS	Vice Chief of the Imperial General Staff
WD	War Diary
WO	War Office

THE POLITICS OF COMMAND

Introduction

Andrew George Latta McNaughton was a gifted soldier-scientist. Prior to the First World War he had pursued postgraduate work in engineering at McGill University, and throughout that war he used his scientific expertise to increase the effectiveness of Canadian artillery. He initially commanded an artillery battery, but by war's end he had taken command of the Heavy Artillery of the Canadian Corps. After the war he remained in the Permanent Force, and by 1926 he was the Vice Chief of the General Staff (VCGS). In 1929, at the age of forty-two, McNaughton was promoted to major-general and Chief of the General Staff (CGS), the highest position in the army. He held that post until 1935, when he stepped aside to take up the prestigious office of President of the National Research Council (NRC) of Canada.

In late 1939, just months after the Second World War began, McNaughton took the 1st Canadian Infantry Division (CID) overseas. By mid-1940 he had been promoted to lieutenant-general and given command of the British VII Corps, which on Christmas Day 1940 was redesignated the Canadian Corps. He supervised the expansion of the Canadian Active Service Force (CASF) from a fledgling organization in 1939 to the powerful First Canadian Army in April 1942. McNaughton took part in the abortive attempt to re-establish the British Expeditionary Force (BEF) in France after Dunkirk in June 1940, authorized the raid on Dieppe in August 1942, and authorized the deployment of the 1st CID and the 1st Canadian Army Tank Brigade to Sicily in July 1943.

McNaughton never led the Canadian Army – his self-proclaimed 'dagger pointed at the heart of Berlin,' – into action. At the end of 1943 he was relieved of his command and entered semi-retirement. Then in October 1944 he was recalled to national service by Prime

Minister William Lyon Mackenzie King, who named him Minister of National Defence at the height of the crisis over sending conscripts overseas. In the end, however, he failed to gain a seat in Parliament: in February 1945 he was defeated in a by-election in Ontario; later that year, he failed to win a seat in the general election. Unbroken by these defeats and by the acrimonious debate over conscription (during which old Tory friends turned on him for daring to join King's Liberal Cabinet), McNaughton went on to serve Canada in various ways after the war, as a UN diplomat, as head of the Canadian Section of the Permanent Joint Board on Defence, and as chair of the Canadian side of the International Joint Commission.

Few Canadians striding the national stage in the first half of the twentieth century had a more diverse, successful, and ultimately controversial career than Andy McNaughton. He was perhaps the most scientifically and intellectually gifted Canadian officer to wear his country's uniform. Canada's foremost military historian, Colonel Charles P. Stacey, considered him remarkable 'by any standard' and declared with no hesitation that of all the senior Canadian military and civilian authorities of the Second World War, 'no other had so magnetic or compelling a personality.'[1] J.L. Granatstein agreed, declaring that McNaughton was 'a great man, highly intelligent, a compelling personality,' and one who 'did well in every single endeavour he took on – except as commander of the First Canadian Army.'[2]

Granatstein's caveat is the subject of this new study of McNaughton.[3] A brilliant man, a highly successful soldier, and a gifted artilleryman, McNaughton was nonetheless deemed incapable of operational command of higher formations in wartime. Perhaps more than anything else, operational command is the acid test for a general officer. This book addresses the three principal arguments put forward by historians to explain McNaughton's apparent failure at the pinnacle of his military career: (1) that he refused to sanction the subdivision of the Canadian Army for use in operations; (2) that he was a poor military trainer and a poor operational commander; and (3) that he possessed an abrasive personality that undermined effective cooperation not only with senior British commanders but also with senior Canadian commanders and government officials.

Historians argue that one of the main reasons for McNaughton's failure was that he refused to see the Canadian Army divided in order to protect and expand the semi-autonomy that General Sir Arthur Currie had achieved while commanding the Canadian Corps during the latter half of the First World War. It is alleged that McNaughton's faith in the 'Currie Doctrine' led him to consciously undermine all efforts – whether they originated in the British War Office or in the National Defence Headquarters (NDHQ) in Ottawa – to employ parts of the Canadian Army under British command in action against the enemy during the early stages of the war.

McNaughton's staunch defence of Canadian national prerogatives, and his single-minded obsession with seeing the entire Canadian Army spearhead the invasion of North-West Europe, brought him into conflict with the Chief of the Imperial General Staff (CIGS), General Sir Alan F. Brooke, the Canadian Minister of National Defence, James Layton Ralston, and the CGS, Lieutenant-General Kenneth Stuart. Grave concerns about the army's lack of battle experience and rising public unrest over the lack of active participation in the war drove Ralston and Stuart to join with Brooke in forcing McNaughton from command.

Historians also argue that McNaughton's military incompetence contributed to his removal. They claim that he was a poor operational commander who lacked the necessary skills to properly train the army – in particular, his formation commanders. To prove this, historians have relied heavily on the testimony of Brooke and General Sir Bernard Toliver Paget, the Commander-in-Chief of Home Forces, regarding McNaughton's performance on SPARTAN, a massive army-level exercise held in March 1943. McNaughton, it is said, devoted too much time and energy to weapons design and to the political infighting required to protect Canadian autonomy, this at the cost of preparing himself and others for the burdens of command in modern war. In essence, Brooke, Paget, and others questioned McNaughton's military professionalism, and so have later historians.

The final factor in McNaughton's removal as army commander was his personality. Historians have consistently portrayed him as abrasive and inflexible, as quick to form judgements, and as totally lacking diplomatic skills. These personality faults, so it is said, led to debilitating feuds with key figures such as Brooke, Ralston, Stuart, and General Sir Bernard Law Montgomery. Indeed, McNaughton has been roundly blamed for these poor relations. Much of this book, then, will be about McNaughton's personal relations with his peers and superiors, both British and Canadian; for as General Sir Archibald Wavell said in his life of Field Marshal Lord Allenby: 'History, and more especially military history, is dry, misleading stuff without a clear understanding of the character and motives of the chief actors.'[4]

In his essay 'Stress Lines and Grey Areas: The Utility of the Historical Method to the Military Profession,' Dominick Graham has advocated greater dialectical inquiry: 'High-ranking officers are required to balance dialectical factors against functional considerations, and the higher their place in the chain of command, the more weight they must give the former.'[5] The *functional* plane of military history corresponds to such things as doctrine and weapons systems in military organizations, whereas the *dialectical* plane encompasses the interplay of people and institutions.

An examination of the dialectical aspects of McNaughton's story is influenced by the fact that important nuances of friendship and friction often disappear in the formalism of official memoranda. Even the private letters of the principal actors evidence some restraint, for various reasons. The documents do not always reveal the facial expressions, the body language, and the tone of voice displayed by McNaughton, Brooke, Montgomery, Ralston, Stuart, and others when they met to discuss contentious issues. Sir Ian Jacob has observed that historians find it hard to recapture the atmosphere of the time and to evaluate 'the pressures bearing at any given moment on the principal actors.'[6] The historian cannot have witnessed the actual meetings and so must attempt to plug the gaps without exceeding the plausible.

Another potential problem when studying personalities is the historian's natural tendency to side with certain actors over others. In the introduction to his life of Augustus, John Buchan noted that 'I am conscious that my interpretation of Augustus is a personal thing, coloured insensibly by my own beliefs.' Michael Howard once said that the historian has an obligation to ensure, 'even within the limits imposed by his own cultural environment, that our view of the past is not distorted by fraud, by evident prejudice or by simple error.'[7] The historian must strive to be objective, to avoid imposing current concepts on past events, and to abide by the rules of evidence. A close study of McNaughton's interactions with Brooke, Paget, Montgomery, Ralston, Stuart, and others will facilitate an assessment of his innate capacity to achieve those goals he considered important.

This fresh study of McNaughton has been framed in terms of the three principal arguments identified above. The first two – regarding McNaughton's unwillingness to see the army divided, and questions about his military professionalism – are inescapably coloured by the third – the effects of personality. This work represents a revisionist approach to the historical problem of why McNaughton failed as commander of First

Canadian Army. It is not my intention to totally dismiss existing interpretations of his failure – indeed, the three arguments identified retain some basic validity. Rather, I hope to show where the existing interpretations diverge from the documentary record, and why.

PART ONE
The Making of Andy McNaughton

1

Early Life and the Crucible of the First World War

I am a Canadian, not a professional soldier.

McNaughton to Currie, 1918

Andrew McNaughton was born on 25 February 1887 (the same year as Montgomery) in the prairie town of Moosomin, thirty-six miles east of Regina, to Robert Duncan McNaughton and Christina Mary Ann Armour. Both of Andrew's parents were descendants of immigrants from Glasgow, Scotland. Had the Canadian Pacific Railway not reached Saskatchewan by 1882, Andrew McNaughton might well have been an American, for his father, fascinated by the American West, had traded in buffalo hides in Chicago in the 1870s. In 1885, Major-General Sir Frederick Middleton, commander of the Canadian Militia, used the railway to defeat Louis Riel's North West Rebellion, thereby securing the prairies for white settlement. That same year Robert McNaughton established himself in Moosomin and started a family.[1]

In Victorian Canada the family was a formidable institution, characterized by unchallenged paternal authority and dominant maternal care. It represented duty and was the principal engine of moral upbringing. In the areas of new settlement, children were the product not only of love but also of the capitalist requirement for labour. At the time of McNaughton's birth western Canada was still a frontier society and the Great Plains were yielding only slowly to the plough. Hard work was the norm. It would be difficult to imagine any other ethic taking root on the bleak prairies. Even as the McNaughtons prospered through hard work, in their daily lives they had to struggle constantly against the raw environment.

The influence of childhood on later development is undisputed, and

Andrew, unlike too many other youths, benefited from loving parents. They frowned on idleness and instilled in him a deep sense of industry. His biographer, John Swettenham, wrote that 'his home life – a gentle young mother for whom he had great affection, a strong, adventurous father and a brother for companionship – was happy.' His brother, Murray William Armour, eighteen months younger, offered ready companionship. Had McNaughton seen fit to write his memoirs, he probably would have echoed General Sir Brian Horrocks's simple comment: 'I had an extremely happy childhood.'[2] Though Andy's youth was less exotic than that of Horrocks – who grew up in Gibraltar with free access to Spain – McNaughton's early life, too, was full of fun and adventure, centred on hunting, fishing, and horses.

The McNaughtons were Anglican and took their religion seriously. Andrew most certainly read the bible as a child (Eisenhower finished it by age twelve), but his specific religious beliefs are not clear. He developed into a formidable scientist, and it would not have been surprising had he encountered difficulty reconciling religion with science. The Anglican Church was the established church of the British Empire, and for many Anglicans faith could exist without serious conflict with the secular world. At any rate, McNaughton was no religious zealot and never attempted to equate military leadership with Christian values the way Montgomery did.[3] And McNaughton was not so 'orange' that it prevented him from falling in love with and marrying Mabel Weir, a Catholic, in 1914. In the event, like the vast majority of senior Canadian – as well as American and British – officers of the Second World War, McNaughton professed the Protestant faith.[4]

In Victorian Canada, to be of British descent and Protestant most often meant being conservative in one's politics. The McNaughtons supported John A. Macdonald's Conservative Party, which dominated Confederation until 1896. Supporting the Tories equated to a strong affinity for Britain and the Empire. Macdonald maintained that he had won the 1891 election in part because of the 'loyalty cry' towards Britain.[5] It was difficult not to be loyal, for even though Canada had become a self-governing Dominion in 1867, it remained dependent on Britain for diplomatic and military protection and financial support. Moreover, Queen Victoria's Diamond Jubilee in 1897 renewed Canadians' satisfaction in the Empire's accomplishments.

Carl Berger has observed that Canadians' affection for imperialism was in fact 'one form of Canadian nationalism' in that the future of the country – and its sense of its own growing power – depended on the

Empire's continued dominance.⁶ Writers like G.M. Grant and George Parkin convinced an untold number of Canadians that the Empire was the greatest secular instrument for the world's betterment that existed at that time. So as the century closed, most English-speaking Canadians adhered to moral and social justifications for that Empire's existence and continuation.⁷ William L. Morton was no doubt correct when he observed that in Victorian Canada, the conscious decision to be British rather than American meant that the 'Canadian imitation often outdid the British imitated.'⁸

McNaughton grew up in the midst of this Canadian enchantment with the Empire and could scarcely have ignored the influence of its wide-ranging accomplishments. At the time of his birth, Queen Victoria had been on the British throne for forty years. In 1800 the British controlled 20,000,000 people and 1,500,000 square miles. By 1900 the Empire had expanded to dominate 390,000,000 people and 11,000,000 square miles.⁹ The British Empire was the dominant military, economic, industrial, and scientific power in the world. The Royal Navy, supreme on the oceans since Trafalgar in 1805, guaranteed British overseas trade. The history books were dominated by the exploits of Captain James Cook in the South Pacific, General James Wolfe on the Plains of Abraham, Lord Clive in India, and the Iron Duke at Waterloo. Even military defeats such as the massacre of the 24th Regiment of Foot at Isandlwana in 1879 and the death of Major-General Charles Gordon at Khartoum in 1885 were portrayed as heroic struggles against barbarians.¹⁰ Everything the British did, be it inventing, proselytizing, or exploring the world, was depicted as heroic and glamorous.

Yet at his core, McNaughton was an urban Scotsman. In Swettenham's view, he had 'inherited a clearly identifiable and tough sense of [Scottish] nationalism which, wholly transferred to the country of his birth, was the origin of his dedication to Canada as a strong and independent nation.'¹¹ That may have been true, but the allure of the Empire was also strong, and even though he would come into conflict with British generals, he always held the British people and their accomplishments in high regard.¹²

In 1900, Robert McNaughton, having prospered through his trading store in Moosomin, decided to send both his boys to a good school. Andrew, who had shown intellectual promise as a child, entered Bishop's College School in Lennoxville, Quebec, with his brother Murray. During his five years at Bishop's, Andrew rose to head prefect. While Montgomery was floundering through the Army Class at St Paul's Day School with mediocre grades, McNaughton was excelling at academics. He also proved to be a good athlete. It was at Bishop's, under former British army sergeant J.H. Harney, that he received his first taste of military training. At the time, he enjoyed it as merely another experience without becoming obsessed with it.

In stark contrast to the other prominent Canadian officers of the Second World War – including Guy Simonds, Eedson Burns, Harry Crerar, and Kenneth Stuart – McNaughton was not a graduate of Kingston's Royal Military College (RMC). He was the most notable exception to the general rule that applied to the key British and Canadian commanders of the Second World War. Alexander and Montgomery, infantrymen, had attended Sandhurst; Brooke, a gunner, had attended the Royal Military Academy at Woolwich. They had done so not from a passing fancy but rather out of a strong desire to follow the profession of arms.[13] McNaughton most probably would have studied engineering had he attended RMC, but the Canadian Society of Civil Engineers did not yet recognize an RMC diploma in engineering. That simple fact may have been what persuaded him to follow a civilian education.[14] In 1905, at the age of eighteen, McNaughton went to McGill University to pursue higher education.

Having chosen to study hydroelectric engineering at McGill, McNaughton found himself under the tutelage of the great Ernest Rutherford, who would later win the Nobel Prize for developing the modern theory of radioactivity. In 1910 McNaughton earned his Bachelor of Science degree with honours. At the prompting of Dr Louis Herdt, he stayed at McGill to pursue postgraduate work; by 1912 he had earned a Master of Science degree. Over the next two years he lectured at McGill and published six research papers. McNaughton was a brilliant engineer, but academic studies, as satisfying as they were for him, never completely extinguished his interest in military pursuits as he first experienced them at Bishop's.

While studying at McGill, McNaughton qualified for a commission in the British Army. Why he chose to do so is unclear, but it would appear that his boyhood reading about the British Army on imperial service made a lasting impression. Perhaps he felt a higher obligation to the country than could be satisfied by an academic career; perhaps he simply craved adventure. When interviewed by Major-General P.H.N. Lake, Canada's first CGS, McNaughton identified the cavalry as his first choice, preferably in the Indian Army.[15] India was exotic, and McNaughton's choice suggests a passion for adventure. His later abandonment of his

Indian dream was the result of a desire to be closer to his ailing mother. McNaughton later reflected that had fate not intervened to preclude him from pursuing his goal of joining the Indian Army, he might have spent the First World War in a Turkish prison.

Instead he settled for the Canadian Army, joining 'B' Battery of the Royal Canadian Artillery at Kingston in 1909. Then in 1910 he took a commission in a Montreal field battery of the Non-Permanent Militia. A reserve commission was well suited to his principal career as a scientist. Thus began the dual professions that would characterize his life. Three years later he was promoted to major and took command of the Montreal battery. During the haphazard First World War mobilization conducted by Sam Hughes, the Minister of Militia, McNaughton's battery became the 4th Battery/2nd Brigade of the Canadian Expeditionary Force (CEF). In France it was redesignated once more as the 7th Battery. McNaughton saw his duty in no uncertain terms and had given up a promising engineering practice to sail with the great armada carrying the first contingent in early October 1914. University had been a turning point in his life, but the test of modern warfare that lay ahead, dominated by trenches and artillery barrages, was to fully define him.

McNaughton's battery underwent a short period of indoctrination shooting near Armentières–River Lys–Ploegsteert Wood in February 1915. His first taste of combat came at Ypres in April during the great German gas attack. He suffered a serious shrapnel wound in the shoulder but stayed with his 18-pounder battery for twelve hours and directed 1,800 rounds on the enemy, many of them over open sights at close range, in support of the 1st Canadian Division. Being wounded during the First World War was a trauma endured by many young officers who went on to higher command in the Second World War. George Patton, Douglas MacArthur, Harold Alexander, Bernard Montgomery, Bernard Paget, Kenneth Anderson, and Oliver Leese, to name but a few, also sustained one or more wounds during the First World War.[16]

The horror of war was burned into McNaughton's brain when he watched the Algerians at Ypres running for their lives ahead of a gas cloud while coughing up their lungs. It was a grim introduction to combat. 'It was a very disturbing, very distressing sight,' he later reflected. However, he possessed the necessary strength of character to shrug off both the gruesome impact of gas warfare and the unnerving effects of German snipers as 'the normal course of war.' Regarding his days on the Somme in 1916 – a grim battle that for Britain became the defining moment of the twentieth century – he declared that 'I have never been happier in my life,' in reference to the special pride he took in his unit's accomplishments.[17] His reaction appears to have been similar to that of Alexander, the future field marshal, who declared two years into the slaughter: 'I wouldn't miss it [the fighting] or be out of it for anything.'[18] McNaughton probably did not have the cavalier, romantic notions of war to the extent that Alexander did (though his wish to join the Indian Army gives some evidence of them). In the event, he was here demonstrating two of the most important attributes combat leaders can possess: physical courage, and steadiness under fire.

After Ypres, McNaughton convalesced at the Royal Victoria Hospital in England, where for several months he wondered whether he would ever regain the use of his left arm. With traumatic injuries, rehabilitation depends almost entirely on the mental fortitude of the individual, and McNaughton possessed this characteristic in great quantity.[19] While he was healing, his scientific mind worked on ideas for improving the accuracy of artillery. After failing two medical boards he was finally declared fit for service again in October 1915. He took command of the 21st Battery/6th Howitzer Brigade in the newly arrived 2nd Canadian Division. In January 1916 he was with the brigade in France; two months later he was called back to England to assume command of the four batteries of the 11th Brigade, Canadian Forces Artillery (CFA).[20] With the new position came promotion to lieutenant-colonel.

After missing the chaotic action at the St Eloi Craters in April 1916 and the tough fighting at Mount Sorrell on Ypres Ridge in June, McNaughton took part in the grim ordeal on the Somme with the 3rd Division Artillery during the summer. His brigade was employed wherever it was needed, supporting British formations and even the Australian–New Zealand Army Corps (ANZAC). He considered it great experience to be marching and fighting all over the front line, though he would later frown on the practice of intentionally separating Canadian units.

McNaughton also missed the first-ever use of tanks at Flers-Courcelette in September 1916. In October, though, he fought in the failed attempts to take Regina and Desire Trenches on the Somme. Major-General David Watson's 4th Division finally captured Regina Trench in early November with the help of McNaughton's brigade. By the end of November the Somme battles had run their course. That the Somme pushed the participants to their mental and physical limits, and beyond, there can be no doubt. McNaughton: 'There wasn't much left of us after the Somme was over, I can tell you.'[21] At the time, he wrote Mabel a rambling letter, the disjointed nature of which suggests he may have been suffering from shell shock or battle fatigue.[22] If he was, he had the good fortune of having the opportunity to recover his wits before assuming an even more challenging command.

In late January 1917, at the age of twenty-nine, McNaughton was appointed to the new position of Counter-Battery Staff Officer (CBSO) for the Canadian Corps. In his new capacity he played a key role at Vimy Ridge, the first time in the war that all four Canadian divisions fought and attacked simultaneously under Canadian Corps Headquarters. Two years earlier the French Tenth Army under General Henri Philippe Pétain had suffered 100,000 casualties trying to take the ridge. In early 1917 the Canadians resolved not to fail, and superb employment of artillery proved to be the key to taking and holding the ridge. Counter-battery work demanded precision, and McNaughton's scientifically attuned mind readily worked the problem. In this he was aided by a newly formed operational research team (predating such efforts in the Second World War).[23] He and his team were so successful, in fact, that his counter-battery organization destroyed 83 per cent of the 212 German guns facing the Canadians at Vimy.[24] The systematic destruction of the German artillery at Vimy led directly to the successful capture of the ridge after two days of hard fighting.

After Vimy, McNaughton's counter-battery organization supported the Canadian Corps at Hill 70 in August. When General Arthur Currie indicated that General Sir Henry Horne, commander of the British First Army, originally wanted the Canadian Corps to attack Lens (instead of Hill 70), McNaughton called it a 'bloody fool operation.' In the end, Horne relented and the Canadians captured and held the high ground north of Lens itself. Of Passchendaele in late October, McNaughton declared: 'I had twenty-eight days of it and I think it was the most dreadful of the lot.'[25] Still, it was another success for the corps, and he had personally organized the defensive fire to make certain the Germans could not retake Passchendaele Ridge after the Canadians captured it in early November.

McNaughton made a significant contribution to the corps' dramatic gains during the last hundred days of the war. He again persuaded Currie that Horne's plan for attacking the Festubert sector was seriously flawed. So McNaughton and Major-General Edward Morrison, the corps' General Officer Commanding (GOC) Royal Artillery (RA), did the artillery planning for Currie's grand assault on Amiens, which began the last '100 Days' and which led General Erich Ludendorff to declare 8 August the 'black day of the German Army.' Fighting side by side with the Australian Corps, the Canadian Corps drove eight miles into the German lines, achieving virtually all its objectives at a cost of 4,000 casualties – a rate of advance far greater at a cost in human terms far less than had been possible only the year before.

From Amiens the Canadian Corps pushed down the Arras–Cambrai highway, where McNaughton employed air force light bomber squadrons to engage targets beyond gun range.[26] The battles for the Drocourt–Queant Line, the Canal du Nord, and the Hindenburg Line followed in rapid succession. McNaughton's last battle was at Valenciennes, where he coordinated heavy-artillery and counter-battery assets. Indeed, for many months towards the end of the war he was the nominal GOC of the Corps Heavy Artillery because Brigadier-General R.H. Massie was ailing. McNaughton reflected that he was running the show 'so far as the fighting was concerned.'[27] During the final month he was promoted to brigadier-general (though not officially confirmed until 25 November, he was ordered to wear the rank), assuming command of the Corps Heavy Artillery on 22 October.

By any measure McNaughton had a successful war. He rose from major and the command of a four-gun battery to brigadier-general and the command of the immensely powerful Heavy Artillery of the Canadian Corps, which comprised 96 officers, 3,300 other ranks, and 92 guns of various calibres. As the CBSO at Hill 70 he had actually commanded more guns than that – 111 – with the authority to call on five more.[28] In terms of promotions, he certainly kept pace with his great rivals of the war to come: Alexander, Montgomery, and Brooke. Alexander rose from platoon commander to become one of the British Army's youngest majors and subsequently one of its youngest lieutenant-colonels before temporarily commanding a brigade. Montgomery, after being wounded as a platoon commander in August 1914, never commanded in the field for the rest of the war, though he rose swiftly to temporary lieutenant-colonel and General Staff Officer (GSO) 1 of the 47th (London) Division. Brooke was principally a staff officer throughout the war and rose to become GSO 1 of First Army.[29]

McNaughton's character had been severely tested during the slaughter along the Western Front, and it had not been found wanting. General Wavell was not alone when he declared that in war, 'character is of greater importance than brains or experience.'[30] Despite sustaining a second serious wound on 5 February 1918, McNaughton remained steady and courageous when executing his duties and responsibilities. That same month he was awarded the Distinguished Service Cross for 'conspicuous gallantry' as CBSO. By war's end he had been mentioned several times in dispatches.

McNaughton had remained sensitive to the sufferings of his men throughout the ordeal on the Western Front. He was constantly on the prowl to ascertain their physical and moral condition. He ate what they ate and endured the same physical hardships, and he knew how to fit in easily with them. Ernest G. Black would later recount how at the Somme, McNaughton 'hopped upon a pile of shells and had a mug of tea with the boys. I do not recall any other officer sharing our mugs of tea.' It was because of this type of leadership, Black added, that McNaughton could push his men hard 'and get cheerful obedience.'[31]

Regarding the technical aspects of gunnery, McNaughton had mastered his craft as it then existed. While both Brooke and Montgomery had proved themselves as exceptional staff officers, McNaughton proved to be an exceptional artillery commander. Harry Crerar, who worked with McNaughton in the counter-battery team, and who was destined to replace him in command of First Canadian Army, had great respect for McNaughton's drive and imagination in developing new techniques.[32] Currie declared after the war that McNaughton was the greatest gunner not just in the Empire, but in the whole world. Even John English, one of McNaughton's sterner critics, would concede that his wartime counter-battery work rated a 'superior performance.' General Sir Frederick Pile, head of Britain's Anti-Aircraft Command during the Second World War, reflected that McNaughton was 'probably the best and most scientific gunner of any army in the world. His ideas were colossal.'[33]

Canadian military historians have never found any cause to question McNaughton's skill as a gunner, but looking ahead to the Second World War, it is important to recognize the validity of Samuel Huntington's point: that command 'is not a skill which can be mastered by simply learning existing techniques.'[34] Gerald W.L. Nicholson has noted that the type of counter-battery work practised by McNaughton could take full advantage of the unique conditions of positional warfare, 'for under such conditions the communications facilities could be as elaborate as the state of the signallers' art permitted.' Most critically, 'there was usually ample time to co-ordinate the intelligence received, before taking action.'[35] Such conditions would not generally be encountered in the fast-paced manoeuvre warfare of the Second World War.

That McNaughton had performed well as a brigade commander in the First World War did not automatically signify that he would handle a much larger Canadian force with the same skill in the next war. In discussing American corps commanders in the Second World War, Robert H. Berlin concluded that the key to their success in high-level leadership positions was 'not necessarily found in World War I battlefield service.'[36]

The value of McNaughton's First World War experience may lie in less tangible areas. Wavell never exercised command at all during the war; instead, he traced his preparation for high command in the Second World War to the lessons he had learned while serving under Allenby in Palestine.[37] Mentorship is a recurring theme in military history. Patton consciously modelled himself on his First World War idol, General John J. 'Black Jack' Pershing, commander of the American Expeditionary Force. McNaughton developed a deep affection for Currie during the war, but whereas Patton mimicked Pershing's bearing and mannerisms, McNaughton came to identify with Currie's vision of a unified, semi-autonomous, and powerful Canadian Corps fighting for achievable objectives. Currie's willingness to stand up to General Sir Douglas Haig when the interests of the Canadian Corps were at stake left an especially deep impression on McNaughton.[38]

From the moment Currie assumed command of the Canadian Corps in early June 1917, he rebelled against the British approach to fighting the war. He made his first stand when Horne, commander of the First Army, ordered him to assault the city of Lens. Currie considered a direct assault on the city suicidal and extracted from Horne permission to take the city by first securing a dominating height known as Hill 70.[39] Even as a brigade commander in the early stages of the war, Currie had exhibited a commonsense approach to operations. His decisions, often taken at some personal risk, were driven by an intense desire to minimize casualties. Brigadier J.E.B. Seely, commander of the Canadian Cavalry Brigade, declared that Currie had 'an almost fanatical hatred of casualties ... I think Currie was the man who took the most care of the lives of his troops.'[40]

Currie demonstrated his commitment to avoiding unnecessary casualties in no uncertain terms when the Canadian Corps moved to Passchendaele in October 1917. After digesting the situation there, he openly defied Haig. 'I carried my protest to the extreme limit,' he recalled, 'which I believe would have resulted in my being sent home had I been other than the Canadian Corps Commander.'[41] Currie operated on the simple principle that the objectives assigned to his corps had to be feasible and worthwhile. Though Haig gave him the 'some day I will tell you why, but Passchendaele must be taken' speech, Currie extracted a promise from the Commander-in-Chief that the Canadian

Corps would not have to attack until all preparations were complete to his satisfaction.[42]

In January 1918, with overall British strength plummeting, Currie resisted pressure from the War Office to reduce the strength of his infantry brigades from four battalions to three. Even though this would have made possible the deployment of a 5th and 6th Division – indeed, the creation of a Canadian army – Currie stayed focused on preserving the combat power of his corps and refused to take that step. His single greatest contribution to Canadian autonomy was that he prevented Haig from using his divisions piecemeal during the great German spring offensive of 1918. 'He lodged a complaint,' fumed Haig, 'when I ordered the Canadian Division to ... support the front and take part in the battle elsewhere. He wishes to fight only as a "Canadian Corps."'[43]

Had the Canadian Corps not evolved into a semi-autonomous force under Currie's guidance, it is unlikely that Prime Minister Sir Robert Borden would have had the opportunity to sit at the League of Nations in Geneva and independently sign the Treaty of Versailles. In this vein, Borden's support was vital to Currie, for without it Currie could not possibly have been as aggressive as he was. As early as May 1915, in a letter to Sir George Perley, Canada's High Commissioner in London, Borden made it clear that 'it can hardly be expected that we shall put 400,000 or 500,000 men in the field and willingly accept the position of having no more voice and receiving no more consideration than if we were toy automata.'[44] Currie's martial skills and Borden's staunch political support guaranteed a powerful formation in the field – one, moreover, that was capable of achieving greater autonomy in military affairs. As Stacey observed, the corps 'left behind it a national legend' and was quite simply 'the greatest thing Canada had ever done.'[45] McNaughton knew from first-hand experience that this was true, and he never forgot it.

There was another lesson from the war that McNaughton never forgot: one did not have to be a professional soldier in order to fight effectively. For proof he needed only look at Currie, the amateur soldier whose pre-war militia training, including attendance on the Militia Staff Course, hardly qualified him for command of a brigade in operations.[46] By comparison, Haig, fourteen years older than Currie, was a professional soldier with a wealth of imperial war experience. Churchill claimed that by the time Haig became Commander-in-Chief he had 'no professional rivals ... and none appeared thereafter.'[47] Montgomery frequently observed that many commanders had a 'ceiling' beyond which they 'should not be allowed to rise,' especially in combat. Though Currie made mistakes early on, he also demonstrated the skill to move to the next level.[48] He had never commanded more than 400 soldiers on any given militia night. Yet later he proved capable of commanding a 4,000-man brigade, then a division, and finally a 100,000-man corps that had the combat power of a small army.

Currie's professional approach to training also left a lasting impression on McNaughton. In the war's last '100 Days' the Canadian Corps put aside the limited-objective, set-piece attacks of the first three years for a 'nearly continuous cycle of rapid pace, coordination intensive attacks' employing tanks, indirect fire, tactical air support, chemical munitions, electronic deception, and command-and-control and intelligence systems, all of which finally crippled the German Army. Haig trusted Currie to such an extent at the Canal du Nord and Cambrai that he allowed him a virtual veto power over General Horne as to the method of attack.[49] Currie, the amateur soldier in a colonial army surrounded by professional British soldiers, turned out to be one of the most 'professional' of all in his approach to the conduct of war.

2

The Road to High Command

I am sure we have made in this, the most important of all choices, the right one.
King on McNaughton to command 1st Canadian Division

McNaughton was thirty-two when the First World War ended. He was prepared to resume his scientific studies and pursue a promising engineering career, but his plans were derailed when Currie, the newly appointed Inspector General of the Militia Forces of Canada, made a personal appeal for him to stay in the army. Currie wanted McNaughton to serve on the Otter Committee, a body charged with incorporating the CEF units into the post-war militia. This was unfortunate, for while others such as Patton were investigating the future employment of tanks (at least for a while), McNaughton was essentially involved in a retrenchment.[1] Nevertheless, after serving on the Otter Committee, McNaughton began a meteoric rise to the top of the military profession in Canada.

Historians have paid little heed to the fact that McNaughton rose far too quickly in the interwar army. By January 1920 he had been appointed to the Permanent Force as Director of Military Training and Staff Duties in Ottawa. His rank was brevet lieutenant-colonel, temporary colonel, and acting brigadier-general. By comparison, Brooke rose to the equivalent position in the British Army only in 1936, though with the substantive rank of major-general.[2] By 1920 Montgomery had been reduced in rank to brevet-major; Patton, a colonel in the First World War, experienced the same setback and only made lieutenant-colonel again in March 1934 after nearly fourteen years as a major.

On his climb to the top McNaughton was afforded various opportunities for professional development. In 1921 Currie, long convinced of McNaughton's bright future, sent the world's greatest gunner off to the Staff College at Camberley. Brooke had attended in 1919; Montgomery had followed him in 1920. Camberley was founded in 1858 to address the lamentable performance of British commanders and staff officers in the Crimean War. Since Britain did not establish a regular General Staff along German lines until 1906, the 'brain' of the army was to be found at Camberley.[3] Horrocks called the Staff College 'the seat of all military knowledge'; Oliver Leese, who commanded Eighth Army in Italy after Montgomery's departure, reflected that it taught one 'how to work.'[4] Through lectures, sand-table exercises, Tactical Exercises Without Troops (TEWTs), and syndicate discussions, Camberley trained officers to fill appointments up to the rank of brigade major or equivalent.

Though Montgomery could not even recall if he had done well at Camberley, Brooke had impressed the Directing Staff, and so, now, did McNaughton as a member of the senior division.[5] The commandant, Major-General Hastings Anderson, concluded that though much of what McNaughton learned at the college 'was new to him,' he nevertheless 'has more than held his own.' He possessed a 'wide general knowledge, and brings a highly trained, scientific mind to bear on all military problems.' As a scientific gunner he was simply 'outstanding' and possessed 'good tactical ability.' His 'immense' powers of concentration and 'great strength' of character were also duly noted. Anderson thought him capable of 'great efficiency as a Commander or Staff Officer under conditions of modern war.'[6] There is little reason to question this fine assessment. The simple fact of the matter is, however, that McNaughton never had the opportunity to practise and expand on the valuable lessons he undoubtedly learned at Camberley in an operational field environment in Canada between the wars. The same was not true for Brooke and Montgomery.

Soon after his time at Staff College, Brooke began a three-year posting to the General Staff of the 50th Northumbrian Division (Territorial Army) while Montgomery went directly to Ireland as Brigade Major of the 17th Infantry Brigade in Cork, where he fought Sinn Féin for almost eighteen months.[7] McNaughton, however, returned to the position of Director of Military Training and Staff Duties in 1922. The following year, only two years removed from Camberley, he became Deputy Chief of the General Staff (DCGS) with the rank of brigadier-general. McNaughton's grooming for further advancement was helped along by

Table 2.1 Advanced training of British and Canadian senior officers, 1919–1939

	Staff College, Camberley	Quetta	IDC
†Wavell, Archibald	1909–10	–	–
†Dill, John**	1913	–	–
#Ironside, Edmund	1913–14	–	–
#Alan F. Brooke,**	1919	–	1927
†O'Connor, Richard*	1919	–	1935
†Pearkes, George R.	**1919**	–	**1937**
†Gort, John*	1919–20	–	–
†Wilson, Henry Maitland*	1919	–	–
†Montgomery, Bernard*	1920	–	–
†Paget, Bernard	1920	–	?
#McNaughton, A.G.L.	**1921**	–	**1927**
#Crerar, Harry D.G.	**1923**	–	**1934**
†Percival, Arthur*	1923–4	–	1935
††Pope, Maurice	**1924–5**	–	–
†Sansom, E.W.	**1925**	–	–
†Alexander, Harold	1926	–	1930
†Slim, William*	–	1926	1937
†Auchinleck, Claude	–	1920	1927
#Cunningham, Alan	1925	–	1937
†Leese, Oliver*	1927–8	–	–
††Stuart, Kenneth	**1927**	–	–
†Dempsey, Miles C.	1927	–	?
^McCreery, Richard	1928	–	–
^Foster, Harry	1928	–	–
††Burns, Eedson L.M.	–	**1928–9**	**1938–9**
†Ritchie, Neil	1929	–	–
†Salmon, H.L.	**1931**	–	–
†Horrocks, Brian*	1931–2	–	–
††Stein, C.R.S.	**1932–3**	–	–
†Vokes, Chris	**1934–5**	–	–
†Keller, Rod	**1936**	–	–
#Simonds, Guy G.	**1936–7**	–	–
†Foulkes, Charles	**1937**	–	–
†Hoffmeister, Bertram M.	–	–	–
^Foster, Harry	1939	–	–

* Indicates later position as Directing Staff at Camberley or Imperial Defence College.
** Indicates Directing Staff at both
\# Artillery   ^ Cavalry
† Infantry   †† Engineer
Bold indicates Canadian citizens.

Major-General James H. MacBrien, the CGS, who considered him the best of the army's younger officers.[8] MacBrien sent him off to the newly established Imperial Defence College (IDC) in January 1927, where his classmates included Brooke, Claude Auchinleck, and other future Second World War senior commanders. The commandant was Admiral Sir Herbert Richmond, the famous historian and naval theorist, and the Chief Instructor was Major-General John Dill, a future field marshal, CIGS, and member of the Combined Chiefs of Staff (CCOS).

Whereas Camberley prepared officers to function on higher staffs, the IDC (founded in 1926) gave senior officers an opportunity to explore the broader issues of war and imperial defence policy from a joint perspective rather than through the narrow lens of their individual services.[9] Comprising just thirty students in the Senior Division, it was an intimate club within which personal and professional ties were established. Like the U.S. Army War College, it also represented that last critical prerequisite for high command. The IDC curriculum was varied, and McNaughton had the opportunity to present papers on various imperial issues.

McNaughton performed at a high level and drew considerable praise from Richmond for leaving 'no stone unturned to add to his knowledge of the military problems of the empire.'[10] The students visited industrial plants and defence establishments. McNaughton briefly visited the Tank and Track Transport Experimental Establishment at Farnborough; unfortunately, he never recorded any opinion of what he saw. Between terms he observed Edmund Ironside, another future Second World War CIGS, manoeuvre his 2nd Division near Thame. This was the first time McNaughton had seen a division in the field since 1918.

In 1928, after completing his IDC studies, McNaughton briefly commanded Military District No. 11 in Victoria before being promoted to major-general at the age of forty-one; then in January 1929 he ascended to CGS over several senior officers to become the professional head of the Canadian Army. That same year, Brooke was a new brigadier commanding the School of Artillery at Larkhill; he would not rise to CIGS with the rank of full general until December 1941, after commanding a brigade, a division, and a corps. Forty-nine-year-old General Douglas MacArthur, a divisional commander in the First World War, was still a year away from becoming Chief of Staff of the U.S. Army. The slow process of promotion usually associated with the interwar American, British, and Canadian armies simply did not apply to McNaughton. J.L. Granatstein has suggested that his 'phenomenal rise' had been achieved on 'ability alone.'[11] However, it was not an ability verified by the successive and incrementally difficult command of units and formations.

As CGS, McNaughton went to great lengths to get more funding for the air force (until 1935 part of the army) at the expense of the navy. He also waged an ultimately fruitless battle to focus all departmental power in the hands of the CGS, believing, as MacBrien had, that the army alone, by virtue of its dominance in the Canadian military establishment, should speak for all three services. MacBrien's vision of a unified voice in the department faded in 1927 when his successor, Major-General H.C. Thacker, voluntarily surrendered any claims to army primacy. McNaughton, loyal to MacBrien and to many of his ideas, never forgave Thacker for deliberately undermining the position of CGS.[12]

McNaughton also fought hard to prevent the deputy ministers in his department, G.J. Desbarat and L.R. Lafleche, from exercising any authority in the realm of policy or doctrine. He often bypassed them completely to discuss issues with the Minister of National Defence; indeed, he sometimes bypassed everyone and took his concerns directly to Prime Minister R.B. Bennett, with whom he enjoyed a particularly beneficial relationship.

In the course of this departmental power struggle, McNaughton convinced Bennett in October 1932 that the army could establish unemployment relief camps to ease the sufferings of single, able-bodied men left homeless and unemployed by the Great Depression. McNaughton's motives were commendable and demonstrated his commitment to the nation; even so, the project diverted scarce resources from the army. As English rightly concluded, the scheme – which quickly proved to be a political liability for Bennett – represented in the end an 'unprofessional focus' on McNaughton's part.[13]

Despite being sidetracked by the unemployment relief camps, McNaughton did succeed in making the army more expeditionary-minded. He fought hard to dismantle Defence Scheme No. 1, which entailed a fifteen-division militia to defend against an American invasion. Colonel James Sutherland 'Buster' Brown, Director of Military Operations and Intelligence (DMO&I), was the principal architect of the scheme. Though McNaughton had actually favoured it at one time, the unilateral British decision to dismantle Royal Navy facilities in the Caribbean and at Halifax in October 1929 had convinced him that defending against an American invasion without British help was impossible.[14]

Much to the distaste of the anti-American Brown, when McNaughton became CGS he prioritized Defence Scheme No. 3, a plan to send an expeditionary force to aid Britain in time of war. In December 1932 McNaughton informed the CIGS, General Sir George Milne, that 'the most serious or important issue for which we ... require to be organized concerns itself with the mobilization and dispatch of a Canadian Expeditionary Force to take part in an Empire War of first magnitude.'[15] McNaughton was certainly right about the need for an expeditionary capability; unfortunately, he gave far less attention to how a Canadian Expeditionary Force would fight after it had been raised and dispatched across the ocean.

McNaughton remained CGS for seven years, for the most part during the Great Depression. In that time he failed to prepare the army for modern war. When he departed in 1935 he summed up his acute failure in a memorandum titled 'The Defence of Canada: A Review of the Present Situation.' In it he observed that the state of modern equipment should be viewed 'only with the greatest concern.'[16] Here was a great irony, for he had always been wary of pushing the government too hard for fear of losing what the army did have. In 1921 he had stated that he was 'no advocate of extensive re-equipment at present. We would have no money left over if we did for training' – a position he was to reiterate eight years later.[17] Moreover, he admitted in the same memorandum that his estimates submitted while CGS had been based on the 'Ten Year Rule' – a planning assumption in effect in the War Office since 1919 and attributed to Churchill, which maintained that there would be no major war for ten years. Ultimately, the rule was placed on a rolling basis. Using the rule as his own yardstick, McNaughton sought only those funds that were 'immediately necessary to the maintenance and training of cadre forces.'[18]

It was in no way certain that McNaughton could have squeezed additional funding from the government through more aggressive lobbying. MacArthur, no shrinking violet, struggled for five years to get more funds for the U.S. Army while serving as Army Chief of Staff; he failed miserably. 'I stormed, begged, ranted, and roared,' he would recount. 'I almost licked the boots of certain gentlemen to get funds for motorization and mechanization and air power.' In every year that he served in the army's top position, he had to watch funding for his service drop.[19] With the approach of war McNaughton's successors, Major-Generals E.C. Ashton and Thomas Victor Anderson, made attempts to rectify the Canadian equipment situation, but the end state was, as Desmond Morton concluded, 'a Canadian army that, taken all in all, was worse in 1939 than it had been in 1914. Imagine that.'[20]

McNaughton was also well aware that during his watch as CGS the Permanent Force never held collective training in the field. It would not do so until 1938, just a year before it went to war. The British Army, on the other hand, managed to get to the field for collective exercises every year throughout the 1930s. Every year from the time McNaughton became CGS until the outbreak of the war, the British Army conducted collective field training at the brigade level or higher. One observer of the 1935 manoeuvres considered them 'probably the best ... since the War' and certainly better than those conducted a decade earlier. The Permanent Force exercised at Camp Borden in the summer of 1938 at the brigade level, but the poor overall quality of pre-exercise training hindered the manoeuvres.[21] For a professional army to remain ignorant of collective training for almost a full decade was unacceptable in any circumstances. It is hard to dispute Stephen Harris's claim that the army was 'never so poorly equipped' or 'so starved of funds,' that it could not even practise 'that body of fundamental professional knowledge applicable or adaptable to most battlefields even during the hostile environment of the inter-war years.'[22]

The German Army, too, was starved of funds and equipment throughout the 1920s, but it succeeded in working through various practical doctrinal problems. Heinz Guderian recalled that the army tested future tank tactics with canvas dummies 'pushed about by men on foot.' Major Eedson Burns, the future commander of I Canadian Corps in Italy during the Second World War, urged the same sort of experimental expediency in a 1935 *Canadian Defence Quarterly* (*CDQ*) article titled 'A Step Towards Modernization.' He suggested using automobiles, visible 'at the door of any armoury on training nights,' to practise manoeuvre 'without impossible expense.'[23]

It is readily conceded that McNaughton was not operating in a political environment favourable to the military in general. Military budgets were at subsistence levels and would only shrink during the Depression. Moreover, the sheer size of the country worked against efforts to bring units together for combined-arms training. That he chose to commit his great intellectual capacity to the political infighting demanded by the position of CGS for the purposes of 'holding the fort' is, in this broad context, understandable, but only to a point. Had he exploited his inventive genius to experiment with combined-arms exercises, even on a rudimentary level, he would have rendered great service to the army and sharpened his own professional skills.

Though he was the professional head of the Canadian Army, Mc-

Naughton demonstrated little enthusiasm for training it. Most critically, he showed almost no interest in mastering the art and science of command at a level beyond that which he had achieved in the First World War. According to James Eayrs, he expended a disproportionate amount of time and energy on political battles because the backward Canadian military of the 1930s offered few challenges for his great intellect. Perhaps McNaughton considered the possibility of commanding again unlikely given his senior position, but hubris also surely played a part. Eayrs added that as CGS, 'there was no one to match his qualifications or to rival his reputation ... His grasp of military matters in their widest sense was sure [and he had] perfect confidence in his own abilities.'[24] Those abilities, scientifically based and entrenched by his gunnery successes in the First World War, profoundly shaped McNaughton's entire approach to 'professionalism' during the interwar years.

McNaughton's faith in a scientific approach to war was most evident in 'The Military Engineer and Canadian Defence,' an article he published in *CDQ* in 1932. In it he wrote: 'The whole question of the right forces to maintain, *like many engineering problems* [emphasis added], is a question of proper balance between conflicting factors ... It is our business as soldiers to know all there is to know about the quantity and quality of our materials of construction – animate as well as inanimate – about the facilities existing and required for training or manufacture; we must appreciate what we can do in a given time starting from a given level of organization; we must be prepared to estimate and advise on the risks to be run.'

McNaughton concluded that Canada had to base its defence organization, not on an appreciation of the doctrine needed to fight a future opponent, but 'on the resources naturally available in the country, either in men or material.' If and when a new war erupted, he wanted Canada to be faced by a problem of 'adaption rather than creation' because 'quality is far more important than quantity.'[25] This suggests the duality of his 'professionalism' – military and scientific – as well as a reason for his failure to act imaginatively before the war.

In 1921 McNaughton observed that 'when the time comes for us to mechanize, we will find that we have great natural advantages to our credit' because the nation's youth 'are used to handling machinery of all kinds.'[26] Eight years later he was still preaching that philosophy, declaring in a *CDQ* article that 'our people took naturally to gunnery' during the First World War. From this he deduced that it should not be any different with mechanized warfare. It is not known if he ever read John F.C. Fuller's 1919 Gold Prize essay on technology and the future battlefield, but Fuller's advocacy of a future army 'led by scientists and fought by mechanics' would have struck a powerful cord with McNaughton.[27]

War, for McNaughton, seems to have been primarily a function of organization and physics. His extensive experience coordinating the Canadian Corps' counter-battery and heavy-artillery assets influenced him to the point that he believed victory was achievable through simple mathematical calculation. His interwar writings in *CDQ* make this abundantly clear. In 'The Development of Artillery in the Great War,' he stated: 'I know of no organization [Canadian Corps] in the history of War which was able to produce such a high ratio of shell to troops, nor any in which the price paid for victory was lower in personnel.' He maintained the same theme in 'The Capture of Valenciennes,' in which he boasted of the weight of shell 'exceeding by several hundreds of tons that fired by the Germans at Jutland.'[28] Such material expenditure ultimately secured victory and spared men's lives.

McNaughton remained committed to the 'weight of shell' doctrine quite simply because it had worked exceptionally well when the enemy never moved and there was plenty of time to prepare massive artillery strikes. His genius for positional artillery and infantry-based warfare might have been of great value in Normandy in 1944, when the fighting degenerated into a virtual stalemate somewhat reminiscent of the First World War trenches. However, the mechanized warfare (transportation-based) and armoured warfare (centred on tanks) that would be practised in the Second World War would inject greater complexity and present commanders with entirely different problems to solve once static fighting could be turned into mobile action.

Though his direct experience with tanks was limited during the First World War, McNaughton most certainly would have discerned that it was the Australians and Canadians who enjoyed the greatest successes with tanks in 1918.[29] The 4th Tank Brigade was attached to the Canadian Corps for the battle of Amiens and proved effective in overcoming machine-gun positions. Yet Shane Schreiber has suggested that McNaughton ignored Currie's 'consistent efforts to incorporate movement and surprise in his attacks, and his repeated experiments with mechanized forces.'[30] Perhaps McNaughton was influenced here by the blatantly obvious fact that at Amiens, eighty of the brigade's ninety-nine tanks were lost to gunfire trying to lead the advance, at which point the remainder were forced to follow the infantry.[31] Perhaps McNaughton deduced from this that the tank never superseded the infantry–artillery team, which was the real backbone of the offence during the war.

Like McNaughton, Brooke clung to the primacy of artillery well into the interwar period. Brooke was definitely not seen as a tank prophet. According to his biographer, David Fraser, he was voicing doubts about tanks as late as 1935. By 1937 he was still considered 'very much the distinguished artilleryman,' known for his well-organized barrages and for advocating infantry-support tanks. So traditional were Brooke's views on armoured warfare that the famous British military theorist, Basil Liddell Hart, tried to block his promotion to command the Mobile Division.[32] McNaughton never really voiced any detailed or innovative opinions on tanks either. In his profile of contemporary military thought published in *CDQ* between 1923 and 1929, James H. Lutz offhandedly mentioned McNaughton's name once, and even then it was not in the context of mechanization; he focused instead on the contributions of Frank F. Worthington and Kenneth Stuart, Eedson Burns and Guy Simonds. McNaughton discussed various issues in *CDQ*, but tanks were not one of them.[33]

While McNaughton published articles on such subjects as air survey and the workings of the Department of National Defence, Montgomery was attempting to work out the actual role of armour in a future war. In 'The Major Tactics of the Encounter Battle,' published in *Army Quarterly* in 1938, Montgomery described the importance of ground, the type of operational instructions required for mechanized warfare, and how divisional commanders should relay orders. He addressed the critical issue of road moves and time and space, arguing that armoured divisions should *not* be strung out; rather, they should be concentrated for heavy blows.[34] The year before, *CDQ* had published a similar article by Montgomery titled 'The Problem of the Encounter Battle as Affected by Modern British War Establishment.' Kenneth Stuart, the journal's editor, declared it 'the most thoughtful and valuable tactical discussion that has appeared in any British service journal for some considerable time.'[35]

There was no shortage of material on mechanization and armour to study. As DCGS and CGS, McNaughton had access to a wealth of imperial doctrinal and strategic material; moreover, the works of Fuller and Liddell Hart – to name but two of the visionaries – were in wide distribution. Fuller's most complete exposition on future mechanized warfare was contained in *Lectures on F.S.R. III* published in 1932.[36] McNaughton had met both Fuller and Liddell Hart during the 1927 manoeuvres on Salisbury Plain, where the first mechanized combat brigade took shape. However, he chose not to study the latest military writings, and he did not leave behind a massive library filled with penciled annotations of his thoughts on military subjects, like Patton did.[37]

Brigadier R.J. Orde reflected that McNaughton was 'undoubtedly the first CGS that really engendered an atmosphere of "Well, this is a professional job and let us make the best we can of it."'[38] But that was most certainly not true. McNaughton did state in his 1921 address to the United Services Institute that 'we must be thinking about the effect of these new weapons ... and preparing our minds for the new conception of tactics which they involve. I am an earnest advocate of moral preparation which is largely thought, and thought costs little.'[39] However, Harris's indictment that he 'argued against the existence of a unique profession of arms if that entailed full-time, life-long, and concentrated study' comes much closer to defining him as CGS. He simply saw no personal value in full-time, lifelong, and concentrated study, even though in the absence of field manoeuvres and actual war, it represented his only recourse for professional development.

McNaughton saw no need for professional study because he believed fundamentally in the militia tradition – after all, the Canadian Corps from Currie down had been a militia army. Almost two-thirds of the Canadian commanders in the First World War with the rank of brigadier or higher were from the militia.[40] Surely this was proof enough that rigorous and formalized military training was not required in order to fight effectively. It was these amateurs who had made such a fine impression on Pershing in 1918. 'The alertness and confidence of these neighbors of ours,' he recorded, 'were admirable' and one 'soon caught the spirit of that superb corps.'[41] What McNaughton had obviously missed, however, was that Currie and the other amateur Canadian commanders had gone about their business in a 'professional' manner. As William A. Stewart noted, the Canadians had to be 'shocked by several sanguinary savagings' before they realized that the 'amateur Militia approach was wildly incompatible' with the realities of the Western Front.[42]

By May 1935 it was clear that Bennett considered McNaughton a political liability because of the unemployment relief camps. The prime minister pressed him to accept a new position: President of the NRC. Formed in 1916, the NRC coordinated the nation's scientific and industrial research. As alluring as a scientific post must have been for him, McNaughton did not vacate his post as CGS without a fight, though he had already received one extension. He told Bennett he would 'much prefer' to remain as CGS because he could 'be of greater service to the state' there. But after consultation with Grote Stirling, the Minister of National Defence, McNaughton relented. At his new post, he quickly established the NRC as a leader in Canadian scientific research. During his tenure he launched several lines of war research that were destined to bear fruit by 1939.[43] According to Lieutenant-General Elliot Rodger, nothing gave McNaughton 'greater pleasure' than working with the scientists. C.J. Mackenzie, who replaced McNaughton at the NRC when the war started, stated that his time on the council was 'the most tranquil period of his life and the most happy.'[44]

When Prime Minister William Lyon Mackenzie King's government decided in September 1939 that one division would be sent overseas to aid Britain, Norman Rogers, the Minister of National Defence, suggested that McNaughton be recalled from the NRC to command it.[45] King welcomed the suggestion, and on 6 October he and Rogers interviewed McNaughton, who quickly indicated that Canada's war effort should focus on production and that 'every effort should be made to arm and equip the troops to spare human lives.' This was precisely what King, who was obsessed with the spectre of another conscription crisis if Canadian casualties soared, wanted to hear. King recorded in his diary that McNaughton's eyes 'filled with tears and the right side of his mouth visibly twitched as he spoke of what the responsibility meant.'[46] This was a not altogether uncommon display for commanders charged with the power of life and death.[47]

King considered the interview 'as deeply moving as any I have witnessed in my public life,' and he judged McNaughton 'the best equipped man for the purpose.' He saw McNaughton again in early December and, though still convinced that he was the best choice for command, hedged his bet somewhat. 'I felt a little concern,' King wrote, 'about his being able to see this war through without a breakdown. I felt he was too far on in years to be taking on so great a job. Having been through the strain of a previous war he and many others like him might find they had not the endurance that they believed they had.'[48]

The other possibilities for division command in 1939 were few, and much has been made of the fact that the Permanent Force consisted of a mere 450 officers at the outbreak of the war, only half of whom were fit for field duty.[49] Only Brigadier-General George R. Pearkes – a Victoria Cross winner, a Permanent Force officer, and commander of Military District No. 13 – would have been a serious alternative to McNaughton in 1939. Many Canadian military historians have little sympathy for Pearkes: they have accepted Montgomery's criticism of him, voiced in 1942, as unassailable evidence of his command deficiencies.[50] This judgement is too harsh and superficial.

Pearkes had attended the IDC in 1937 and while there had been intimately involved with the month-long army manoeuvres in East Anglia between the 1st and 2nd Infantry Divisions. While McNaughton was labouring at the NRC, Pearkes was gaining first-hand experience of the possibilities and problems of divisional command. Reginald Roy has pointed out that by the end of the summer of 1937 Pearkes's experience with large bodies of troops on manoeuvre 'was far greater than [that of] any other senior officer in the Canadian permanent or non-permanent militia.'[51] Moreover, he had been to the field to oversee militia training at the platoon and company levels in the summer of 1938.

It is a sad fact, however, that none of the senior officers of the 1st CID that sailed for England in late 1939 had exercised so much as a battalion on manoeuvre since 1918, including McNaughton: a period of twenty-one years.[52] Contrast this with Alexander, who had commanded the Irish Guards from 1928 to 1930, the Nowshera Brigade of Northern Command in India from 1934 to 1938, and the 1st Division from 1938 to 1940. Montgomery had commanded the 1st Battalion, Royal Warwickshires, from 1931 to 1934, the 9th Infantry Brigade from 1937 to 1938, and the 8th Infantry Division from 1938 to 1939. Brooke had commanded the 8th Infantry Brigade for eighteen months and led it on several manoeuvres in 1934, and in 1937 he had assumed command of the Mobile Division on Salisbury Plain.[53]

Before the United States entered the war, Patton (and many other senior American commanders) had taken part in the large-scale Tennessee, Louisiana, and Carolina Manoeuvres of 1940 and 1941. When his 2nd Armoured Division left Fort Benning, Georgia, to commence the Tennessee exercise, it was in two columns each sixty miles long, and he practised moving it in the dark, without lights, and on radio silence.[54] Being able to move large bodies of men and machines efficiently was to be a professional minimum in the Second World War. Burns himself knew this, stating in a 1935 article: 'The mechanics of military leadership is the calculation of time and space.'[55]

Other scientifically gifted commanders of the past, such as Napoleon and Robert E. Lee, had chosen not to rest on their natural mathematical and engineering abilities. It appears that this is precisely what McNaughton *did* do after 1918. He was a man of great conviction and was certain that a citizen army backed by the latest technology would be sufficient to win the next war. Yet for all his intellectual brilliance he allowed himself to be blinded to the fact that the technological changes of the interwar years would not permit a citizen army the necessary time to 'figure out' the battlefield of the next war, especially if it was thrown into the breach from the outset.

Table 2.2 McNaughton's relative interwar command experience

	Bn	Reg	Bde	Div	Corps
Alexander, Harold	1922	1927	1934	1938	1940
*Anderson, Kenneth	1929	–	1937	1940	1941
*Auchinleck, Claude	1929	–	1933	–	1940
Bradley, Omar N.	–	–	–	1942	1943
Brook, Alan F.	–	–	1934	1937	1939
Crerar, Harry D.G.	–	–	–	1943	1943
*Cunningham, A.G.	–	–	1937	1938	1941
Dempsey, Miles C.	1938	–	1939	1941	1943
Eisenhower, D.D.	1940	–	1919	–	–
Gort, John S.V.	–	–	1930	–	–
Hodges, Courtney	–	–	–	–	–
Ironside, William E.	–	–	–	1926	–
Leese, Oliver	1936	–	–	1941	1942
McCreery, Richard	–	1935	1940	1940	1943
*McNaughton, A.	–	–	–	1939	1940
Montgomery, B.L.	1931	–	1937	1938	1940
Paget, Bernard C.T.	–	–	1936–37	1940	–
Patch, Alexander M.	–	1942	–	1942	1942
Patton Jr., George S.	1920	1938	1919/1940	1941	1942
Ritchie, Neil	1938	–	–	1940	–
Slim, William J.	1938	–	1939	1941	1942
Simpson, William H.	–	1925	–	1941	1942
*Wavell, Archibald	–	–	1930	1935	1938
Wilson, Henry M.	1927	–	1936	1937	1939

* Indicates those relieved of army command during the war.

At various times during the interwar years McNaughton hinted that he understood the requirement for the type of serious doctrinal inquiry exemplified in the writings of Burns, Simonds, Montgomery, and others. But there is little direct evidence that he successfully made the mental leap from a gunner's perspective to visualizing how armoured units in the next war would actually move, attack, and defend and cooperate with the other arms at the tactical and operational level. McNaughton, though he wore an officer's uniform, was an amateur commander on the verge of taking the Canadian army into a professionals' war on the continent of Europe.

PART TWO

The Problem of Deploying the Army

3

A Willingness to Fight, 1940–1941

The tasks of the Canadian field army developed in a manner which no one foresaw.

Charles P. Stacey

With the exception of Hong Kong and Dieppe the Canadian Army saw no combat during the first three years of the Second World War. A division and an army tank brigade fought in Sicily and Italy in 1943, but the bulk of the army was not committed to battle until 1944. By that time the British, Australians, Indians, South Africans, New Zealanders, and even the Americans had been heavily engaged for years. The blame for this unfortunate turn of events has consistently fallen on McNaughton. J.L. Granatstein, one of Canada's most prominent historians, has perpetuated this school of thought, arguing that McNaughton was a 'prickly nationalist' who intentionally hoarded the army in anticipation of the cross-Channel invasion in order to satisfy his deep-seated urge for autonomy. Granatstein is not alone in his view.[1] Yet even a superficial examination of the war's course suggests that there were deeper reasons for the long delay in committing the Canadian Army to battle. McNaughton was not the main impediment to deploying the army (or parts thereof) on operations with the British Army between 1940 and 1943. Current thinking on this issue grossly misrepresents the complex circumstances in which he commanded the army during the early years of the war. His decisions need to be placed in the context of the multilayered personal, political, and strategic calculations predominating at the time within the War Office and in Ottawa.

McNaughton's perspective on employing the army was fundamentally different from Mackenzie King's. The idea of once again raising great expeditionary forces to fight in France terrified King, to such an extent that he penned a 'deep sadness' at the prospect of the 1st Canadian Division moving into a 'fiery furnace to be devoured whole.' Heavy casualties on the continent might easily trigger conscription – a policy decision that had divided the country during the First World War.[2] This is why King jumped at Prime Minister Sir Neville Chamberlain's proposal for a joint pilot training plan in Canada in late September 1939, and why he expressed profound disappointment that the proposal had not been made sooner so that Canada would not have to 'head so strongly into expeditionary forces at the start.'[3]

McNaughton, for his part, was totally committed to the idea of fighting in Europe, even though the CGS at the time, Major-General Thomas Victor Anderson, along with the other service chiefs, had stated in a memorandum of 29 August 1939 titled 'Canada's National Effort (Armed Forces) in the Early Stages of a Major War' that the idea of fighting beside other imperial forces had been given 'a secondary and incidental consideration' in Canadian military planning.[4] McNaughton went so far as to warn Norman Rogers, the Minister of National Defence, that if the Canadian government pursued a policy that did not specifically involve 'a fighting unit in the line of battle' it would 'have difficulty remaining in office.'[5] Yet McNaughton also cherished the virtues of a volunteer army and told King so during their October 1939 meeting. Even so, he sought to mitigate the need for conscription, not by minimizing the army's exposure to combat, but by effectively applying science to reduce casualties. Churchill held a similar conviction, maintaining that the war could only be won through 'scientific leadership.'[6]

McNaughton wanted to fight, but he also worried – for good reason – that Canadian units and formations might be siphoned off piecemeal by British commanders. There was a strong possibility that colonial units and formations would be overused. This might well lead to heavy casualties. He had been in England for only a short time when he informed Major-General Sir Archibald E. Nye, the Director of Staff Duties (DSD) at the War Office, that the chief requirement from a Canadian perspective was to make the Canadian contingent a coordinated organization rather than a collection of individual units.[7] McNaughton's strong sympathy for Currie's struggles during the First World War suggests that his position was inevitable; that said, he was also reflecting the position of King and the Cabinet War Committee (CWC), which would declare a few months later that Canadian forces overseas 'should be as autonomous as possible.'[8] McNaughton's initial problem was finding feasible military operations to participate in with the rapidly expanding British Army. There was no easy solution, and the dilemma would persist for the next three years.

As early as the German invasion of Norway in early April 1940, McNaughton demonstrated a sincere willingness to consider any and all requests from the War Office for Canadian forces to join the fighting. On 16 April, Major-General Richard H. Dewing, the Director of Military Operations (DMO) at the War Office, formally asked McNaughton to provide Canadian troops for Operation MAURICE, a hurriedly planned operation against Trondheim scheduled to take place only nine days later.[9] At that time, almost all trained British troops were in France or were already earmarked for Norway. Under the stipulations of the Visiting Forces Act, McNaughton placed the 2nd Infantry Brigade 'in combination' with British forces for participation in the operation, soon renamed HAMMER and commanded by Major-General Sir Adrian Carton de Wiart, V.C.[10]

Fortunately the War Office cancelled the operation on 20 April because of the operational risks involved. The Canadians might have been tasked to neutralize the Trondheim forts by frontal assault, and the CIGS, General William E. Ironside, frowned on the idea. There was 'no possible landing place,' he observed, because of 'mountainous cliffs running straight down into the water.' Churchill, First Lord of the Admiralty at the time, referred to the entire Norwegian campaign as a 'ramshackle' affair and only grudgingly supported HAMMER. Carton de Wiart later reflected: 'I felt in my bones that the campaign was unlikely to be either long or successful.'[11] Charles P. Stacey, who idolized McNaughton, was correct when he observed that his willingness to commit to 'so desperate a venture' as HAMMER proved that 'the long period in which the Canadians took no part in active operations was not the result of any reluctance to embark upon dangerous projects.'[12]

It has been suggested that McNaughton embraced HAMMER out of a pure fit of enthusiasm.[13] Yet it is difficult to see how he could have refused this initial request to help the British. Granted, he probably downplayed the risks involved; the point, however, is that refusal most surely would have soured relations with the War Office at the very beginning of the joint Anglo-Canadian venture against the Germans. McNaughton believed firmly that the real 'acid test of sovereignty' was the control of one's own armed forces. He has generally been portrayed as an ultra-constitutionalist at this time; even so, he was no obstructionist once the shooting started. It is almost inconceivable that he would have refused to participate in HAMMER.

Early in the war McNaughton complained to Lester B. Pearson of the High Commission in London: 'How can I fight the war if I have to worry about the Statute of Westminster?' Pearson believed that McNaughton was 'as anxious as any nationalist politician to maintain the Canadian identity of his army, and to ensure that its control would remain Canadian [but he] always and rightly made the reservation of military necessity.'[14] Ironically, as the war progressed with the Canadians sitting in Britain, it would become more – not less – difficult for McNaughton to refuse risky operations.

When McNaughton first arrived in England in late 1939, Pearson noted: 'He'll be stepping on lots of toes before long in his restless zeal, his driving imagination, and his insistence on efficiency rather than "spit and polish."'[15] Yet his 'can do' attitude earned him the respect of the senior British leaders during the Norwegian campaign. Oliver Stanley, the Secretary of State for War, praised him for the 'splendid way' in which he had 'responded to our invitation to take part in this operation,' while Dewing recorded in his diary that 'Andy played up magnificently.'[16]

Praise also came from Ironside, who penned a brief note of gratitude to McNaughton in which he expressed his conviction that the experience 'augurs well for co-operation between British and Canadian Forces in the future. I feel that it is owing to your personal interest in the matter that we were able, without getting involved in legal arguments, to come to a swift decision.' McNaughton replied: 'With common sense, mutual confidence, and prompt decision, the legalistic problems involved in close co-operation can be solved.'[17] As early as December 1939, McNaughton had expressed his sincere faith that the Canadians and the British would be able to work together. He told Eric Hutton of *Star Weekly* that since Canadian officers had regularly attended British staff colleges, 'we have gained the priceless advantage of knowing each other well, of organizing our forces the same way, of writing our orders in identical manner.'[18] Ironside was a gunner like McNaughton and had been GSO 1 in Major-General David Watson's 4th Canadian Division during the First World War. Ironside had been somewhat confounded early in 1940 by McNaughton's views on command, control, and training, but their friendship from the previous war facilitated understanding. McNaughton later maintained: 'We were very close personal friends and he certainly knew that he'd get action from me, as he did.'[19]

No such bonds of sympathy existed between McNaughton and James

L. Ralston, the Canadian Minister of Finance, during the Norwegian crisis. Ralston, the Acting Prime Minister while King was in the United States visiting President Franklin Delano Roosevelt, used this opportunity to rekindle his long-running personal and professional quarrel with McNaughton. Ralston took great exception to the fact that McNaughton had authorized Canadian participation in HAMMER on his own initiative without first getting permission from Ottawa.[20]

McNaughton was certain that he had acted legally because he had solicited the opinion of the Deputy Judge Advocate General, Brigadier-General Price Montague, before authorizing participation.[21] McNaughton was furious at Ralston and exploded to Pearson about politicians 'trying to run the war while 3000 miles away.' H.A. 'Sandy' Dyde, who accompanied Rogers to England, reflected: 'It was obvious to several of us ... [that] Ralston was clearly the object of his [McNaughton's] resentment.'[22]

Rogers, the Canadian Defence Minister, had only just arrived in Britain. He and Vincent Massey, the Canadian High Commissioner in London, firmly supported McNaughton's position against Ralston's claim that McNaughton had no legal basis to authorize deployment beyond the British Isles. In his diary of 20 April, Rogers recorded: 'It was quite evident from the recital of the events which had occurred before my arrival in London that the matter was of extreme urgency and if Canadian troops had not been made available immediately the War Office would have been obliged to seek troops elsewhere ... The movement itself had to be carried out with the utmost secrecy ... This fact, no doubt, made it more difficult to reproduce through cable dispatch the actual situation under which General McNaughton was obliged to accept or reject the proposal made by the War Office.'

Massey noted in his diary that the government 'is wrong in thinking that the matter could have waited until the Cabinet deliberated. Military measures must take place with a speed the Cabinets know not.' On 22 April, Rogers replied to Ralston that there were 'dynamic features in present military situation which argue against too rigid limitation.'[23]

There had been a time when Ralston and McNaughton seemed well disposed to like each other, considering their common First World War experiences. Ralston, six years older than McNaughton, had served in the war as a major and adjutant in the Nova Scotia Highlanders and had fought on the Somme as well as at Vimy Ridge, Hill 70, and Passchendaele. In June 1917 he was wounded and awarded the Distinguished Service Order (DSO). In April 1918 he took command of the 85th Battalion, suffered another wound at Amiens, and was recommended for the Victoria Cross (he did not receive it).[24] It had been Ralston, during his first tenure as Minister of National Defence in the late 1920s, who recommended McNaughton for the position of CGS. But even before McNaughton took up that appointment in January 1929, Ralston had strongly censured him for exceeding his authority on policy matters.[25] Soon after, they butted heads so sharply over McNaughton's advocacy of a Dominion Arsenal that McNaughton considered resigning and probably would have had the 1930 federal election not removed Mackenzie King's Liberals, and Ralston, from power.

Canadian journalist Grant Dexter observed at the time that Ralston felt inferior in the presence of military men, despite his own military experience. If this is true, it does much to explain his tendency to micromanage.[26] King, himself a stubborn man, noted: 'There is something inhumanly determined about [Ralston's] getting his own way, regardless of what the effects may be on all others.'[27] Ralston found it all too easy to immerse himself in administrative detail, and this translated into a failure to delegate authority. This could only lead to trouble with McNaughton, who was too self-confident to tolerate close oversight.

There was little love lost between McNaughton and Ralston, and fate quickly conspired to bring the two into even closer contact when on 10 June 1940 Rogers died in a plane crash near Newcastle, Ontario, and Ralston became Minister of National Defence. McNaughton was saddened at Rogers's death: 'There was a feeling of mutual confidence between us. I had no confidence in Ralston [who] was very, very small potatoes and I had suffered from him.'[28] After the war, Major-General Daniel C. Spry, McNaughton's one-time Personal Assistant (PA), stated that after Norway McNaughton and Ralston were like 'two dogs sniffing at one another'; Massey described the two as 'oil and water.'[29] Pearson recorded in his diary at the end of December 1939 that McNaughton was 'explosive' in his criticism of several individuals on both sides of the Atlantic and 'damns people like Ralston.'[30] The personal animosity did not work to McNaughton's advantage as tensions mounted over the army's idleness.

Many of the problems that eventually afflicted both McNaughton and the army might have been mitigated had he taken his division to France to serve under a British corps as originally planned.[31] The precipitous collapse of the French Army in May 1940 reduced McNaughton's role to that of a rover looking for a suitable place (at Ironside's request) where the Canadians could help secure the lines of communication of the

British Expeditionary Force (BEF). After conducting a personal reconnaissance in France over 23 and 24 May, McNaughton returned to London and made a painful judgement: the single Canadian brigade group selected to take part could make no real difference in the larger picture. So he argued. When Lord Gort, the BEF's hard-pressed Commander-in-Chief (C-in-C), requested Canadian reinforcements at Dunkirk a few days later, McNaughton squared his jaw and made it clear that he would do whatever was asked of him. He demanded only that he be allowed to take his artillery. In the event, ANGEL MOVE – the deployment of the Canadians to France – was proposed and dropped three times, highlighting the chaotic predicament of the BEF's withdrawal to Dunkirk.[32] Dewing described McNaughton's situation as 'extraordinarily difficult' but thanked him for being 'absolutely determined to do thoroughly whatever might be asked of you.'[33]

Operation DYNAMO, the evacuation of the BEF from Dunkirk, should have been the end of the calamitous battle for France. However, McNaughton was quickly drawn into Churchill's wild, politically driven scheme of establishing a bridgehead in Brittany to keep the French government fighting. A second BEF, built around the British II Corps and commanded by Lieutenant-General Alan F. Brooke, made its way across the Channel in mid-June. Brooke, deeply sceptical of the operation's feasibility, quickly decided it was a lost cause and ordered his forces – which included a brigade from McNaughton's division – back to Britain. From McNaughton's perspective the entire 'Brittany Redoubt' operation was a farce. After the war he would declare that had he known of its political impetus 'I would have put my foot down and classified this expedition as *not* a practical operation of war.'[34] Like Lieutenant-General Bernard C. Freyberg, General Officer Commanding (GOC) of the New Zealand Expeditionary Force, and like the other Dominion military leaders, McNaughton possessed the authority to intervene should he feel that actions taken by British commanders unnecessarily risked the lives of Canadian troops.[35] McNaughton took that authority seriously and did not wield it lightly.

The new CIGS, Major-General Sir John Dill (who had replaced Ironside on 27 May), expressed his gratitude for McNaughton's help in Brittany on 21 June. 'I cannot tell you,' he wrote, 'how much I regret all the disappointment you have had ... It is all rather a sorry tale but I am sure in the circumstances you will realize what our difficulties were and will forgive us for all the inconvenience you have been caused.'[36] This was the language of friends, and indeed, McNaughton counted Dill (as he had Ironside) among his close associates from the First World War. In the first half of 1940, McNaughton had done all that had been asked of him; only the fortunes of war had thwarted his good intentions.

Britain's haphazard forays against the enemy in the first half of 1940 soon gave way to preparations to defend against a German invasion. Dewing told McNaughton on 27 May, even before the Brittany operation, that opportunities to employ the Canadians would come 'soon enough, but it must come in circumstances in which it can play a sound military role.'[37] That was the crux of the matter; only four British divisions would fight against the Germans in the two years following Dunkirk.[38] Dewing was merely trying to sooth McNaughton's frustrations, however, for according to Colonel Maurice Pope, Crerar's senior staff officer at Canadian Military Headquarters (CMHQ), Dewing also declared in the fall of 1940 that it would be impossible to return to the continent 'for a very long time to come.'[39] This simple and inescapable fact places the other options available for Canadian deployment after June 1940 in their proper context.

In exploring the issue of employment for the Canadians, one is immediately struck by one historian's observation that a 'strange paralysis of will regarding the use of the Canadian Army seems to have gripped Canadian military and political leaders' after the fall of France. Other historians have asked why Canadian formations were not sent to the Middle East or North Africa between 1940 and 1942.[40] At first this query seems logical enough, given that Churchill had no choice but to pursue a peripheral strategy after the fall of France and Italy's declaration of war on 10 June. In a 3 September 1940 memorandum he stated categorically: 'The only major theatre of war which can be foreseen in 1940/41 is the Middle East.'[41] Traditionally the 'Middle East' denoted the area bounded to the west by the Mediterranean and the Red Sea, to the southeast by the Arabian Sea and the frontiers of India, and to the north by the frontiers of Turkey and Russia. The 'Near East' delineated the lands bordering the eastern Mediterranean, including Egypt and those lands that in 1914 had constituted the Ottoman Empire.[42] However, in August 1942 Brooke would divide the Middle East Command into a Middle East that included Egypt and the Mediterranean and a Near East that encompassed Palestine and Persia. McNaughton has been blamed for not deploying the army to the Middle East, but the facts simply do not support such an accusation. The Canadian Army never left Britain between the middle of 1940 and the middle of 1943 for a multitude of complex reasons. The best way to understand the issue is by breaking it down into two distinct periods: the defence against invasion of 1940 and 1941; and the assault preparations of 1942 and 1943. The second period will be discussed in the next two chapters.

The most obvious reason – and it was a compelling one – why the Canadians did not see the shores of the Mediterranean in late 1940 was that Churchill and the British Chiefs of Staff (COS) wanted and needed the Canadians to defend Britain. Clearly, the British Army's first priority was the defence of the British Isles.[43] After Dunkirk the Canadians were a highly valued commodity, and by late July, McNaughton had taken command of the British VII Corps, charged with repelling an invasion. It seems that at the time he was content with the command of VII Corps, for it allowed him to sidestep Ralston. He later reflected: 'I had none of this trouble about having to work around and having to get permission to blow my nose from Ralston and that psychological thing, you couldn't tolerate it, you see, it got beyond endurance ... That was the real reason why they set up a British Corps.' In this, McNaughton was only partly right, for Crerar had been urging his friend, Lieutenant-General Dick Haining, Vice Chief of the Imperial General Staff (VCIGS), to place McNaughton in command of a combined Anglo-Canadian corps.[44]

By September 1940 the 2nd Canadian Infantry Division (CID) had assembled in Britain to bolster the defences. Fears of invasion were acute at the time, and they never completely disappeared, not even after ULTRA intelligence indicated that the Germans were about to invade the Soviet Union.[45] Moreover, even while the threat of invasion remained, the British Army was still trying to deploy more of its own divisions to active theatres. Further evidence of the predicament faced by McNaughton and the Canadians comes from November 1940, when Ralston and Crerar came to Britain to discuss the army's future role and composition. Prior to their arrival a confidential memorandum had circulated through the War Office touching on the subject of how to employ the Canadians. In turn, it generated a new memorandum titled 'Possible Employment for Canadian Formations.' This one identified four possible courses of action: (1) send them to the Middle East, (2) designate the Canadians as a cross-Channel invasion corps, (3) use them in irregular operations, and (4) Home Defence.[46] The logistics involved in the first option undermined it from the start, and the possibility of crossing the Channel seemed remote in late 1940. The idea of employing the Canadians in irregular operations was summarily dismissed. That left Home Defence.

Though defending England was the most logical thing for the Canadians to do, McNaughton remained open to other possibilities. At the end of June 1941, Colonel Eedson Burns, Brigadier General Staff (BGS) of the Canadian Corps, presented him with the idea of deploying perhaps a division to North Africa. According to Burns, McNaughton was 'very interested' and sent him to the War Office to test the waters, but General John Noble Kennedy told him it was 'not too encouraging' because there were 'a good many formations ahead of you in the queue.' Burns added: 'Nothing came of this rather informal and confidential negotiation.' However, he also spoke with Brigadier Aubertin W.S. Mallaby, Deputy Director of Military Operations (DDMO), who was under the impression that Canadian troops were only available for service in Britain.[47] Burns quickly set him straight; what is not readily apparent is whether Mallaby's conception was the product of McNaughton's position or the recommendation of the War Office memorandum circulated in late 1940. Certainly, at no time between 1940 and 1942 did the British COS or Churchill show any real interest in seeing the Canadians leave Britain.

In mid-August 1941, King travelled to Britain, where Churchill informed him that though the Canadians 'would be having fighting before the war was over ... his whole point of view was the security which their presence would give to the island.' According to King, Churchill hoped the Canadians 'would be able to hold out in England [and] really wanted them for the British Isles.'[48] A few days later King told Canadian troops at Aldershot that it was the War Office, not Ottawa, that was keeping them in Britain. On 27 August *The Times* quoted him as saying that the Canadians could be employed anywhere. Every time the issue was raised, however, the British demurred.[49]

Though the British enjoyed a few notable successes up to the time of King's arrival in Britain – the Royal Air Force (RAF) had won the Battle of Britain by mid-September 1940, and the Royal Navy had exacted revenge for the sinking of *HMS Hood* by sinking the *Bismarck* on 27 May 1941 – British fortunes were still at a low ebb. General of Panzer Troops Erwin Rommel and his soon to be famous Afrika Korps began pushing back the British in North Africa in March 1941. By June the British Army had been pushed out of Greece and had lost the island of Crete. Even the drawing off of German might into the Russian vastness beginning on 22 June 1941 did little to change British views about what to do with the Canadians. On 6 November 1941 Ralston reiterated King's position to David Margesson, the Secretary of State for War; Margesson replied that the Canadians were needed to defend Britain.[50] A pattern was readily apparent by this time: Canadian officials would routinely declare that Canadian troops could be used anywhere; the British authorities would smile, say thank you, and then declare that the Canadians were needed in Britain.[51] This consistent British position cannot be attributed exclusively to McNaughton's lobbying to keep the army united.

Another important factor working against Canadian deployment to the Mediterranean was Churchill's acute concern about the ratio of British to imperial divisions engaged against the Axis in North Africa. On 17 September 1941 he told Lieutenant-General Sir Claude Auchinleck, commander of Eighth Army: 'I have long feared the dangerous reactions on Australian and world opinion of our seeming to fight all our battles in the Middle East only with Dominion troops.' During Operation CRUSADER in late 1941, Lieutenant-General Alan Cunningham's order of battle included the British 7th Armoured Division, the 70th Division, the 1st Army Tank Brigade, and the 22nd Guards (Motor) Brigade, but also the New Zealand Division, the 4th Indian Division, the 1st and 2nd South African Divisions, the 6th South African Armoured Car Regiment, and the Polish Carpathian Infantry Brigade Group.[52] From the above it is clear that Churchill's concerns about perception were entirely justified. He was keenly aware that he needed to free himself 'from the imputation, however unjust, of always using other people's troops and blood.' So he pressed the COS to send only British divisions to the Mediterranean.[53]

A third impediment to sending Canadian formations to North Africa was the simple fact that there was never any compelling operational requirement for Canadian forces there. Between December 1940 and February 1941, during Operation COMPASS, the first British offensive of the war, Lieutenant-General Sir Richard O'Connor's 30,000-man Western Desert Force had defeated Marshal Rodolpho Graziani's 100,000-man Italian Tenth Army.[54] British fortunes then plummeted with the arrival of Rommel and the Afrika Korps in February, but for most of 1941 and 1942 there were never more than three German divisions in North Africa, and ULTRA revealed serious German logistical limitations.[55] Churchill's great trump card, however, was the firm support of Roosevelt. 'I know of your determination to win on that front,' the President cabled Churchill in early May 1941, 'and we shall do everything that we possibly can to help you do it.'[56] American material support was more urgently sought after than reinforcement by Canadian troops. Churchill was content with the size of the British force in the Mediterranean but risked Operations TIGER and TIGER No. 2 to urgently ship tank reinforcements to North Africa in April and July 1941. Physically getting a Canadian division to North Africa was not a simple task.

The reasons so far cited as to why the Canadians did not go to the Middle East were ultimately overshadowed by King's profound reluctance to send the army anywhere near a battle zone. While he stated publicly that there were no restrictions on its deployment, such pronouncements did not reflect his true position. As early as 1 October 1940 he recorded that Major-General Harry Crerar, sent back by McNaughton to be CGS in July, wanted a Canadian army fighting in the Middle East, Africa, or elsewhere. 'I shall have to watch this particularly,' King added, and he did.[57] That same month, Britain's Foreign Secretary, Anthony Eden, raised with Mr James G. Gardiner (Canadian Minister of Agriculture and National War Services) the possibility of Canadians deploying to Egypt during Gardiner's trip to Britain.[58] This prompted Ralston, before heading off to Britain for his own visit, to seek the CWC's approval to approach the War Office about the possibility of Canadian troops serving in Egypt. The CWC – meaning ultimately King – declined to take the carrot.

Crerar does not seem to have been aware of the CWC's position, for on 4 December he asked Dill what the chances were of a Canadian formation going to the Near East. Dill gave the standard War Office reply: the Canadians were to be employed 'nearer home.' Crerar then offered that as far as he was aware, Ottawa was open to any possibilities 'no matter where the theatre might be.'[59] This was simply not true, for two days later King had a telegram sent to Ralston that specifically stated: 'If troops are being sent to the Near East they should be sent from the parts of the Commonwealth which control policy in the Near East or which are more geographically concerned with the Near East.' King reflected: 'I thought the logical thing was to have Canadians continue to defend Britain, our position being that we were at the side of Britain, and not to begin to play the role of those who want Empire war.'[60] In subsequent conversations with Churchill in December, Ralston impressed on him King's true position – that any request for Canadian troops to serve outside Britain would have to originate in Canada. Churchill accepted this condition.[61]

When King said that requests had to originate in Canada, he really meant that they had to come from him. This was apparent in May 1941 when Charles G. 'Chubby' Power, the Minister of National Defence for Air, once again suggested sending a brigade to Egypt. King immediately cut him off at the knees: 'I would not countenance anything of the kind ... I do not think we should interfere with the disposition of troops, when our policy was that of allowing the High Command to make whatever disposition was thought most effective.' King had in effect established a litmus test: Canadians were not to be sent to Egypt (and it must be assumed that his thinking extended to other possible theatres) 'unless we could be shown that the argument there *far* [emphasis added] outweighed their remaining in Britain.'[62]

King was determined to undermine all efforts to deploy even a small force such as a brigade. Meanwhile, McNaughton found himself in a curious predicament. Opportunities for deployment were few and far between, yet the Canadian forces were still growing. From the beginning McNaughton had adopted a slow approach to building up first the Canadian Corps and then the army – slower, in fact, than the War Office and Canadian officials such as Rogers and Crerar would have liked.[63] McNaughton told Rogers in April that he had 'at no time pressed for the formation of a Canadian corps,' even though he thought that a corps was the 'smallest organization through which the Canadian forces in the field could be effectively administered and fought.' He made it perfectly clear that it was 'much more effective to have a smaller formation promptly maintained to full establishment than a larger formation under strength.'[64] The influence of Currie's obstinate refusal to add more divisions to the Canadian order of battle during the First World War merely to have an army (which would also have diluted the strength of the individual divisions) is unmistakable in McNaughton's attitude. Even during the discussions that led to the establishment of the Canadian Corps on Christmas Day 1940 (based on VII Corps, which McNaughton had commanded since July), he made sound arguments against forming it at the time.[65]

It is now beyond doubt that it was Crerar who pushed the idea of a big army. Crerar feared minimal post-war political influence if only small expeditionary ground forces were raised; also, he genuinely believed that large ground forces were required to defeat Germany. In determining the actual size of the Canadian Army he relied on the War Office's strategic plan to raise fifty-five divisions, fourteen of them from the Dominions.[66] Even before the Canadian Corps was formed, however, Crerar probed McNaughton on the idea of forming an army when there was no compelling need for it. On 9 September he told McNaughton: 'Shortly after the Canadian Corps is formed, and a going concern, your elevation to Army Command would, I believe, give similar satisfaction to Canada, if that is of any interest to you.'[67] McNaughton never replied, nor was he inclined to give Crerar's renewed proposal of 11 August 1941 any considerable attention.

After assuring McNaughton that there was sufficient manpower to field and sustain eight divisions (six deployed overseas) for more than six years, Crerar asked: 'Have you ever considered the pros and cons of a Canadian Army comprising 2 Corps each of 2 Divisions and an Armoured Division? I fully admit that this is a pretty ambitious proposal because the necessary increase in Corps, etc., troops will be fairly heavy. At the same time, I do not think that the picture is an impossible one.'[68] Perhaps McNaughton did not think it impossible, but he was far from keen on the idea. He did not respond to the specifics of Crerar's argument because he was anticipating a fuller elaboration of the army's position in 1942 from the War Office. He told a *Toronto Globe and Mail* reporter on 29 September 1941 that the possibility of forming an army of two corps 'cannot be considered until it is known precisely how many men are available.'[69]

In July 1941, McNaughton again reiterated his understandable concern for matching Canadian resources to tasks already committed to.[70] By the end of the year there were 124,472 Canadians in Britain comprising three infantry divisions, an armoured division, and an army tank brigade. Clearly, McNaughton was correctly thinking long-term towards the eventual cross-Channel invasion; but in the short term an obvious question was, where was this growing force supposed to go? A better question is how many divisions would have had to deploy somewhere to protect McNaughton from the charge that he would not countenance the breaking up of his army? If he sent off one division, the bulk of the army would be left sitting in England.[71] Finding a legitimate role for an entire corps was not easy. In 1944 the British even baulked at the idea of a Canadian Corps in Italy at a time when there were far more German divisions to deal with there than had been the case in North Africa. On 29 September 1941, McNaughton told the *Globe and Mail* that the Canadian Corps 'is a dagger pointed at the heart of Berlin – make no mistake about this.' It was great copy, but the Canadian divisions were certainly not headed for Berlin any time soon. Neither he nor Crerar had sufficient strategic information to predict with any accuracy where the army might fight or what the environmental requirements might be.[72]

To summarize, the Canadian Army never deployed beyond the shores of Britain before 1942 for four main reasons: (1) Churchill clearly wanted the Canadians to defend Britain, and the COS concurred; (2) he did not want them in North Africa at a time when British forces there were already outnumbered by imperial formations; (3) there was no compelling operational requirement for the Canadians to reinforce the Allies there, and in any event it would not have been easy to fit Canadian formations into existing British operations; and (4) perhaps most decisively, King was working hard to undermine any attempt to get the Canadians into action. Thus there is plenty of evidence that throughout 1940 and 1941, McNaughton was hardly the biggest obstacle to deployment of the army, notwithstanding what many historians have contended.

4

From *ROUNDUP* to *TORCH*

> I thought I must have more complete and timely information of plans as they developed.
>
> McNaughton, 17 September 1942

The prospects for getting the Canadian Army into action did not significantly improve during 1942. The entry of the United States into the war, brought about by the Japanese attack on Pearl Harbor on 7 December – and Hitler's declaration of war three days later – was a welcome addition to the Allied cause, but it threw the proverbial 'monkey wrench' into strategic calculations as far as the Canadian Army was concerned. Under the RAINBOW 5 plan, the United States was committed to a 'Germany first' strategy, but exactly when and how the Western Allies would come to grips with the Germans became a giant game of chess, and unfortunately the Canadians were assigned the role of pawns in the short term. Without the Americans' entry, however, McNaughton's 'dagger pointed at the heart of Berlin' would never have been unsheathed in France.

The CCOS was established at the ARCADIA Conference in Washington in late December 1941, but King was not informed about this meeting until well after the fact.[1] Entry into this exclusive club was virtually impossible – all the more so because King had already decided to leave the war's strategic direction to Churchill.[2] In this new strategic context McNaughton had to rely on a top-down flow of information. This was the normal method of military command and control, but the flow of information was inconsistent and often delayed. As a consequence, his ability to forecast operational requirements for the Canadian Army suffered.[3]

Perhaps the most important change at the beginning of 1942 was that Brooke had replaced Dill as CIGS (unofficially on 1 December, officially on 25 December 1941) and now held the most powerful military post in Britain. McNaughton reflected that the 'kindly, co-operative, mutual confidence that had existed under Dill became a thing of the past.'[4] First Rogers had been replaced by Ralston; now Dill had been succeeded by another long-time rival. The tension between McNaughton and Brooke had grown throughout the preceding two years, and it is important to trace this development in some detail because it led to a misperception on Brooke's part regarding McNaughton's willingness to see the army divided.

The personal friction between McNaughton and Brooke dates back to their association in the Canadian Corps during the First World War, though its origins are much better documented from McNaughton's perspective than from Brooke's.[5] There is evidence on the Canadian side that they clashed over professional issues. For his success in designing the 4th Canadian Division's artillery plan for Vimy Ridge, Brooke later noted: 'I was named "The Barrage King" or "the man who put the B in Barrage!"'[6] McNaughton was a superb counter-battery officer, and it is apparent that both he and Brooke considered themselves great gunners. Yet just as clearly they had different views about how artillery ought to be employed. Sir Julian Byng had brought Brooke into Canadian Corps Headquarters from the 18th Division to be Staff Officer, Royal Artillery, and assigned him the responsibility for sighting the field guns of the reinforcing artillery at Vimy. When Currie saw Brooke's dispositions, he was not happy. McNaughton suggested pulling the guns back a few thousand yards to avoid hitting the Canadian troops. Brigadier R.J. Leach recalled that McNaughton and Brooke also argued over the proper command and control of brigades engaged in counter-battery firing.[7] It is not difficult to surmise that the professional dispute between McNaughton and Brooke developed into personal animosity.

The first major clash between McNaughton and Brooke during the Second World War happened in July 1940. McNaughton's promotion to lieutenant-general and command of the newly created British VII Corps coincided with Sir Alan Brooke's promotion to full general two days earlier (he had been knighted for services in France on 11 June). With the promotion Brooke assumed the responsibilities of C-in-C Home Forces from Ironside and became McNaughton's immediate superior. Brooke quickly confronted McNaughton regarding comments the latter had made about the aborted Brittany operation. It seems that during his first visit to the Canadians as C-in-C Home Forces, Brooke declared: 'Andy, I hear you're very disconcerted and you've been very critical of the way I've handled the [Canadian] division in France.'[8] McNaughton conceded that Brooke had been in an 'extraordinary difficult position' in Brittany, then added, 'Brook[e]y, I certainly am. I'm not going to hide it, I was very upset about it. We had an arrangement.'[9] McNaughton's version of the encounter has it that Brooke remarked: 'I want you to know that you're under my orders now and I'll fight the division as I see fit.' McNaughton strongly objected: 'Well, we'll see!' After the war he reflected: 'What can you do with a man like that ... I let him know that ... I wouldn't have it.'[10]

Canadian military historians have been far too willing to place virtually all the blame on McNaughton for his poor personal relations with Brooke. McNaughton is always characterized as the 'prickly nationalist' or as possessing an 'abrasive' personality. There is some truth to that assertion. He also had what General W.A.B. Anderson called 'an aura about him, a flamboyance without trying.' Elliot Rodger, McNaughton's PA, noted that he could 'inspire the most extraordinary loyalty in all the people below him.'[11] That flamboyance, combined with his considerable self-confidence, probably bordered on arrogance and was interpreted as such by others. From the very beginning he was not impressed with how the British were running the war. Later he would state: 'I had the confidence of the War Office because of the way I had been getting things done for them in the defence of Britain when everybody else was sitting around sucking their tongues.' He even admitted that 'I was never inclined to stick too much to the normal proprieties of role' – in other words, he would say or do whatever was necessary to get the job done.[12]

McNaughton clearly had idiosyncrasies, but so did Brooke. 'Brookie,' as Montgomery said, 'is not an easy person to get to know.' He carried deep personal scars that undoubtedly caused him to raise emotional barriers.[13] Brooke has been described as 'impatient, peppery, quick tempered, frank to the point of brutality [and] intolerant of stupidity.' General Sir Leslie Hollis, the Assistant Secretary to the War Cabinet and the COS Committee, described him as 'resolute, volatile, vibrant, versatile and sharp tempered.' Professionally, Brooke was highly intimidating, willing to go toe to toe with the prime minister. Churchill once said that when he thumped the table and got in Brooke's face, 'what does he do? Thumps the table harder and glares back at me – I know these Brookes – stiff-necked Ulstermen and there's no one worse to deal with than that.'[14] Walter Bedell Smith, Eisenhower's Chief of Staff, also clashed with Brooke. When it came to finding qualified staff officers for the invasion of Northwest Europe, Smith naturally looked to the Mediterranean for experience, but Brooke resisted. Smith told him, 'You are not being very helpful.' Brooke replied, 'You'll get nothing this way.' In the end, Bedell Smith concluded that Brooke was 'tricky' and could do 'a lot of double-talking. He was a slick Irishman.'[15] There is little doubt that McNaughton could be difficult to work with, but so was Brooke, perhaps even more so, and historians need to recognize this fact.

It was not long before McNaughton and Brooke collided again over the control of Canadian forces. In January 1941, Major-General George R. Pearkes's 1st CID took part in Exercise VICTOR, a large TEWT. At Brooke's direction large portions of the division were taken from Pearkes's command in spite of his protests and placed under British formations to meet exercise contingencies. It seemed that Brooke was doing precisely what he had told McNaughton he would do – that is, command the Canadians as he saw fit. McNaughton responded immediately, calling in Charles P. Stacey, the army's historian, to tell him that a grave constitutional crisis was brewing that might result in the removal of Canadian forces from British control within twenty-four hours. Stacey recorded that McNaughton 'took a serious view of the matter, and obviously proposed to make an issue of it.' McNaughton's reaction and his portrayal of the incident to Stacey reflected a tendency on his part – demonstrated in the past – to overdramatize events for effect, but after clashing with Brooke over Brittany it was probably inevitable. McNaughton told Stacey there was 'nothing personal in his attitude ... it was not egocentric but purely objective' because he had a 'responsibility to the Government and people of Canada for the lives of the men entrusted to his command.'[16]

On the last day of January McNaughton and Brooke discussed the issue. Brooke noted: 'With a little talk we settled the matter quite amicably and all is now well.' But he added that McNaughton's constitutional arguments did not make things easier and actually rendered the use of Canadian troops 'even more difficult than that of allies!'[17] Brooke was simply wrong here, for McNaughton had always accepted the principle of military necessity and had clearly demonstrated this acceptance throughout the Norway and Brittany misadventures.

In a letter penned the next day McNaughton thanked Brooke for his willingness to listen the day before, but added:

> From our conversation and from your own experience years ago with the Canadian Corps and more recently with your own Corps in France, I feel sure you recognize the military advantage of employing the present Canadian Corps as a whole ... If not so used and units or subordinate formations were detached ... then a very heavy price would have been paid in military efficiency and effectiveness against the enemy.[18]

'I am naturally most anxious,' McNaughton continued, 'on grounds both of military advantage and of constitutional propriety, that the Canadian Corps should be kept together.' Circumstances might demand separation, but he left Brooke with no misperceptions:

> I recall your promise that before any 'instructions' to make a detachment are issued, that the alternatives will be carefully weighed, and I confirm that under this condition I will accept your judgment at the time; it being definitely understood that a Canadian Division is not to be subdivided except with the consent of its commander and that it will be returned to the Canadian Corps at the first practicable moment; the fact that a Canadian Division is detached will not interfere in any way with the normal system of Canadian administration nor with my right and duty to intervene should the situation so require.

On 5 February, Brooke responded favourably: 'I agree with all you say and have forwarded a copy of your letter to the Army Commanders of Southern and South Eastern Commands.'[19] It must be remembered that McNaughton was making this case for unified divisions within a corps during the defence of Britain. Unfortunately, it seems that Brooke was developing the idea that McNaughton would not countenance the use of individual divisions outside Britain in any circumstances other than an absolute emergency. However, Brooke did have larger issues to deal with, and his heavy responsibilities meant in the last instance that he could not always give every problem equal attention.

When Brooke took over as CIGS he wasted no time in establishing his priorities. The continued securing of Britain and its communications remained paramount and actually argued in favour of the Canadians staying put there. Next was the defence of Singapore and communications through the Indian Ocean. The Middle East came third.[20] Brooke's strategic calculations, however, disintegrated as quickly as he generated them. On 10 December the Japanese sank *HMS Repulse* and *HMS Prince of Wales*, a blow that crippled the Royal Navy's control of the seas from the east coast of Africa through the Indian Ocean. Hong Kong, defended by a token force including two Canadian battalions, fell on Christmas Eve. The domino effect of events in the Far East complicated Brooke's responsibilities beyond measure: he had to devote considerable attention to keeping the nervous Australians 'quiet' because they were 'fretting about [sending Australian] reinforcements to Singapore.'[21]

Even before the final defence of Hong Kong played out, Brooke instinctively knew that 'we shall suffer many more losses in the Far East.'[22] Though possessing superior forces, Lieutenant-General Sir Arthur Percival surrendered the British force at Singapore, some 130,000 men, including an entire Australian division, on 15 February 1942. If McNaughton feared the inappropriate use of Canadian troops by British commanders, for cause he had only to look at the unnecessary disaster that had fallen on the Australian troops at Singapore. The loss of Singapore was a body blow to Brooke – indeed, Churchill himself declared in the House of Commons that it was 'the greatest disaster to British arms which our history records.'[23] Buffeted by these heavy blows, Brooke recalculated. By default, the Middle East now assumed greater importance. North Africa quickly and literally became the 'only show in town' as well as the only place where the Canadians could have joined the British Army in an active theatre against the Germans. Yet this small window of opportunity did not coincide well with the larger and growing Canadian force sitting in Britain.

On Christmas Day of 1941, McNaughton finally yielded to Crerar's desire for a larger force and informed Lieutenant-General Bernard Paget, Brooke's replacement as C-in-C Home Forces, that an army headquarters with two corps headquarters would be necessary to control the Canadian forces envisioned by the already approved Canadian Army Programme for 1942–3.[24] Crerar drafted the program while he was CGS and presented it to Ralston on 18 November 1941. Crerar declared that a 'very definite need exists for a maximum expansion during 1942,' and to that end he requested a second armoured division, an additional army tank brigade, and some eighty-three additional corps and army-level units.[25]

On 26 January 1942, King announced that the Canadians would be establishing an army headquarters. Yet two weeks earlier Brooke had told McNaughton 'it was his wish to continue to employ them [Canadians] in the defence of this country until such time as operations on the Continent of Europe became practicable.'[26] This was precisely what McNaughton told Roosevelt, General George C. Marshall, the Army Chief of Staff, and Brigadier-General Dwight D. Eisenhower, Acting Chief of the War

Plans Division of the War Department, during his visit to Washington a few days later.[27] Brooke supported the Canadian Army's expansion even while arguing for its retention in Britain; thus he bears considerable responsibility for shaping how the Canadian Army finally got into action. Inadvertently, the arrival of the Americans influenced it as well.

On 9 April 1942, Marshall presented the American strategy for defeating Germany to Churchill and the COS during discussions in London. The plan called for a massive American build-up in Britain, code-named BOLERO, in anticipation of a fully fledged cross-Channel invasion in the spring of 1943, soon to be known as ROUNDUP (a code name originally used by the British Joint Planning Staff in December 1941 for a possible twelve-division assault on France). In the event that the Russians appeared on the verge of defeat, Marshall also proposed an emergency landing in France, no earlier than September 1942, with whatever strength the Allies could muster.[28] Since early March 1942 the COS had also been exploring diversionary operations to help the Russians that year. The British plan was called SLEDGEHAMMER, and this code name quickly came to represent the American *and* British emergency schemes. SLEDGEHAMMER envisioned invading and holding the Cherbourg peninsula. A few days later Churchill and the COS accepted Marshall's proposal for offensive action in Northwest Europe. Brooke recorded that they were anticipating action 'perhaps' in 1942 and 'for certain' in 1943.[29] On 5 May Churchill gave King an overview of this new strategic reality.

Towards the end of June, Major-General Eisenhower arrived in London as Commanding General, European Theater of Operations, United States Army (ETOUSA), with orders to prepare for SLEDGEHAMMER and ROUNDUP, the latter tentatively scheduled for 1 May 1943. At this juncture Churchill was allowing the Americans to think that he and the COS were on-board with the cross-Channel operations, even though he never had any intention of sacrificing nine British and Canadian divisions (the butcher's bill for SLEDGEHAMMER) in an attempt to help the Russians.[30] Nor was Churchill inclined to throw away more of the British Army executing ROUNDUP. Later he would maintain that ROUNDUP would have precipitated 'a bloody defeat of the first magnitude.' Even Eisenhower later admitted that he had misjudged the difficulties involved.[31]

Much to Marshall's astonishment and anger, Churchill informed Roosevelt on 8 July that the British would not be taking part in SLEDGEHAMMER.[32] British planners had begun to consider the problems associated with a return to the Continent soon after Dunkirk.[33] The series of defeats suffered by the British Army at the hands of the Wehrmacht was strongly influencing Churchill's psyche at the time. The surrender of the Tobruk garrison only two weeks earlier on 20 June had been the most recent blow. He was not at all convinced that the British Army was up to the task of facing the German Army. This fear helped kill SLEDGE-HAMMER.[34]

With SLEDGEHAMMER stillborn, Churchill proposed two other projects that had been consuming his attention for several months. One was Operation JUPITER, an invasion of Norway; the other was Operation GYMNAST, an invasion of French North Africa. By late July, under direct orders from Roosevelt to find some accommodation with the British for offensive action in 1942 (to respond to heavy American domestic pressure for action that year, to prevent a 'Japan First' surge, and to help the Russians), Marshall committed the United States to Operation SUPER-GYMNAST, a combined Anglo-American descent on the Mediterranean.[35]

On 24 July, Marshall informed the COS that the American commitment to GYMNAST 'renders ROUND-UP, in all probability impracticable of successful execution in 1943 and therefore ... we have definitely accepted a defensive, encircling line of action for the CONTINENTAL EUROPEAN THEATER.'[36] The COS readily agreed, for Marshall's decision was perfectly in sync with their long-held strategic views. As preparations for GYMNAST – rechristened TORCH by Churchill that day – got under way, First Canadian Army, stood up on 6 April, was pointedly left out of the operation's order of battle. Historians have blamed McNaughton for this omission, but it is apparent that other factors were at play.

In June, Paget had indicated that McNaughton would soon be brought 'inside' a newly established planning committee for expeditionary operations. This was the Combined Commanders, but McNaughton was never actually invited to participate as an equal. As far as he was concerned, Brooke intentionally 'kept me apart from Eisenhower as far as he could [in order to avoid the] complication of another nationality.'[37] McNaughton was not even told of the decision to mount TORCH until at least September, and Brooke certainly never briefed him on it, even informally, until 17 October.[38] Stacey suggested that McNaughton was not told because the British government was 'well pleased to have the Canadian formations remain in the United Kingdom.'[39] Evidence in support of Stacey's contention came from Brooke, who told Ralston in late November 1943 that Paget 'had refused to agree to weaken the U.K.

and also that they did not want to have to undertake this first joint show [TORCH] with the U.S. and have not only two but (if the Canadians come in) three different lots of troops to look after and service, with problems of reinforcements, equipment, supply and administration.'[40]

Paget's argument was reasonable enough, but it must not be forgotten that the War Office reaped real dividends by having the entire First Canadian Army stay put and out of TORCH, in that it freed British formations for deployment overseas. Churchill exploited American divisions in Britain in the same way and with equal effect.[41] As of 26 May 1942, Eighth Army had six divisions, three of which (and two army tank brigades) were British. As David Fraser has observed, this was 'still only a tiny proportion of the British Army being raised.' When General Sir Brian Horrocks went off to the desert in mid-August 1942 he noted that he was to command 'one of the only two [British] corps which were actually fighting.' The British desperately needed to 'blood' many of their inexperienced divisions; they also had a new army headquarters – First Army Headquarters (established 6 July 1942) – to command the divisions that had been assigned for TORCH.[42] Moreover, British First Army needed battle experience as much as First Canadian Army.

Since McNaughton knew nothing about TORCH, he remained focused on ROUNDUP, which from his perspective was the only other operation that seemed to have any permanency. He did not know that with the decision to mount TORCH, British and American planning for ROUNDUP had come to a virtual standstill.[43] Throughout the summer of 1942 there were hints that changes to ROUNDUP had been made. However, hints were all they were, for as Paget told Eisenhower in late June: 'We constantly go over the same ground and no real progress has been made.'[44] Paget had indicated on 8 July that First Canadian Army should remain intact. This suggested that it was to be concentrated for a ROUNDUP-type operation. In early August, however, Major-General Sir Archibald Nye, the ACIGS, told McNaughton and Stuart that such an operation would be possible in 1943 only if German morale cracked, but they had to prepare for it in any event.[45] This was not the first indication from Nye that cross-Channel operations in 1943 were problematic. Back in February he had told Burns that even with American assistance, critical shipping shortages would delay a massive offensive on the Continent well into 1943. Brooke's attitude towards the shipping dilemma was even more negative: '[It] may well be our undoing!!'[46] Prospects for active Canadian participation in the fighting, therefore, looked grim, with one haunting exception.

Well before the decision to mount TORCH, the 2nd CID had been identified for participation in Operation RUTTER, the raid on Dieppe. During 1941, McNaughton had argued for and obtained greater discretionary power to authorize raids on his own. His expanded authority paid dividends in August when the 2nd Canadian Infantry Brigade (CIB) took part in GAUNTLET, the raid on the island of Spitzbergen.[47] Yet it was Crerar who pushed for Canadian inclusion in RUTTER.[48] It was cancelled because of weather on 7 July, and debate continues to this day regarding why it was pushed forward again.

McNaughton would thereafter always insist that the ultimate responsibility for allowing the Canadians to take part in the raid 'was mine and nobody else's.' He refused to shift any responsibility to the British even when an American journalist, Quentin Reynolds, blamed Canadian commanders for the debacle in *Dress Rehearsal: The Story of the Dieppe Raid* (1943).[49] Yet it is clear that on 14 July, Combined Operations Headquarters (COHQ) decided on its own to revive the raid. McNaughton's War Diary for that date specifically states that Crerar had sent a message indicating that the 2nd CID would probably recommence its special amphibious training. The entry is for 0900, so it is clear that Crerar received word of the COHQ decision faster than McNaughton did.[50] McNaughton believed that he knew precisely why Churchill resurrected RUTTER after its cancellation. On 25 July he recorded: 'It appears that Stalin had cabled the Prime Minister asking what was being done to distract the Germans by raiding. The Prime Minister had been very pleased to be able to reply indicating action was in hand and in consequence he had approved the highest priority in preparation for JUBILEE.'[51]

This justification probably did not surprise McNaughton. The Brittany operation in 1940 had had a similar political imperative – to keep the French fighting if possible. Also, during his visit to Washington in March, Roosevelt had told him about the pressure for a second front, suggesting that 'military considerations might well have to give way to the overriding broad political factors' when it came to helping the Russians. In July, Marshall had presented the COS with a memorandum from Roosevelt strongly recommending that SLEDGEHAMMER proceed 'with utmost vigor [and] whether or not the Russian collapse becomes imminent.'[52] Marshall's own view was that raiding was key to the 'establishment of a preliminary active front' and would – it was hoped – 'be of some help' to the Russians.[53] McNaughton clearly grasped at least some of the higher political intent, for on 16 July, in the presence of Mountbatten, he ordered Major-General James Hamilton Roberts, commander of the 2nd CID, to execute JUBILEE notwithstanding not only Roberts's own reservations but also those of Montgomery.[54]

There is no question that Churchill was under enormous pressure to immediately appease Russian demands for a second front. He was also in the unenviable position of having to simultaneously achieve two things. First, he had to display token interest in SLEDGEHAMMER in order to prevent the Americans from shifting their emphasis to the Pacific. Marshall had actually proposed this shift to Roosevelt on 10 July and was still contemplating it on 11 August. Second, Churchill had to demonstrate to the Americans in concrete terms just how suicidal SLEDGEHAMMER actually was.[55]

On 13 July, Marshall told Eisenhower to investigate how SLEDGEHAMMER might be executed. Four days later Eisenhower recommended that it be kept alive until 1 September.[56] In this context JUBILEE became a modified SLEDGEHAMMER – perhaps even a psychological substitute – initiated by Churchill to address the convergence of several nasty political issues that threatened the 'Grand Alliance.'[57] Sholto Douglas, Commander-in-Chief of Fighter Command, was correct in his view that 'irrational pressures were brought to bear – from the Americans as well as the Russians.'[58] Even after TORCH replaced SLEDGEHAMMER, the need to assist the Russians remained. The raid on Dieppe would serve as a catalyst for drawing the Luftwaffe into battle over the Channel.

Luring the Luftwaffe into an attritional battle was one of the fundamental premises of Allied strategy, and Brooke's diary contains many references to the fact that it was the only tangible way to help the Russians at the time. Eisenhower informed Marshall on 11 July: 'In lieu of completely preparing for SLEDGEHAMMER under the conditions laid down by the Prime Minister, the [British] Staff will utilize assembled shipping for the purposes of deception and, whenever possible, for the conduct of raids on an increasing scale in size and intensity. They hope to thus bring on air battles and to keep the enemy upset.'[59]

Bear in mind that raiding operations IMPERATOR and SESAME were also considered. On 8 June 1942, Churchill very clearly minuted a COS document that these were 'to be regarded as a "bait" to draw the German fighters into combat.' He asked: 'Would they be wise to make this sacrifice?'[60] By this logic the Canadian ground troops assigned to JUBILEE were merely bait, fixing the enemy's attention while the RAF manoeuvred to strike the real objective overhead.

Dieppe served yet another purpose: to hold German attention in northern France for the benefit of TORCH. Proof of this comes from

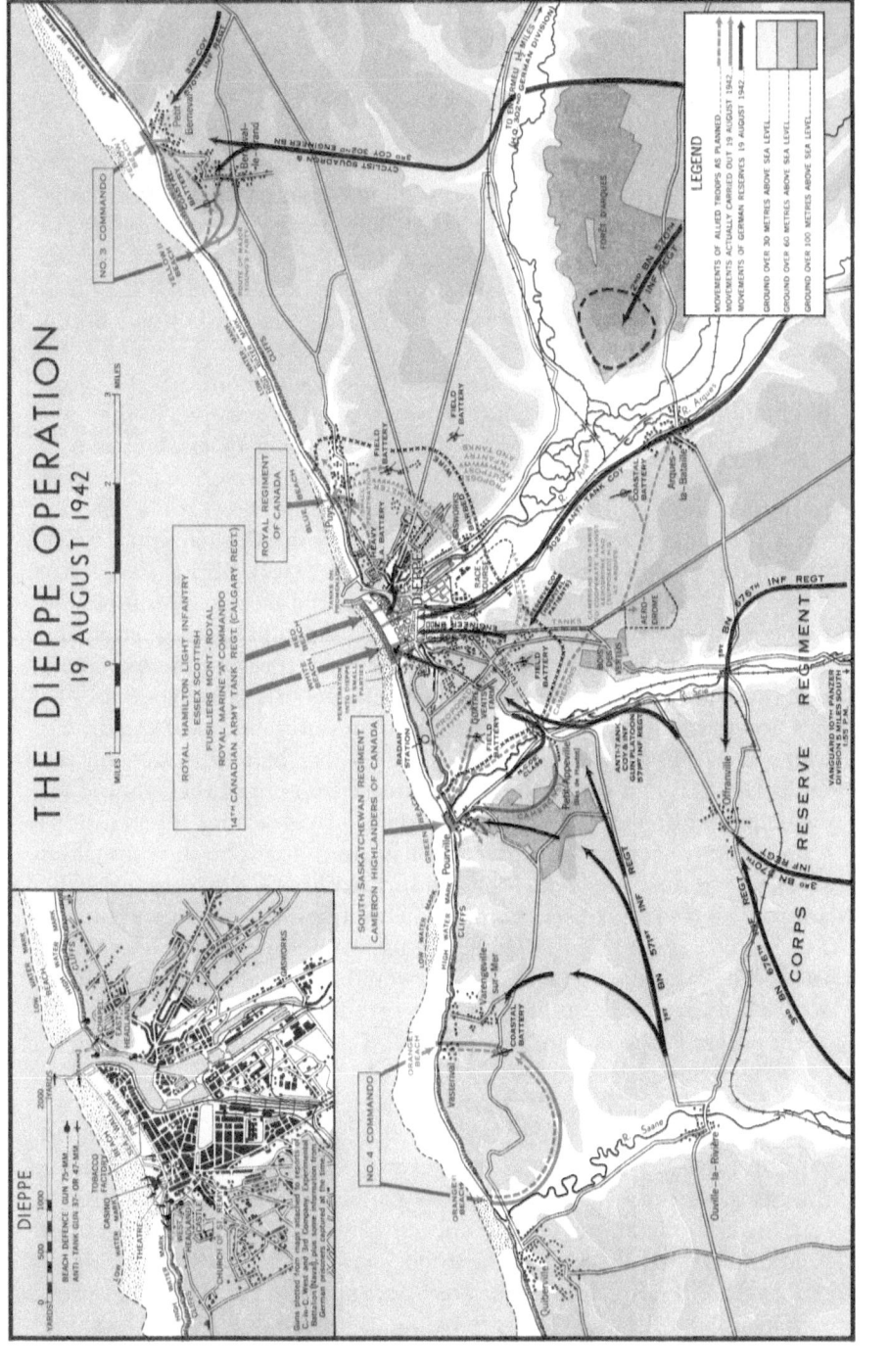

Major-General John Noble Kennedy, the DMO, who recalled Churchill's comments after visiting Stalin. Churchill told him that Stalin had remarked that 'Dieppe will be explained by Torch.' This, he added, indicated Stalin's 'quickness of mind in that he connected the two things.'[61] Indeed, Churchill had specifically assigned Mountbatten the task of dissuading Roosevelt from pushing SLEDGEHAMMER. He also told Stalin that the most ambitious action being planned was for a large-scale raid on Dieppe. The idea put forward by many historians that JUBILEE was resurrected solely to gain lessons for an at-that-time distant invasion is simply too neat and tidy.[62]

The raid on Dieppe, carried out by Roberts's division on 19 August, did in fact spark a great air battle in which the RAF sustained its greatest single-day loss of the war – some 106 aircraft, against German losses of 48 aircraft destroyed and 24 damaged. The benefits were hardly worth the cost. Colonel Oliver Stanley, head of the London Controlling Station (LCS), the organization established to deceive the Germans about British intentions, was appalled when he learned that JUBILEE was to proceed. Potentially sacrificing 5,000 loyal imperial troops was simply too high a price to pay to keep the Russians fighting, and he resigned.[63] The Canadians suffered horrendous casualties. Forced to execute the operation without heavy bombers or capital ships, Roberts carried out his duty in a no-win situation and brought his shattered formation back to Britain with only 2,211 men of the original 4,963 that had embarked on the raid. JUBILEE had taken one Canadian division out of the running for participation in TORCH. More important, it had serious repercussions for attempts to get the army into action again. The frightening casualties sustained in but a few hours only confirmed King's conviction that the army needed to be conserved 'until the moment when it was absolutely necessary to attempt invasion.'[64] After the disaster on the beaches of Dieppe it was inconceivable that he would have consented, without a considerable struggle, to part of the army going to North Africa to fight a still potent German panzer force that at that time was threatening to drive the British across the Suez Canal.

Had Churchill had his way the rest of First Canadian Army would have been committed to yet another diversionary operation, JUPITER, to capture Luftwaffe aerodromes in northern Norway and thereby relieve the heavy pressure on the Murmansk convoys.[65] The COS had already rejected the operation as unsound, but Churchill simply would not let it go and wanted the Canadians to do it. He reflected: 'I thought that this operation would give a glorious opportunity to the Canadian Army, which had now for two years been eating its heart out in Britain, awaiting the invader.'[66] On 9 July, McNaughton attended the COS meeting where he was invited, at the suggestion of Clement Attlee, the Deputy Prime Minister, to assess the feasibility of JUPITER. Brooke was incensed at McNaughton's inclusion, and there is evidence that the COS as a body was not impressed either.[67] Brooke told McNaughton privately 'how matters stood as I did not want him afterwards to imagine that we were suggesting that the Canadians should undertake an operation which we [COS] considered impracticable.'[68]

In early August, McNaughton, based on Guy Simonds's analysis of JUPITER, reported to the COS that JUPITER was 'extremely hazardous' but politically acceptable if the results 'were judged to be of the highest importance.' The COS praised Simonds's analysis as 'one of the clearest and most ably worked out appreciations which they had ever had before them.' That should have been the end of it, but Churchill went so far as to suggest that McNaughton go to Moscow for staff talks to make JUPITER work. King, however, sensed danger and intervened to stop it.[69] Brooke later commented on his original diary entry at the time that McNaughton 'did not rise in my estimation as a result of these transactions; he seemed incapable of telling Winston that an operation was impracticable but always handed the baby over to Mackenzie King.'[70] Yet Brooke, of all people, knew from first-hand experience how difficult it was to divert Churchill's attention from pet projects.

Ultimately, it is difficult to discern Churchill's real motives in advocating JUPITER. There is some evidence that at first he considered it a legitimate operation; later, however, he gave indications that it was a deception measure for TORCH. Indeed, he told Roosevelt on 27 July: 'Secrecy can only be maintained by deception. For this purpose I am running JUPITER ... The Canadian Army here will be fitted for arctic service. Thus shall we be able to keep the enemy in doubt till the last moment.'[71] There is no evidence that McNaughton knew of this. What is beyond doubt, however, is that the success of Churchill's deception scheme was directly linked to the realism of McNaughton's planning, which in turn depended on the Canadian Army (scheduled to make up the bulk of the JUPITER force) remaining in Britain.

In one sense, McNaughton was already involved in TORCH. By early September he had agreed to undertake OVERTHROW, the formal, authorized deception component of TORCH.[72] Even so, as the summer turned into the fall of 1942, McNaughton was like the captain of a rudderless ship. He made attempts to set a clearer course, finally telling Brooke that he was 'under some considerable embarrassment through the lack of stability in plans for operations in which the Cdn Army had been invited to participate.' He was especially frustrated that ROUNDUP had been postponed for TORCH 'and other schemes of which so far I had only the vaguest particulars.' At this stage he was operating under the assumption that the army was to play a principal role in ROUNDUP, scheduled for 1 May 1943. 'Then suddenly, and without warning,' he added, 'a change of plans had been made, which had not been communicated to me until later.'[73] He told Brooke: 'In fact, very grave questions of Cdn policy had been involved, which might have had most serious affects on our war effort. I suggested that it was most unwise to leave this sort of condition to chance and yet, while I did not wish to interject myself into the British direction of the war (as I had all I could undertake in getting the Cdn Army ready for a particular theatre of operations) nevertheless, I thought I must have more complete and timely information of plans as they developed.' Brooke's solution – to meet with McNaughton at least once every two or three weeks – was 'the only possible way to keep me informed,' McNaughton observed, because the CIGS's position was 'on such an unstable basis that one day a plan was on and the next day off.'[74]

On 3 October, McNaughton, Ralston, and Stuart met to discuss future operations. McNaughton explained SLEDGEHAMMER, ROUNDUP, WETBOB (a variant of SLEDGEHAMMER), and TORCH but only in outline. He told Ralston that because of the 'unsatisfactory' flow of information, he was 'unable adequately to advise the Minister of National Defence.' It was probably because of this acute uncertainty in operational planning that McNaughton pointed out that he was against any splitting up of the Canadian Army, 'just as the commander of the Cdn Corps in the last war had been and the meeting *generally* [emphasis added] agreed.' SLEDGEHAMMER and WETBOB had been cancelled, and there was no invitation for TORCH. From this McNaughton concluded that ROUNDUP was the army's future task, and he made this known to Ralston and Stuart.[75]

The years 1941 and 1942 were difficult for McNaughton and his army. Feasible operations were scarce, and his ability to forecast future requirements was hindered by inadequate information. So he had little choice but to consider and agree to risky operations such as JUBILEE; had Brooke not intervened, he might even have succumbed to Churchill's advocacy of JUPITER. McNaughton was frustrated by the lack of sound information on which to plan the operations of the Canadian Army throughout 1941 and into late 1942 and his frustration would only continue in 1943.

McNaughton, (second from left front row) and Brooke (second from right front row) on the staff of the Canadian Corps in 1917. (LAC)

McNaughton as a Brigadier-General in 1918. (LAC, PA-034150)

Lieutenant-General Andrew G.L. McNaughton, commander First Canadian Army. (LAC, PA-164285)

McNaughton and the indomitable Winston Churchill. (LAC, PA-119399)

Field Marshal Sir Alan F. Brooke, Chief of the Imperial General Staff, and McNaughton's principal British nemesis. (LAC)

General Sir Bernard Paget, Commander-in-Chief Home Forces. (LAC, PA-140144)

Lieutenant-General Bernard L. Montgomery inspects training with Lieutenant-General Harold Alexander in 1941. Montgomery exercised a heavy influence on the training of the Canadian Army. (LAC)

Major-General Guy Turner, McNaughton's confidant and Brigadier General Staff. (LAC, PA-140890)

Henry Duncan Graham Crerar, pictured here as Chief of the General Staff in 1940. He replaced McNaughton as commander of First Canadian Army. (LAC, PA-116458)

Crerar, McNaughton and Turner at Headley Court, April 1942.
(LAC, PA-034127)

McNaughton's principal Canadian nemesis, Minister of National Defence, James Layton Ralston, with Harry Crerar on the right. Major-General Victor Wentworth Odlum is between them. (LAC, C-064025)

McNaughton's other Canadian nemesis, Lieutenant-General Kenneth Stuart, Chief of the General Staff, on the right. Ralston is in the middle. (LAC)

McNaughton with Vincent Massey, Canadian High Commissioner in London. (LAC, PA-150935)

McNaughton's ally, for a time, Prime Minister Mackenzie King.
(LAC, PA-164304)

The price of fighting for the sake of fighting, Dieppe, August 19, 1942. (LAC, C-014160)

Guy Simonds, right, was McNaughton's BGS during SPARTAN. Montgomery, however, was Simonds' true mentor. (LAC, PA-142272)

Major-General Ernest W. Sansom, commander II Canadian Corps. McNaughton defended his performance on Exercise SPARTAN too long and undermined his relationship with Paget as a result. (LAC, PA-188991)

McNaughton finally gets to see his Canadian troops in Sicily, August 1943 after being refused entry to the island by Montgomery the previous month. (LAC, PA-136200)

5

Practical Operations of War

> I was prepared to consider any proposition on its merits.
>
> McNaughton, 19 November 1942

While McNaughton continued to stick by his (and King's) policy of looking to the COS to recommend Canadian deployment, Ralston and Stuart were disheartened by the 3 October meeting and immediately began to increase pressure on the War Office. Three days later Brooke recorded that Stuart 'is anxious for Canadian forces to be given a more active role.' A little more than a week later, in conversation with Massey, Ralston, and Stuart, Brooke noted that it was not an 'easy matter to lock them into any of our proposed offensives.'[1] Indeed, it never had been, but Ralston kept up the pressure, simultaneously working Churchill and Sir James Grigg, the Secretary of State for War, to find employment for the Canadians as soon as possible, stressing that there were 'no strings' and that Ottawa was ready to 'consider any proposals.'[2] Ralston was entirely within his rights to do this, and McNaughton ultimately worked for him. The problem was that Brooke was entirely correct: there were scant opportunities to deploy a large enough portion of the army to satisfy all critics, and that was not McNaughton's fault.

The pressure exerted by Ralston and Stuart quickly bore fruit, for on 17 October Brooke asked McNaughton if he was willing to undertake TONIC, an operation to seize the Canary Islands for use as an alternative airfield in case Spain occupied Gibraltar.[3] Brooke made it clear that this offer was the direct result of actions by Ralston; he even called McNaughton back later in the day to reiterate that TONIC was not a certainty but was being offered to the Canadians as a result of Ralston's representations. According to McNaughton, Brooke was 'anxious that any Canadian force employed might be so used that it could be returned to the Canadian Army at an early date as he attached great importance to keeping ... a well-balanced, self-contained organization for Home Defence and eventual employment on the Continent.'[4] This was entirely consistent with what Brooke had told him in January 1942.[5]

McNaughton once again told Brooke that the army was for use 'in whole or in part and [he] would give most careful consideration to any project,' but he added that he wanted to make a real contribution and was not interested in 'fighting for its own sake or glamour.' This was not a unique McNaughton idiosyncrasy: during the flurry of proposed operations in the Mediterranean in mid-1941 the COS had refused to consider 'action for its own sake.'[6] It took McNaughton only one day to decide that TONIC was a feasible operation, and by 23 October he had informed Brooke that Crerar's I Canadian Corps had been tasked to undertake the operation.[7] TONIC, however, was nothing more than a sop to Ralston and Stuart. It also represented in every way McNaughton's fear of 'fighting for its own sake.' As he later stated, he and Brooke both knew it was a 'very remote contingency.'[8] It certainly would not have satisfied the cries for battle experience, for there were no German troops on the Canary Islands.

The sudden Canadian demand for immediate deployment only heightened Brooke's frustration. McNaughton clearly perceived it during their 23 October meeting, noting that he was 'very intense and dramatic and ... was overwrought and very tired and harassed; when I departed at 1545 hrs he seemed very loath to let me go.' The TONIC sop aside, they had at least agreed that it was of 'paramount importance that there should continue to be the frankest and closest relations between ourselves and that if any cloud should seem to arise we would meet at once to clear the air.'[9] If Brooke was prepared to do this, it represented far more openness than what existed between McNaughton and Ralston.

McNaughton resented the fact that neither Ralston nor Stuart had seen fit to let him know that they intended to push for immediate deployment. McNaughton also criticized Crerar for advocating the same thing behind his back. On 19 October, Crerar wrote to McNaughton that it was not his habit to 'talk out of turn, tell stories out of school or knife people behind their backs.' Yet this was precisely what Crerar was doing and what he *had* been doing since returning to Britain as a corps commander in November 1941. Guy Simonds, no fan of Crerar's to be sure, told Stacey after the war that Crerar was constantly disloyal to McNaughton and that he actively lobbied for McNaughton's job.[10] The evidence in support of this charge has been well documented by Canadian military historians.[11]

It has been suggested that by the fall of 1942 Crerar was completely at odds with McNaughton with regard to how to get the army into action and that he bitterly resented being left out of higher policy decisions.[12] Yet McNaughton had authorized every feasible and not so feasible operation that came his way, including JUBILEE. He had not been told of TORCH in time to make a contribution, but he had agreed to OVERTHROW and TONIC. Moreover, Crerar had pushed for the creation of an army headquarters in part as a means to free the corps commanders to focus exclusively on training the army's formations. In his resentment of McNaughton's dominance of higher policy, Crerar acted as if he was still the CGS with a right to interfere in McNaughton's prerogatives. McNaughton was perfectly correct in censuring Crerar for his actions. No leader would tolerate such a breach of the chain of command. Corps commanders do not make policy.

A month later, on 19 November, McNaughton and Brooke had one of their most important meetings, the former describing it as 'long and intimate.' It was here that McNaughton made a shocking proposal: 'I told General Brooke that while I, perhaps naturally, thought that Canada's contribution could be most effective if given in the form of a Cdn Army substantially self-contained, yet the thing which weighed most with me was that the contribution should be the maximum towards winning the war and that if this, having regard to time and space, required the use of individual divisions separately or *even the breaking up of divisions* [emphasis added] then I was prepared to consider any proposition on its merits and to report thereon to Canada.' This statement perfectly reflected Ralston's own position (and perhaps a degree of loyalty on McNaughton's part). Yet Brooke 'expressed himself as very much against breaking up any of our formations,' speaking particularly of the 'importance of maintaining the possibility of operating as a self-contained Cdn Army which he said would give the best results.'[13]

To determine whether breaking up the army was called for, McNaughton asked Brooke for as much detailed information as possible. After a lengthy discussion of the strategic situation as it then stood, they agreed that the Canadian program should be framed to provide for the following:

By April 1943 – Large scale raids of limited scope and duration on the U Boat Bases in Bay of Biscay ports. The 3 Inf and 2 Armd Divs should be up to strength with reasonable reinforcements available in the U.K.

By 1 August 1943 – We should be ready to go on the Continent in strength to stay there holding a bridgehead of limited depth from the coast, should a definite crack in German morale be evident. We need not have full Army, L. of C., etc. tps, as under the conditions envisaged, these might be extemporized; nor should we need under these conditions a large scale of reinforcements.

By 1 October 1943 – He [Brooke] would like to see the full structure of the Cdn Army completed.[14]

This meeting placed McNaughton squarely on record as consenting to the possible break-up of the army. Two days later he told Stuart precisely the same thing, but added: 'It is very definitely General Brooke's opinion that the project for a Cdn Army should be maintained.'[15] Paget held a similar conviction. As 1942 drew to a close he offered McNaughton encouragement for the future: 'Looking ahead to the possibilities of 1943, I derive high hopes from all I see and know of the Cdn Army. When the great day comes, it will play a decisive part in the decisive battle, and results will certainly compensate for the long period of waiting for the opportunity to prove its worth.'[16] Yet the situation continued to look bleak as the new year approached. Mercifully, the COS cancelled TONIC on 19 December; even so, considerable staff effort had been wasted, as it had been for JUPITER.

On the last day of 1942, Brooke offered McNaughton another improbable operation, noting that the COS 'thought that the Cdns would not like to be left out.' This was BRIMSTONE, an attack on Sardinia to be executed by a British corps. Brooke had little time for the operation. Three days before offering it to McNaughton he had noted in his diary that he had turned down a 'very bad plan' for Sardinia offered by Eisenhower.[17] McNaughton nevertheless consented to study it, just as he had done for the past tentative operations. He also scheduled the 1st Division to commence combined training in Scotland. But under pressure from Brooke and the COS, BRIMSTONE (like TONIC) soon joined the graveyard of ill-conceived British operations.

McNaughton's 31 December meeting with Brooke created additional confusion. Brooke stated that with only twelve British, five Canadian, and up to twelve American divisions scheduled to be in Britain by the autumn of 1943, a cross-Channel invasion was not practicable that year.[18]

There was little McNaughton could say to this. Two days later, however, on 2 January 1943, he informed Stuart that Paget had selected him to command the 'British' side of a large-scale exercise named SPARTAN to study an attack on the Continent from an established bridgehead 'which is role contemplated for Cdn Army.'[19] The army's role had been consistently defined along these lines by McNaughton, with the full support of the War Office, but the timetable remained obscure. In any event, McNaughton was not oblivious to the fact that the Canadian Army could not wait until 1944 to see action.

Proof of McNaughton's willingness to do what was necessary to address the cry for action is found in the minutes of his 10 January meeting with Crerar and other senior Canadian officers regarding the reorganization of the army. McNaughton declared that since invasion was impossible in 1943, the army needed to prepare to 'take its place in close concert with British formations in smaller operations' against the belly of the enemy. Specifically, he told those assembled that divisions 'may be used separately.' A few weeks later he declared to the British press that the army was available, in whole or in part, for operations anywhere.[20] McNaughton's words fell on deaf ears at the War Office. The COS continued to plan operations without the Canadians in mind, and the passage of information to McNaughton remained less than optimal.

At the Casablanca Conference (code-named SYMBOL) in January 1943 the CCOS selected Sicily as the next objective after the final clearance of North Africa. King received a briefing memo about future operations from Churchill on 30 January, but it did not specifically mention Sicily. King was not even aware of the invasion, code-named HUSKY, until 14 May – a situation that he described as 'simply absurd.'[21] The War Office had no intention of including the Canadians in HUSKY, for on 9 February Brooke told McNaughton and Stuart that they would be taking part in a Northern Task Force assaulting Northwest Europe to capture the Cherbourg Peninsula.[22]

The following day Crerar again went behind McNaughton's back to pressure Brooke for deployment. 'I fully see his point and his difficulties,' Brooke noted, 'but I wish he could see mine and all their complexity!!!'[23] Brooke added that 'the main factor that had up to date mitigated against their use in Africa was the stipulation made by Canadian Government that the Canadian Army must not be split up and must only be used as a whole. A Conception McNaughton had always upheld with the greatest tenacity. Crerar realized this concept must be broken down.' This was absolutely false: Brooke had consistently advocated a unified

Canadian Army for cross-Channel operations, and McNaughton had specifically told him on two recent occasions (19 November 1942 and 10 January 1943) that divisions could be used separately. Brooke could not convince Crerar that 'there would still be opportunities in operations connected with the re-entry into France [because he] looked upon such conceptions as castles in the air.'[24]

Crerar's scepticism looked deceptively groundless in the context of Paget's brief to McNaughton on 10 February. McNaughton recorded Paget's views as follows:

> Gen Paget said his mind had long been clear that the Cdn formations should be used to exploit from the Bridgehead and that we should organize for this mobile role. He said it had already been decided that the initial attack would be under comd First Cdn Army. That for the purpose of the landing and assault to secure the Bridgehead he would place 1 (British) Corps (Lt-gen Morgan) – three inf divs – at my disposal. The 1 and 2 Cdn Corps would follow through and exploit. As soon as certain progress had been made an army under Lt-gen Swayne would land on our left. Another army would also be brought in a later phase under Lt-gen Gammell.[25]

Had this scenario actually transpired, McNaughton's 'dagger pointed at the heart of Berlin' would have become a reality. Yet, as Paget had noted in mid-1942, such operations would be feasible in 1943 only if there was a serious deterioration in German morale. In the end, Stacey's conclusion was more than justified when he concluded that 'looking back on the development of this period, it is evident that the plans being considered at the War Office for action in France were extremely vague and fluid.'[26] McNaughton's actions must be considered in this context.

At the beginning of March 1943, Ralston returned to Ottawa and told King that the army was to stay in Britain. King stated: 'It is Churchill's and Brooke's action ... that keeps our men in England and is keeping the two Corps intact.' In pure King fantasy he added that they could have gone to North Africa 'if we had had our way and had not been over-ruled.' Actually, King was fully prepared to see the army sit in Britain another year to ensure complete success in the eventual invasion. When Churchill on 15 March requested that Canada give up its allotted shipping (to complete the army) for American aircrews, King noted: 'It really was a relief to my mind. I have been afraid of a second Dieppe.'[27] Even so, the continual deferral of cross-Channel operations and the lack of Canadian Army activity anywhere were issues that King could not ignore any more.

Throughout 1941 and 1942, King faced increasing political troubles at home, the most serious of which was the ever-present issue of conscription. A November 1941 Gallup poll indicated that 61 per cent of the country was satisfied with the war effort; but at the same time, 60 per cent demanded conscription. A national plebiscite on the issue held on 27 April 1942 freed King from the unequivocal stance he had taken against conscription at the beginning of the war, but it had serious repercussions, including the resignation of his ranking cabinet minister from Quebec, Pierre Cardin. King was left with a country more divided than ever over compulsory military service (English Canada versus Quebec), and with a Cabinet divided over essentially the same issue.[28]

King's political capital was being eroded by the perception that Canada was doing too little to help win the war. By the end of 1942 the perception was strengthening that the tide had turned against the Axis. The U.S. Navy had reversed the disastrous course of events in the Pacific with a stunning defeat of the Imperial Japanese Navy at Midway in June 1942; by early February 1943 the U.S. Army had defeated the Japanese on Guadalcanal in the Solomon Islands. In Europe the German Sixth Army was trapped at Stalingrad; it would surrender at the end of January 1943. Meanwhile, Axis forces in North Africa were trapped between Montgomery's Eighth Army and the Anglo-American forces that had invaded French North Africa on 8 November 1942.

Canadians were playing no conspicuous role in any of these great Allied victories, and King's government was not publicly exploiting the tremendous contributions that Canada was making in other areas to the extent it could have. After spending two months in Britain, C.A. Bowman, editor of the *Ottawa Citizen*, told King in September 1942 that Canada's war effort 'was magnificent – so regarded everywhere in England, but not known in Canada.'[29] The problem was simple: when it came to headlines, nothing trumped the dramatic effect of forces in direct, visible action against the enemy. Thus on 23 December 1942, Ralston sat in front of his Cabinet colleagues and declared: 'If the war ended now we would have to hang our heads in shame.'[30] At the same time, Canadian newspapers were becoming disenchanted with the lack of achievements on the battlefield. The *Winnipeg Free Press* contended that Canada should 'send a full division to some theatre of war' immediately; the *Vancouver Sun* published the views of a Great War veteran who ranted that 'soldiers sitting idle in England [were] the greatest disgrace of the war.'[31]

In the early months of 1943, King faced a profound predicament regarding the deployment of the army. If he blocked all attempts to get it into action because he truly feared 'another Dieppe,' he would be highly susceptible to growing agitation in English Canada for a total war effort in support of Britain. Conversely, if he loosed the reins and allowed the army to fight, the inevitable casualties might well force conscription to the fore even more dramatically than in mid-1942, driven once more by English Canada's desire for a total war effort. King could not vacillate forever and found himself compelled to make a decision. Though desperately afraid of the U-boat menace, he decided to release some of the pressure, not only from the public but from Ralston and Stuart as well, by personally pressing Churchill for greater Canadian activity.

On 17 March, King cabled Churchill: 'The strong considerations with which you are familiar in favour of employment of Canadian troops in North Africa appear to require earnest re-examination.' The following day Stuart wrote to McNaughton that if the invasion was impossible in 1943, 'we should urge re-examination for one *and perhaps two* [emphasis added] divisions going as early as possible to an active theatre.'[32] On 20 March both McNaughton and Churchill responded, the latter stating: 'I fully realize and appreciate the anxiety of your fine troops to take an active part in operations and you may be assured that I am keeping this very much in mind.' Churchill added that only one more division was to go to North Africa and that plans for HUSKY were too far along to include the Canadians. King eagerly accepted this explanation, declaring: 'The record, therefore, shows that it is quite clear that the Chiefs of Staff Committee in England and the Prime Minister feel that they [Canadian formations] should be kept where they are.'[33] McNaughton's own response to Stuart's suggestion was as follows:

> My view remains: (1) that Canadian forces *in whole or in part* [emphasis added] should be used where and when they can make the best contribution to winning the war (2) that we should continue to recognize that the strategical situations can only be brought to a focus in Chiefs of Staff Committee (3) that proposals for use of Cdn Forces should initiate with this committee (4) that on receipt of these proposals I should examine them objectively and report thereon to you with recommendations. I do not repeat NOT recommend that we should press for employment merely to satisfy a desire for activity or for representation in particular theatres however much myself and all here may desire this from our own narrow point of view.[34]

When King saw this cable he was 'well satisfied' with McNaughton's logic and 'felt more convinced than ever that my duty was to back up

McNaughton rather than Ralston ... There is a feeling between them which I think warps Ralston's judgment a bit.'[35]

Confronted by King's renewed obstinacy, Ralston and Stuart changed tactics. Stuart told King that morale simply could not hold any longer and that McNaughton 'had become far too removed from the troops.' Ralston backed up Stuart's charge. Even so, McNaughton's determination to avoid heavy casualties continued to strike a powerful chord with King, who noted in his diary: 'The Defence Department can make its own representations to McNaughton, but ... I will not let my name or office be used to further their [Stuart and Ralston's] point of view as against McNaughton's.' King must have felt fully justified in his position when, on 31 March, Anthony Eden 'made it clear ... that he felt that McNaughton and the British would wish to keep the Canadian Army intact for a final blow.'[36]

As March gave way to April, Brooke continued to provide McNaughton with conflicting assessments of future operations. On 5 April he suggested that there might be 'an essay against the continent in the fall of 1943 with a view to testing the enemy resistance,' with First Canadian Army advancing through an established bridgehead as practised during SPARTAN, which had taken place in early March.[37] Yet on 23 April, Brooke again informed him that there was little chance of operations against the Continent in 1943. That same day Brooke also put forward to McNaughton the first concrete proposal for a Canadian formation to take part in an active theatre. Because of the 'insistent' demands of Ralston and Stuart, Churchill had told Brooke to arrange Canadian participation in the next operation.

Brooke asked McNaughton to agree to the participation of one Canadian infantry division and one Canadian army tank brigade in HUSKY, the invasion of Sicily. Brooke indicated that the British 3rd Division would be 'very disappointed' to be replaced and that he and his staff at the War Office were against such a move, which would be made 'at great inconvenience to the British Army.'[38] Brooke then revealed Churchill's minute approving the inclusion of the Canadians subject to the proviso that British command arrangements be accepted completely. At this point McNaughton stated that he did not wish to see the Canadian formations broken up unnecessarily; however, if operations demanded such action, 'this might have to be accepted' because using the fully trained Canadian formations for a specific task such as the one proposed met the established Canadian principle of serving the general interest.[39] Brooke noted in his diary: 'Although [the] Canadians have continually been asking for this I received no sign of gratitude from McNaughton.'[40] McNaughton's lack of gratitude was perhaps an indication of his concern for the fate of the Canadian formations once HUSKY was completed.

Though given only two days to study the outline plans and to make a decision, McNaughton gave his provisional approval for the operation. He nominated the 1st CID and the 1st Canadian Army Tank Brigade.[41] From Brooke, he understood clearly that once the division had attained some battle experience in Sicily it would be brought back to Britain for the cross-Channel invasion. This was a standard provision that McNaughton had come to expect, for when BRIMSTONE was still in play in early January, he had been told that if it went forward, the Canadian division would be brought back in time for the invasion.[42] Perhaps the most satisfying element of agreeing to HUSKY was that after three-and-a-half years of war, Canadian forces were finally being committed to an operation that Brooke and the COS actually believed in. The long wait has traditionally been framed in terms of McNaughton's obsession with fielding a fully unified and autonomous force for the invasion of France.[43] However, it is now quite evident that this perception is wrong.

Not including the obvious obligation to fight in France in 1940, McNaughton had consented to five significant operations involving a brigade or more: HAMMER, JUBILEE, SLEDGEHAMMER, TONIC, and BRIMSTONE. He had also willingly explored the possibility of executing JUPITER if conditions were right. Of these major operations, only JUBILEE had been executed, but he had also authorized two other raids: GAUNTLET and ABERCROMBIE. McNaughton was forced to balance, as best he could, the considerable risks inherent in some of these operations with his earnest desire to help. Neither HAMMER nor JUPITER were sound operations of war, and JUBILEE proved to be a disaster that crippled the 2nd CID for a long time.[44] Thus the actual record shows that McNaughton was willing to consider and act on the various operations proposed by the War Office. The perception that McNaughton was the principal obstacle to the deployment of the army outside Britain has long been based on an uncritical acceptance of Brooke's position.

After the war Brooke described his dispute with McNaughton over deploying the Canadian Army in the following terms:

> The employment of the Canadian Forces ... presented many difficulties. McNaughton always informed me that the Canadian Government was

Table 5.1 McNaughton's willingess to employ Canadian troops, 1940–1943

Date	Location	Operation and unit
Jan 1940	Northern Scandinavia	5,000–7,000 men of 1st Div
Apr 1940	Norway	HAMMER (2 CIB)
Jun 1940	France (Original BEF)	1 CID
	France (2nd BEF)	1 CID*
Aug 1941	Spitzbergen	GAUNTLET (2 CIB)*
April 1942	Hardelot	ABERCROMBIE *
Aug 1942	Dieppe	JUBILEE (2nd CID (-)*)
Sept 1942	Norway	JUPITER (x 2 Divs) (willing to consider)
Sept 1942	Pas de Calais	SLEDGEHAMMER (up to 3 divs)
Oct 1942	Canary Islands	TONIC (2 Divs)
Dec 1942	Sardinia	BRIMSTONE (1 Div)
Jul 1943	Sicily	HUSKY 1 CID/1 Cdn Army Tank Bde*

* Operations actually carried out.

strongly opposed to the use of Canadian Divisions detached from the force as a whole. In fact such a use of Canadian troops could only be contemplated in an emergency ... The force as a whole was too large to embody easily in the formations we had in the field. I pointed out these facts to Ralston who entirely agreed with me that we must be allowed to employ individual Divisions in order to ensure that they had opportunities of gaining war experience.[45]

Brooke consistently identified McNaughton as the principal obstacle to deploying the army, but this is inaccurate. In 1940 and 1941 many factors beyond McNaughton's control kept the army in Britain. The Canadians were needed for the defence of Britain. Furthermore, Churchill did not want them to go to the Mediterranean, they were not particularly needed there, and King certainly did not want them to go despite his political statements about the army's availability to serve anywhere.

Brooke was undoubtedly correct that the Canadian Army was far too big to fit into any British operations between 1940 and 1943. Only a small portion of the British Army was in combat – and in a single, peripheral theatre in the larger European war at that. Moreover, the one time that an individual Canadian division was employed – at Dieppe – it had been destroyed within hours, limping back to Britain with a level of 'battle experience' hardly commensurate with the cost. McNaughton could have made all kinds of suggestions for employment, but that did not mean they would have been realistic or that they would have been supported by Brooke and the COS. Indeed, there is no evidence whatsoever to suggest they would have been supported. Also, such suggestions would have represented an ad hoc method for Canadian participation – something which McNaughton had refused to consider. Ralston, Stuart, and King had all supported him in this.

McNaughton's consistent position regarding the deployment of the army, as expressed to Ismay on 27 April 1943, was that 'all through our stay in England we had looked to the Chiefs of Staff to indicate our tasks. We had accepted those offered, had prepared for them with energy and dispatch and when they had not come to fruition we had expressed no discontent. In each case we knew that the decision not to proceed had been sound. We knew our importance as a general strategic reserve and that it might be fatal to dissipate it prematurely.'[46]

Brooke knew this well enough, for he consistently supported the idea and reality of a unified Canadian Army. In the absence of King's willingness to press for stronger influence over strategic decision making, McNaughton's policy was logical and inevitable. He attempted to make sense of the war's strategic direction through an almost bewildering series of formal meetings and backchannel conversations with numerous high-level individuals at the War Office.

McNaughton cannot be completely absolved of all criticism in terms of getting the army into action: it is clear that in his heart, he wanted to see the entire Canadian force cross the Channel together. The idea of Canada's first ever field army – the equivalent of six divisions strong – storming through Northwest Europe was almost intoxicating. Vimy II was in France, the principal theatre of war; it was not in North Africa, Sicily, or elsewhere in the Mediterranean. However, as the above analysis indicates, the blame for not getting the army into action sooner was not so cut and dried, and the long-held perception that McNaughton was unwilling to fight except under ideal circumstances favourable to his own hubris now needs to be revised by historians.

PART THREE
McNaughton as Military Commander and Trainer

6

The Difficulty of Training in 1940

Training was one of his major preoccupations.

John Swettenham

Between 1940 and 1943 McNaughton and Brooke waged a private battle not only over physical control of Canadian formations, but also over when and where those formations would fight. At times, such as after the Brittany operation in 1940 and Exercise VICTOR in early 1941, the tension between the two boiled over into sharp confrontations. Yet there was a parallel fault line between them. Soon after he became C-in-C Home Forces in July 1940, Brooke became concerned about the state of training in the 1st Canadian Division and later the Canadian Corps. Indeed, even after four-and-a-half years of preparation in England, it was generally believed that the Canadian Army was poorly trained when it entered combat in Normandy in 1944. Brooke blamed McNaughton for not properly training the army in Britain during 1940 and 1941.

After the war Brooke reflected: 'The more I saw of the Canadian Corps at that time the more convinced I became that Andy McNaughton had not got the required qualities of command. He did not know his subordinate commanders properly and was lacking in tactical outlook. It stood out clearly that he would have to be relieved of his command.' Brooke's apprehensions about McNaughton's professional shortcomings were so acute that he declared: 'I felt that I could not accept the responsibility of allowing the Canadian Army to go into action under his orders.'[1] Brooke's perspective was supported by other high-ranking Canadian officers such as Guy Simonds and Harry Crerar at the time and has been echoed consistently by Canadian military historians.

Charles P. Stacey, who personally observed much of the army's preparation in Britain, placed the responsibility for the training deficiency squarely at the feet of the unit commanders, maintaining that the army suffered from a 'proportion of regimental officers whose attitude towards training was casual and haphazard rather than urgent and scientific.'[2] This was a clear reflection of the amateur militia mentality. John English, however, disagreed with Stacey, arguing that blaming regimental officers 'clouds the issue that the Canadian high command did not know how to train them properly and, but for Montgomery's intervention, would have likely left them to learn the profession of arms through a process of osmosis.' English believed that McNaughton neglected operational and tactical training and 'proved incapable of training his division commanders.' As a result the army in Britain for all intents and purposes 'wasted its time.'[3] This is a serious indictment and warrants considerable investigation.

Military training as practised by Anglo-Canadian forces in Britain consisted of two fundamental types, 'individual' and 'collective.' Once individual soldiers were instructed in basic skills such as weapons handling and drill, they were collected into single-arm groups – an arm being infantry, artillery, armour, or engineers – to begin learning very basic or 'minor' tactics. A sub-subunit (single-arm) – an armoured troop, infantry platoon, or artillery troop – represented the level at which collective training truly began. A subunit was an armoured squadron, an artillery battery, or an infantry company, each comprising three or more sub-subunits. A unit was a regiment (actually battalion-sized) of three or more subunits; a formation, which contained elements from more than one arm, was a brigade or division and conducted what was termed all-arms collective training.

It must be remembered here that formation training in the field for brigade commanders did not start as soon as one was 'off the boat.' Conditions had to be met before the building-block process of training could make any real headway. Roughly 50 per cent of the soldiers making up the division in 1939 had had no military training whatsoever.[4] At an absolute minimum, officers and non-commissioned officers (NCOs) had to be trained and qualified in a broad range of skills by attending a multitude of courses before they could properly set about training their own men. Officer training needed to be concurrent with the progressive, bottom-up training cycle from individual to collective. The whole process was maximized when commanders at the lower levels received some training in the essentials before taking their subunits and units to the field and evaluating their subordinate commanders. Company commanders (major's rank) trained their platoon commanders (lieutenant's rank); regimental commanders (lieutenant-colonel's rank) – called commanding officers (COs) – trained their company commanders; and brigade commanders trained their regimental commanders. At the end of this long but necessary building process was McNaughton, who was responsible for training his three infantry brigade commanders (brigadier rank) and his Commander Royal Artillery (CRA).

McNaughton's grasp of progressive training was demonstrated in his first two training instructions, dated 26 December 1939 and 9 March 1940. Both were drafted by his General Staff Officer (GSO) 2, Major Guy G. Simonds, described as the 'workhorse of the G Branch staff.'[5] The CIGS, General William E. Ironside, anticipated that the division would join the British forces on the Continent at some point in May 1940, and McNaughton's training program seems to have been built around that time line. He anticipated achieving enough with individual training by March 1940 that platoons, companies, and batteries would be able to begin collective training. After a month of practice at this level, collective training at the regimental level was to commence, to be completed by the end of April. Brigade commanders would then assemble their regiments and conduct formation exercises between 28 April and 5 June. The cycle was to have been capped off by the division exercising as a full formation with artillery and engineers starting on 6 June. However, this sound training plan went awry almost immediately for multiple reasons beyond McNaughton's control.

On assuming command of the CASF in late 1939, McNaughton undertook the monumental task of raising, equipping, and training a volunteer force capable of taking on a professional German army that had been conditioned for war since 1933. In 1939, waging war effectively required large quantities of all the myriad types of modern mechanized equipment. Yet when he took the 1st CID to England in December 1939, McNaughton was immediately confronted with the debilitating problem of widespread equipment shortages. He later noted that in both the First World War and the Second, 'our deployment in ... personnel was far ahead of the possibility of their equipment with efficient weapons.'[6]

Stacey was in Britain at the time, engaged in writing reports on all aspects of Canada's war effort. He observed that the pace of mechanization and the increased complexity of modern equipment made the equipment problem 'more difficult than in the last war.' The artillery field regiments of the 1st Division had arrived in Britain without guns and were only slowly being equipped with outdated 18-pounders. The infantry units had disembarked with First World War–era Lewis machine guns instead of the more modern Brens. The equipment deficiencies undermined Canadian training efforts early in the war – a situation that could have been avoided, Stacey concluded, had McNaughton's plan for a Dominion Arsenal plant been acted on in the 1930s.[7]

The most critical material shortage was in motor transport. A report at the end of February 1940 noted that though the War Office was giving the 1st Division preferential treatment in training equipment, 'limited issue particularly motor transport hampers progress generally.' Another report a month later maintained that unit collective training was proceeding 'but now reaching stage where progress is definitely limited by shortages of mechanical transport.' The total loss of the 1st Canadian Infantry Brigade Group's transport in France only exacerbated the problem. In mid-July, Brooke noted in his diary that the equipment situation in Britain was 'appalling'; and in early November, Major-General Archibald E. Nye, the Director of Staff Duties (DSD), lamented the 'large deficiencies everywhere.'[8]

McNaughton was so concerned about Canadian equipment shortfalls that he advised CMHQ in late August that 'if no definite assurances of supply of equipment can be given by the War Office, we should consider advising Canada against the despatch of any additional troops.'[9] Yet by October the 2nd Division had assembled in Britain and there were two army field artillery regiments, one medium artillery regiment, and a host of other units. All of them needed equipment with which to train. As an expedient McNaughton planned to withdraw one-third of the 1st Division's weapons and equipment to allow the 2nd Division to train once the invasion season passed in late 1940.[10] Back in Ottawa the new CGS, Crerar, was also deeply concerned about outfitting the army: 'At this stage nothing in the way of military training should be allowed to interfere with production.'[11] Even Simonds acknowledged in a post-war interview that the 'terrible equipment shortages made it almost a pre-eminent requirement to get equipment for the troops.'[12] McNaughton had little choice but to devote disproportionate time and energy throughout 1940 – particularly the first half of that year – to the fundamental problem of equipping the army. He later reflected: 'I had been very deeply concerned with that.'[13]

Another problem McNaughton had to confront immediately was the serious shortage of commanders and staff officers capable of training the expanding army and of functioning at the increased tempo of modern war. The Americans faced the same problem. Their official historians observed: 'The basic training of soldiers, the advanced training of many officers of all grades, and the tactical training of units of all sizes ... had to be carried on simultaneously, with officers and men in every degree of proficiency or lack of it and with only a thin line of Regular Army officers and noncommissioned officers to take the lead.'[14]

General George C. Marshall observed in 1940 – at a time when the Americans, too, were struggling to find adequate senior commanders – that the leadership difficulties of the First World War 'have been enormously multiplied today by the increased mobility and firepower of modern armies,' with the result that the need for 'vigorous commanders is greater now than it has ever been.'[15] On 8 August 1940, McNaughton told Crerar, the newly appointed CGS: 'You will appreciate that number of officers with command and staff experience is limited.'[16] The cadre of Permanent Force officers totalled a meagre 450, and a significant percentage of them were unfit for field duty. Supposedly bolstering their thin ranks were the amateur officers of the Non-Permanent Active Militia (NPAM). The Defence Scheme No. 3 mobilization plans of the CGS in 1939 included an expeditionary force of two mobile divisions plus ancillary troops, but this was a paper exercise, because there were insufficient staff officers to flesh out the large formations. The stark reality was that the talent pool for demanding positions of unit officers and senior staff officers was not deep. This scarcity of expertise at the vitally important lower levels seriously undermined training as well.

Lieutenant-General Howard Graham, commander of the 1st CIB in Sicily and Italy, observed that in the early years there were many officers who were 'not to the standard of training required to prepare men for battle.' Major-General Chris Vokes, who would go on to command the 1st CID in Italy, was even more critical, judging the military competence of most officers in July 1940 to be 'extremely poor.'[17] Proof of Vokes's assertion is provided by Major-General Bert Hoffmeister, commander of the 5th Canadian Armoured Division (CAD) in Italy. In 1940 he was a major in the Seaforth Highlanders; he would later admit that at that time 'I hadn't a clue as to what an operation order looked like, or how to write one.' An operations order (Op O) was the basic method a commander employed to convey his intent for the actions of his unit or formation. He added: 'It disturbed me to no end to realize that here I was, a company commander ... and I didn't know how to look after myself in battle.'[18] Hoffmeister's situation was probably a microcosm of the broader lack of military expertise in the Canadian Army in 1940.

If there were company commanders who had never seen an Op O, then there was a very good chance that the staff officers available to McNaughton in 1939 and 1940 had experienced considerable 'skill fade' since first learning about such things at Camberley. Major-General George Kitching, a staff officer early in the war, stated that even by the end of 1940 there was hardly 'anything like the number of officers trained to staff and maintain the Corps Headquarters and two Divisions.' Only thirty-six staff officers were available in 1939 to fill Grades One, Two, and Three, and there was no system in place to produce more.[19] When the war started, the War Office cancelled the year-long staff courses at Camberley and Quetta, replacing them with staff courses of ten to seventeen weeks. On the first Camberley war course only five positions were reserved for the Canadian Army. The army's expansion only compounded the problem. By the summer of 1942 there was a shortage of 3,500 commissioned officers.[20]

McNaughton and the division staff organized as much 'in-house' training for officers as was possible, given the limited facilities and the dearth of qualified instructors. Division Headquarters planned and directed several map and sand-table exercises and outdoor TEWTs in order to demonstrate to the lower levels the fundamental principles of attack and defence. Also, full advantage was taken of the various courses offered by the British authorities, and each brigade had a British officer attached to assist in tactical training. McNaughton also directed that all officers arriving as reinforcements from Canada attend the Canadian Officer's Training Wing 'so that they may be ... instructed in the latest developments before joining their units.'[21] As far as Stacey was concerned, officer training at this early stage 'got due attention' in the midst of all the other pressing requirements. Yet McNaughton has been heavily criticized for not doing more to train young officers in 1940 – especially for waiting until late that year to establish the Canadian Junior War Staff Course (CJWSC).[22]

McNaughton established the CJWSC at Ford Manor in Surrey in the fall of 1940 with Simonds, by then promoted to lieutenant-colonel, in command. It seems that McNaughton recognized the need for a specific Canadian staff course much earlier but ran into resistance from Canada. He would recall: 'When I proposed the Staff College to the people back home they said, "You mustn't bother with that." Then I remembered telegrams that passed in which I told Crerar who was back in Canada by then that I felt it essential to do it ... I don't know why they opposed me in wanting to start it but I insisted on it and did it and I put some pretty good fellows down there.'[23]

It is also apparent that McNaughton and Crerar did not see eye to eye on relocating the CJWSC to Canada. McNaughton wanted staff training to remain in Britain. On 6 October 1940 he explained to Crerar that there was a lack of qualified instructors in Canada, that too much time was wasted travelling to and from Canada, that senior officers were on hand in Britain for lecturing, that formations were at war establishment, that the British staff colleges were close by, and that there was an ease in maintaining standards with British courses.[24] On 29 November he again laid it out to Crerar: 'It is obvious that officers who have been training in Canada cannot have the same experience on account of the smaller concentrations of troops and lower scales of equipment. It has been the policy to appoint selected regimental officers as "Staff Learners" on the staffs of various Formations and Headquarters of the C.A.S.F. Overseas in order not only to determine their aptitude for staff employment but also to give them practical experience in the field. From such officers the candidates for the staff course are selected.'[25] McNaughton's position on the CJWSC was correct: it is apparent that the relocation back to Canada did not raise the quality of instruction, nor did it sharpen staff training.[26]

Simonds ran a shorter course than the one at Camberley. At McNaughton's insistence he concentrated on elementary staff duties. It seems that McNaughton's focus was on achieving standardization. 'I wanted to train those fellows under my own hand,' he later said. 'I wanted to control what they were going to be taught.'[27] The result, according to English, was that 'the handling of large formations had to be learned on British courses, exercises, or in actual operations.'[28] Even the condensed instruction offered on the CJWSC did not immediately produce the required number of staff officers.

It is clear that the learning curve for officers at all levels was steep in 1940. When Sherwood Lett became the Brigade Major (BM) of the 6th CIB in late May he was suddenly responsible for the training of three battalions, though as the former CO of the Irish Fusiliers before the war he had never exercised a full battalion in the field. He later admitted that he had had to 'soak up an immense amount of knowledge in a short time, as did everyone in the Canadian Corps.' Basil Price, commander of the 1st Brigade, echoed Lett: 'In that very early stage we were all learning so much. A good many of us got away with a good deal perhaps that we wouldn't have a year later, if you know what I mean?'[29] The simple

Table 6.1 Command and staff element, 1st Canadian Division, January 1940

Division Commander:	A.G.L. McNaughton
GSO 1:	Lieutenant-Colonel Guy R. Turner
AA&QMG:	Lieutenant-Colonel E.W. Sansom
DA&QMG:	Major A.E. Walford
DAAG:	Major H.V.D. Laing
GSO 2:	Major Guy G. Simonds
GSO 3:	Captain M.P. Bogert
GSO 3 (I):	Captain C.C. Mann
CRA:	Brigadier J.C. Stewart
BMRA:	Major H.O.N. Brownfield
CRE:	Lieutenant-Colonel C.S.L. Hertzberg
1 Brigade:	Brigadier Armand A. Smith
BM:	Major Harry W. Foster
2 Brigade:	Brigadier George R. Pearkes
BM:	Major R.L.F. Keller
3 Brigade:	Brigadier Basil Price
BM:	Major Charles Foulkes

truth was that there was no quick fix to the problem of learning military skills. It is possible that McNaughton actually inhibited the building-block process by delegating too much of the lower training to his GSO 1, Lieutenant-Colonel Guy R. Turner.

As GSO 1, Turner was the division's Senior Staff Officer. He and McNaughton had struck up a close friendship in 1932 during the days of the unemployment relief camps. According to Elliot Rodger, Turner was a McNaughton loyalist who clung to his coat-tails. Military commanders are quite rightly judged in part by the quality of the men they surround themselves with. Selecting good subordinates requires good judgement, and McNaughton's decision to entrust Turner with considerable responsibility for the critical task of training was questionable.

How McNaughton and Turner actually worked together is not entirely clear. Turner never reflected on their relationship the way William Morgan did regarding his time as Alexander's GSO 1 in France. Morgan recalled how Alexander would constantly 'give me little expositions ... He loved tactics and ... when he wanted to illustrate a thing he took a pad out and drew sketches.' Together they roamed their area all day 'going round looking at training.' One obvious difference between Turner and Morgan was that the latter always knew that a key function of a senior staff officer was to spare his division commander 'as much detail as possible' in the day-to-day routine of command.[30]

After the war Basil Price made an important observation to Reginald Roy about the relationship between a division commander and his GSO 1: 'The training isn't solely the business of the G.O.C. His G.1 is the man who really goes after that you know, and of course he needs guidance and inspiration from his G.O.C. but you can have even with a man whose interests as G.O.C. are somewhat diversified, you can still have a G.1 who'll pound away at the training.'[31] But Turner never gained a good reputation for training while GSO 1 of the 1st CID. Stacey, who routinely saw Turner in the course of his duties as Historical Officer, questioned his capacity and judgement for the important position of GSO 1. Turner was something of a tyrant and seldom initiated action, choosing instead to allow Simonds to 'run with the ball.' As far as Captain M.P. Bogert, the GSO 3, was concerned, Turner was totally unsuited for the position of senior staff officer.[32]

The three fundamental deficiencies described above – in modern equipment, experienced commanders, and sufficient numbers of good staff officers – were not the only impediments to training. Even basic outdoor training was negatively affected by the unusually harsh winter at Aldershot, where the Canadians were concentrated. An endless string of ceremonial parades for dignitaries curious to get a look at the Canadians further curtailed elementary training efforts. Lieutenant-Colonel Ernest W. Sansom, the division's Assistant Adjutant and Quartermaster General (AA&QMG), wrote to his wife on 17 January 1940: 'We have so many visits to arrange programmes for that it is very difficult to get down to the real work of organizing and training the Division.'[33] Finally, when it is considered that the 2nd Brigade was also caught up in preparations for Operation HAMMER in April and that the entire division prepared for the Dunkirk and Brittany operations in May and June, it is easy to understand how progressive training broke down and would continue to do so every time McNaughton agreed to take part in an operation.

Despite the urgency of the equipment problem and the lack of commanders and staff officers, Simonds criticized McNaughton for focusing too much on these issues, concluding that he 'definitely neglected the operational aspects' in the early years.[34] McNaughton did spend considerable time and energy on non–fighting-related issues, and in this regard his role as Senior Combatant Officer Overseas, with its inherent bipolarism – military and political – did not help. Many observers at the time complained that McNaughton spent far too much time on weapons development. General Raymond E. Lee, the U.S. Military Attaché in London and an impartial observer, got the impression during their first encounter in October 1940 that McNaughton was 'more eminent on the technical than on the command side.'[35]

During the crisis over his command in November 1943, McNaughton told Ralston: 'I had myself done very little technical work since the war ... My own part in this had been largely the business of organization which was a proper function. I expressed the view that the provision of the best in weapons and equipments was very much the concern of an Army Comd, and I did not think I had spent undue time on this.'[36] Yet it seems clear that McNaughton underestimated the amount of time he devoted to such pursuits. An examination of his War Diary reveals a consistent and time-consuming dedication to weapons development. Burns noticed this, and McNaughton's one-time PA, Elliot Rodger, has confirmed it. One logical explanation for McNaughton's perspective as given to Ralston was that he was so good at it that it did not register as any burden. Moreover, there was an upside.

McNaughton's efforts to provide the Canadian Army with the best possible equipment and weaponry were praiseworthy, and he possessed the skill to do this well. Commander R.H. Stokes-Rees, working on guns in the Ministry of Supply, reflected that McNaughton made a 'very great contribution' to weapons and equipment. Brooke was so impressed with McNaughton's anti-tank obstacles that in early August 1940 he noted in his diary that he planned on pushing the idea.[37] Equipment and weapons development was vital because as Heinz Guderian observed in his 1937 study, *Achtung Panzer!* modern technical development had 'acquired storm force.'[38] Leveraging technology also saved lives. That cannot be denied – indeed, it was (and remains) the key element in the Anglo-Canadian (and American) approach to waging war. It must not be forgotten that the Anglo-Canadian armies throughout the war would pay a tremendous price in lives trying to carry out many combat tasks with inferior weaponry.[39] Brilliance at conducting operations mattered little in the end if one was consistently murdered by superior firepower at the 'sharp end.'

Still, Simonds and English have a valid point. Knowledge of how to conduct operations *was* important, especially since the 'how' had become far more complex. The 1939 panzer division, with more than 3,000 motor vehicles, required seventy miles of road space or eighty trains of fifty-five cars each to move. This mechanized phalanx bore little resemblance to the seven German armies that had invaded Belgium and France in 1914 with a combined strength of five hundred motor trucks.[40] This sort of mechanization had a dramatic impact on the time and space calculations at the heart of war. Major Ferdinand O. Miksche suggested in 1941 that it had increased the speed of development of any tactical situation by a factor of six or seven over the First World War. British military theorist Basil Liddell Hart maintained that the speed of operations had increased from 3 to 30 mph.[41]

An important caveat is that McNaughton and most other senior commanders in 1940 – including Lord Gort, Commander-in-Chief of the BEF in France – anticipated some form of static fighting in the early stages of a German attack on France. This made sense. Paddy Griffith perceived a 'great continuity of doctrine' from the First World War to the Second that 'argues against any supposed "watershed" in tactical affairs around 1940.' Indeed, the attacker had just as much difficulty penetrating consolidated defensive positions between 1941 and 1945 as his predecessor did between 1914 and 1918.[42] McNaughton certainly did not have a deep appreciation for the rotten nature of French defensive doctrine in 1940, but then again, neither did anyone else. Poland was overrun in three weeks, yes, but France was not Poland.

In anticipation of some type of static fighting, the War Office saw a need for trench training. In conformance, McNaughton had Canadian units rotate through the Trench Warfare Training and Experimental Centre (TWTEC) in Pirbright in early April. He has been criticized for allowing such training to take place, but it looks relatively modern when one considers that at the same time, in Georgia and Louisiana, Adna Chaffee's 7th Mechanized Cavalry Brigade was manoeuvring against Kenyon A. Joyce's traditional horse cavalry.[43] It would be grossly inaccurate to perpetuate the idea that McNaughton did not, or could not, see beyond the trench mentality of the First World War. On the contrary, as he told his assembled officers on 9 June 1940: 'Tactics that had been right and proper in the days of linear defence were no longer applicable. The characteristics of a defensive role were entirely altered. Linear trench and tank obstacles were no longer effective. General Weygand is now using zonal defences with strong points disposed in depth. These contain the anti-tank weapons and the theory was to draw the enemy into the gaps, have the artillery deal with their A.F.Vs. and then counter-attack. The fluid defence requires the highest form of training, discipline and determination.'[44] Three weeks later he informed Ottawa that the need for new tactics 'which differ radically from the old requires that Divisional Commanders be selected from men having current experience.'[45]

Though McNaughton had no opportunity to practise commanding the division as a complete formation in the field, and despite all the distractions and limitations, he still could have done more to pass on his concept of how he intended to fight the division in France to his brigade commanders, Armand A. Smith of the 1st Brigade, George Randolph Pearkes of the 2nd Brigade, and Basil Price of the 3rd Brigade. Perhaps he intended to do so once the division moved to France. There he could have set up the division in an actual operational setting and integrated his brigades properly. But he also could have gathered Smith, Pearkes, and Price (and the CRA, Brigadier James C. Stewart) for discussions and TEWTs while in Britain. McNaughton had always preached that mental preparation 'costs little.'

On 8 June 1940, McNaughton reported to Mackenzie King that the Canadian troops in Britain were 'battle worthy.' With invasion imminent he could hardly have declared them otherwise. However, he had conducted inspections of training throughout March and April, focusing on individual soldiers' skills, and was generally satisfied with what he saw, considering the problems he had encountered during the initial training period.[46] The following day he told the principal commanders and staff officers of the 1st Division that since mobilization he and his staff had 'not been looking for things to criticize, but to gain the confidence of the Commanding Officers and to be able to help them and dispose of their difficulties.' He felt that the period of training had been 'spent to the best advantage.'[47]

McNaughton's efforts in the first seven months of the war, coinciding with the 'phoney war,' must be viewed as an act of creation fuelled by an act of will. Helmuth von Moltke the Elder once said that an army 'cannot be improvised in weeks or months.'[48] Lieutenant-General Leslie J. McNair, Commander of Army Ground Forces, did not think that 'finished' troops could be trained in a single year.[49] McNaughton had been forced to improvise an army: to forge the 1st Division and create a viable military instrument out of nothing. Historians have probably expected far too much of the first six months of 1940. The old Canadian Corps did not suddenly materialize ready to take on the best of the Germans with a snap of the fingers.

It is safe to say that in the first half of 1940, McNaughton believed that he could wait on the higher training of his brigade commanders until such time as the 'act of creation' was complete. McNair himself recognized that there were too many old officers of dubious ability, but he doubted the wisdom of replacing them before new ones were assured. 'Officers of moderate capacity had to be kept on pending the training of better ones,' he observed, and 'new officers had to be trained along with the new troops whom they were eventually to lead in battle but meanwhile they could not exercise mature leadership in training.'[50] This was a growing pain that the Canadian Army would have to endure in 1940. There was no other viable option.

The cold, hard fact was that in the early months McNaughton had little artillery, outdated machine guns, a serious shortage of motor transport, and few facilities. He had few instructors, too few good commanders to supervise training, and even fewer qualified staff officers who might have filled the void. Visitors of all sorts were constantly demanding his attention, the harsh weather was driving much of the important outdoor training inside, and hastily conceived operations by the War Office involving Canadians were repeatedly interrupting the normal training cycle of some units and formations. It was not an optimal training environment.

7

The Politics of Training

> He could not see the deficiency in training and was no judge of the qualities required by a commander.
> Brooke, post-war comments on diary entry, 10 April 1941

> A great deal of the troubles in the Canadian Army really stemmed from the fact that Andy McNaughton would not focus his mind on training and operational problems, and for a very long time we were adrift.
> Guy Simonds to Reginald Roy, 29 December 1972

While McNaughton struggled mightily throughout the first half of 1940 to address the Canadian Army's materiel deficiencies, he fell behind in the actual exercise of command. After Dunkirk, his division returned to General Headquarters (GHQ) Reserve and the 'Canadian Force' came back into being. On 20 June, McNaughton told his senior commanders that the division now had to be prepared to possibly counter-attack in all directions. Therefore he decentralized the division into battalion groups and issued instructions that they were to be trained as complete combined-arms units.[1] The threat of German airborne drops was a major concern at the time, and McNaughton's approach to the problem made sense. He also devoted his time and energy to confirming routes for mobile counter-attacks and overseeing the construction of defences and stop lines.[2] As the new defensive arrangements for Britain continued, the C-in-C of Home Forces, Ironside, wanted a mobile reserve of corps strength south of the Thames. He indicated that McNaughton should command it.

On 21 July – the same day that Brooke replaced Ironside as C-in-C of Home Forces – after little more than six months in command of the 1st Division, McNaughton was promoted to lieutenant-general and assumed command of VII Corps. He was hesitant to take command for a few reasons, later reflecting that 'I didn't particularly want the job.' It meant taking on several 'derelict [*sic*]' formations from France and firing 'a lot of officers.' He did not want to fire British officers on his own, feeling that this was Brooke's responsibility. Moreover, as a result of becoming corps commander he actually functioned as GOC, Aldershot Command, for an extended period of time, which demanded 'a hell of a lot of administrative work ... which I didn't want to be bothered with, dealings with the Civil Power and the Home Guard and all these kinds of people.'[3]

The VII Corps was a powerful composite formation. It comprised Major-General George R. Pearkes's 1st CID, Major-General Roger Evans's 1st Armoured Division, elements of the 2nd New Zealand Division, and the British 1st Army Tank Brigade. Montgomery correctly believed that a corps commander had to be capable of handling '*any combination of divisions* – armoured or unarmoured,' but McNaughton had absolutely no experience with armoured divisions and none commanding his own Canadian division.[4] McNaughton was being pushed up too fast and never should have been given VII Corps.

A corps was the largest formation that fought tactical battles. Major-General Wade H. Haislip, commander of the American XV Corps, observed that a corps commander was the farthest commander to the rear who directed fire on the enemy and that he had to give purpose and direction to his divisions. Corps command required special abilities: to handle large numbers of troops, perhaps upwards of 80,000 (including corps troops); to judge relative capabilities of different types of divisions; and to anticipate operations days in advance.[5] Matthew B. Ridgway, commander of the XVIII Airborne Corps, observed that a corps commander must be 'a man of great flexibility of mind, for he may be fighting six divisions one day and one division the next as his higher commanders transfer divisions to and from his corps.' Moreover, he had to anticipate where the hardest fighting was going to occur, and be there in person, 'ready to help his division commanders in any way he can.'[6]

Though a corps commander conducted operations, the 'operational' level of war as understood today was not recognized by the Anglo-Canadian armies in the Second World War. It has even been suggested that the Anglo-American armies did not wage war at the operational level because logistical constraints had reduced war to a purely tactical endeavour.⁷ This is going too far: the general idea of 'grand tactics' – the movement of formations to the point where the fighting was to occur – did in fact exist.⁸ But it was army commanders, not corps commanders, who decided where, when, and for what purpose fighting was to take place. Army commanders gave corps commanders missions and provided them with divisions and other army-level assets such as heavy artillery to accomplish those missions. The same was true in the Wehrmacht.⁹ By way of simple example, General Sir Brian Horrocks recalled that Montgomery selected XXX Corps to lead the advance into Belgium and decided on the number of divisions it should have under command.¹⁰

The level of corps command had given both General George C. Marshall and Lieutenant-General Leslie J. McNair the greatest anxiety during the 1941 Louisiana manoeuvres. Recalling his First World War experience, Marshall stated: 'I saw the unfortunate results of corps command by individuals who had never commanded divisions in actual operations.' He subsequently insisted that a commander possess a 'sound basic knowledge of divisional requirements and operations' before moving up to take over a corps.¹¹ Under Marshall's system, McNaughton would not have been given VII Corps.

Since VII Corps was a composite formation, its staff was composite as well. McNaughton promoted Lieutenant-Colonel Guy R. Turner to BGS, but only for Canadian administrative issues (Staff Duties).¹² Brigadier-General Miles Christopher Dempsey, former commander of the 13th Infantry Brigade in France (and future GOC, Second British Army, in Normandy), was selected by McNaughton for BGS (Operations). Lieutenant-Colonel W.L. Laurie, Officer Commanding (OC) of I Canadian Corps Signals, told Stacey that Dempsey was 'exceptionally competent.' Brigadier-General Charles Alexander Phipps Murison, Assistant Quartermaster of the BEF, assumed the important position of Deputy Assistant and Quartermaster General (DA&QMG). Dempsey and Murison were British officers and quickly gained McNaughton's respect in their new positions.¹³

Throughout the summer and early fall of 1940, McNaughton was busy trying to establish a new corps headquarters; he was also continuing to work on equipment projects. In August he identified an 'absolute lack' of anti-aircraft guns in the Canadian formations; in September, General Headquarters (GHQ) asked him to organize two temporary super-heavy batteries of 9.2-inch guns on rail mountings to defend the Strait of Dover.¹⁴ Hurried defensive preparations coincided with the Battle of Britain that month, and in October the Canadian brigades began rotating

Table 7.1 Principal staff element, VII Corps and 1st Canadian Division, July 1940

VII Corps:	Lieutenant-General A.G.L. McNaughton
BGS (Admin):	Brigadier Guy R. Turner
BGS (Ops):	Brigadier Miles C. Dempsey
DA&QMG:	Brigadier C.A.P. Murison
CCRA:	Brigadier J.C. Stewart
CE:	Brigadier C.S.L. Hertzberg
GSO 2(I):	Major C.C. Mann
1 Division:	Major-General George R. Pearkes
GSO 1:	Lieutenant-Colonel R.L.F. Keller
AA&QMG:	Lieutenant-Colonel A.E. Walford
GSO 2:	Major Guy G. Simonds (attached for temporary duty)
GSO 3:	Major M.P. Bogert
GSO 3 (Moves):	Not filled
GSO 3 (I):	Captain D.K. Robertson
CRA:	Lieutenant-Colonel James H. Roberts
CE:	Lieutenant-Colonel J.H. Kennedy
1 Brigade:	Brigadier Armand A. Smith
BM:	Major Harry W. Foster
2 Brigade:	Brigadier A.E. Potts
BM:	Major A.C. Gostling
3 Brigade:	Brigadier Basil Price
BM:	Major Charles Foulkes

through static defences on the Sussex coast. The presence of the 1st Armoured Division in VII Corps should have been a golden opportunity for McNaughton to work with large armoured formations, but the division required considerable rehabilitation when it returned from France. McNaughton would recall:

> The first thing that we had to do was to get them so that they could run their bloody tanks. You see, we had to get them trained, the drivers and so on, and we didn't want to lose all the use of the tanks as strong points and help so we associated them with local infantry in suitable places and they were little packets worked in with the infantry while this individual training and so on was going. Well, as the divisional commanders began to get through that phase of training again, retraining if you'd like to call it, then we put them in and raised their business, first of all to companies of tanks that were in central positions that could be called on and finally the division was able to operate as a division again.[15]

Dempsey stayed with McNaughton after VII Corps was transformed on Christmas Day 1940 into the Canadian Corps. He also helped that corps plan anti-invasion exercises in its area of responsibility on the Sussex coast. As these rehearsals got under way McNaughton assured Brooke that an 'aggressive spirit' existed in all elements of the corps. However, certain fundamental problems quickly arose. Skill at road movement was especially lacking. A road move had three main elements: the approach to the Start Point (SP); from the SP to the Dispersal Point (DP); and from the DP to the bivouac area. The first and last were generally the responsibility of brigade and unit staffs; the actual move was a division responsibility. Calculations of time and space were based on the speed and density of vehicles and availability of roads. The road movement speed of infantry divisions was expressed in miles covered in one hour (m.i.1h), whereas the speed of armoured divisions was expressed in miles covered in two hours (m.i.2h). Density was expressed in terms of the number of vehicles per mile (v.t.m.). Regulating headquarters were small headquarters established by formations to enable the commander to control the movement through sectors, a sector being a series of routes eight to thirty miles long.[16]

Staff officers preparing road moves at any level had to consider vehicle density, speed, distance between vehicles, orders of march, time intervals between vehicle packets, and actions for a host of different scenarios, including accidents. Something as seemingly trivial as a single missed timing or the bunching up of vehicles could start a chain reaction leading to the paralysis of larger formations. Lieutenant-Colonels Rod Keller and Chris Vokes, recently appointed GSO 1 and AA&QMG respectively of the 1st CID, were ultimately responsible for making road movement work in the division by ensuring that sound traffic plans, based on the fundamental principles of road movement, were disseminated to lower formations.

Exercise FOX, conducted from 11 to 13 February 1941, stressed road moves to a concentration area and gaining contact with the enemy during an advance to reinforce the Home Guard on the Ashford pillbox line. Stacey observed portions of the exercise and noted: 'We ran into a very bad traffic jam, with the guns and vehicles of the 3rd Field Regiment, R.C.A., along the road with no intervals between them.' He was told at the time that the problem had been caused by the artillery not following their assigned routes. Pearkes actually found himself caught in the jam's epicentre. He later reflected that nobody could be blamed for it; even so, he wondered how carefully the corps staff had reconnoitred the ground beforehand so as to realize the difficulties. He concluded

Table 7.2 Command and staff element, Canadian Corps, January 1941

Corps:	Lieutenant-General A.G.L. Naughton
BGS (Admin):	Brigadier Guy R. Turner
BGS (Ops):	Brigadier Miles C. Dempsey*
CCRA:	Brigadier J.C. Stewart
CRE:	Brigadier C.S.L. Hertzberg
DA&QMG:	Brigadier C.A.P. Murison†
DAAG:	Lieutenant-Colonel C.S. Booth
CSO:	Lieutenant-Colonel J.C. Genet
GSO 1:	Captain Lord Tweedsmuir

*Until 10 June 1941
†Until July 1941

that the exercise 'had been badly planned and it was inevitable, in those roads, that there would be confusion.'[17]

If the exercise was badly planned at the corps level, then Dempsey and Murison were to blame. Division commanders did not do the staffing for road moves. Dempsey was responsible for overseeing the sound practices of the rest of the staff. The Corps GSO 3 and the AA&QMG would have worked together under the GSO 1 – all supervised by Dempsey – to prepare the corps-level road movement plan, which would then have been disseminated to the divisions. Surely Dempsey would have quickly identified any gross violation of road-move principles had he seen it, especially since he was considered an exceptional staff officer with a great capacity for detail.[18] Yet Dempsey had gone from commanding a brigade to supervising the movements of a large corps, and perhaps he too was learning. McNaughton suggested that Dempsey's staff style was quite informal and that 'there was a laugh and a joke about everything that went on.' Moreover, Dempsey functioned with a minimum of issued orders, preferring instead to pass on the commander's intent orally: 'This is what the boss wants.' As far as McNaughton was concerned: 'That's the way it should be.'[19] Perhaps the staff work was not as sharp as it should have been to execute precise road movements.

Down at Pearkes's level the division GSO 1 (lieutenant-colonel) oversaw the 'G' (operations and intelligence) Section of the headquarters and designed training plans in conjunction with the division commander. The GSO 2 (major) carried out the orders and instructions of the GSO 1 and, assisted by the DA&QMG, planned and executed divisional road moves according to the staff march tables. A GSO 3 (captain) understudied the GSO 2, supervised the 'G' draftsman in their preparation of charts and sketch maps, prepared nightly location reports (LOC STATS), circulated situation reports (SITREPS), and helped prepare and execute road moves. The GSO 3 I (Intelligence) prepared situation maps, updated the division commander's battle map, and made deductions from incoming intelligence.[20]

An anti-invasion exercise such as FOX was a defensive rehearsal, a special type of military exercise designed to perfect a set pattern of movement to a predetermined defensive area. The goal was to identify the best possible time to get into position. With this critical information in hand, the commander could react to enemy action with a high degree of certainty. Good road-movement skill was therefore critical. McNaughton admitted:

> I resented the big schemes ... First of all, the troops were not in a position to understand cooperation. We had to supply that from above if it were necessary. [The mission was to] hit the unorganized [invasion] detachment before they could do a thing about it and then gradually build up from that. If it comes to formal operations of war, you have the element of time on your side ... The first thing that became obvious was that the speed into action was the criterion which was of the greatest importance, and having regard to the roads ... it became obvious to me that we had to organize primarily on a battalion basis. The battalion commander became the key kingpin of the whole thing. From the central place he would get his general instructions and, mind you, they had to be very general, that there had been a landing ... Your mission is ... to go hell-for-leather and knock them out.[21]

One incident during the exercise aptly captures the problems inherent in sorting out a traffic mess once a road-movement plan unravels. When Keller and Vokes met Lieutenant-Colonel Howard Graham of the Hastings & Prince Edward Regiment at a crossroads, apparently they could do little more than tell the OC to 'get the hell home the best you can.' McNaughton was technically responsible for the traffic mess. That is the nature of command, but he depended heavily on the effectiveness, energy, and drive of his staff officers when it came to making a sound plan and on his subordinate commanders when it came to making sure the plan was properly executed. This is precisely why *Military Training Pamphlet* No. 47, *Movement by Road*, maintained that a 'great deal' of the success of any road move 'depends on the efforts of regimental officers.'[22] In this sense, Stacey was perfectly correct to blame the regimental commanders.

At the conclusion of FOX, McNaughton, who had acted as the exercise director, addressed various issues in a conference setting. Appropriately enough, his main finding related to traffic control, which in his view was now as important as the artillery barrage map had been in the First World War. The artillery's failure to get forward to assist the infantry owing to the poor assignment of routes 'almost justified corporal punishment.' He pointed out that unit and formation commanders needed to keep their signals officers close by so that they could advise on the signals situation before complicated tactical orders were issued. This was a constructive observation and one that Montgomery would make a year later.[23]

Many of the problems that arose during FOX resurfaced on Exercise DOG, held between 26 and 28 February. DOG was a repetition of FOX, but with Major-General Victor Wentworth Odlum's 2nd CID, in Britain since September 1940, replacing Pearkes's division. The sixty-year-old Odlum had fought in the Boer War and had commanded the 11th Infantry Brigade at Vimy Ridge. Norman Rogers, not McNaughton, had selected Odlum to command the division, though McNaughton might have intervened had he really wanted to do so. Rogers's choice had provoked considerable outrage among senior Canadian officers; his excuse was that he had made it to offset the feeling in Canada that too many professional soldiers were getting too many senior appointments.[24]

DOG was Odlum's first chance to command the full division on exercise, and Stacey and McNaughton were on hand to witness one of his initial Orders Groups. Stacey observed that he issued 'brief and incisive orders for the attack' to his brigadiers and the CRA. There were, however, problems with road movement, partly owing to breakdowns, civilian vehicles, and poor procedures.[25] During the post-exercise conference McNaughton again harped on the poor selection of routes and urged all officers not to be 'complacent' about fundamental staffing requirements. It is not known whether this comment was in reference to the corps staff or the lower staffs, but he was especially hot about a brigade taking a route different than the one ordered by corps. He called everyone's attention to the Field Service pamphlet *Mechanized Movement by Road*, urging careful study of its contents as a means to avoid traffic congestion.

After the war McNaughton reflected:

> We certainly found out how difficult it is to try to manage a division on the roads of Kent, the narrow twisty roads of Kent. We had some terrific jams …

We had to learn it, you know, you can't expect to be a master of roads and movement overnight. [Such confusion was inevitable] until we learned how to do it, the staff officers learned how to do it ... Those exercises were very useful because they failed ... We hadn't had a chance to really have our drivers trained and we hadn't our staff officers trained in the art of handling mechanized forces, motorized forces. We didn't know about the kind of road spaces we had to use and all that sort of thing and we got this whole division into one God-awful mess, I have never seen the like of it. It took three days to get it disentangled ... It was the finest thing we ever had and it let these fellows [staff officers] know that they just had to get their noses down in the ground and learn the art of movement, particularly the art of being able to move lorries at night.[26]

To ensure that the right information reached the right people in a timely manner, McNaughton referred his audience to two General Service (GS) publications of 1939, *Divisional Standing Orders* and *Infantry Brigade Standing Orders*. He stressed that commanders had to ask questions in Orders Groups and that units had to report their arrival times at new locations immediately; and he cautioned against radio silence without permission from the next higher headquarters. Finally, he emphasized that there must always be an officer present in formation headquarters authorized to act in the commander's absence. He may not have preached to his subordinates with Montgomery's messianic fervour but McNaughton nevertheless did what commanders were supposed to do: identify weaknesses and take measures to rectify them. Dempsey was ideally placed to comment on McNaughton's command technique and training ability during this period, but has left nothing for the historian to evaluate.

Brooke seems to have formed a negative opinion of McNaughton during the six months he struggled to equip the division, address myriad personnel and administrative problems, and gear up for hastily conceived operations by the War Office.[27] Brooke formed his negative opinion in large part while watching McNaughton's subordinate commanders on the anti-invasion exercises. In early April 1941 he watched Pearkes's 1st Division practise its counter-attack role in Exercise HARE. McNaughton was ill for a portion of the exercise, so Dempsey escorted Brooke, who recorded that he was 'depressed at the standard of training and efficiency of Canadian Divisional and Brigade Commanders' and that it was a 'great pity to see such excellent material as the Canadian men controlled by such indifferent commanders.' This was a direct indictment of Pearkes and his brigadiers, James H. Roberts, Arthur E. Potts and Hardy Ganong.[28]

Pearkes, a Permanent Force officer, was hardly 'indifferent.' He was a Victoria Cross winner, which meant that he knew what it took to 'win.' Potts felt that Pearkes was 'definitely on the job as Divisional Commander the whole time ... No other interest; single-minded, I'd say.'[29] Major-General Harry Crerar called Pearkes 'an excellent trainer of men and a forceful commander,' while Vokes considered him a 'great teacher of the military arts,' a 'born leader,' and an 'outstanding tactician.'[30] Pearkes seems to have done fairly well on HARE. The exercise report noted that proper reconnaissance had been done before moves and that road moves to the concentration and assembly areas had been 'well carried out. There was no congestion, and, as reported by 400 Sqn. RCAF who were operating on the enemy's side, there were few, if any, good targets presented to the air.' Indeed, the report concluded that the division had now 'mastered' road movement.[31]

The only conceivable criticism of Pearkes would have related to certain timings. He had issued a warning order on 10 April at 0800 hrs and had given confirmatory orders at 2000 hrs with H-Hr at 0445 hrs 11 April. Perhaps this was somewhat too compressed for the artillery preparation. The attack itself, launched on a one-brigade front with all divisional artillery in support, went somewhat astray when some units lost direction. As the narrative stated, this would most likely not have happened had the guns actually been firing. Brooke, however, had little time for either Pearkes or Odlum; after their first encounter in May 1940 he described the latter as nothing more than a 'political general.' In early September 1941, Brooke would tell McNaughton that Odlum was 'too old' and that he viewed the possibility that he might succeed to command of the corps 'with grave apprehension.'[32] Brooke filled his diary with damning comments about the abilities of many commanders, British and Canadian, and it is necessary to understand his motivation.

Brooke was one of the best generals produced by Britain during the war. Montgomery considered him the 'best soldier that *any* nation has produced for very many years.'[33] At first glance Brooke's limited combat experience in the Second World War – he commanded II Corps for little more than a month during the Battle of France – does not seem sufficient grounds for the decapitating judgements he passed on so many commanders. Yet as Nigel Hamilton has argued, he 'towered above his colleagues' in France because he was the only one who had 'prepared himself in mind and deed' for the ordeal. His difficult blocking action on the Ypres–Comines Canal in late May proved sufficient to mark his 'greatness as a field commander.'[34] Brooke's reputation as a commander is undoubtedly exaggerated, but his professionalism is beyond dispute.

The defeat in France was a body blow to Brooke and led him to seriously question the professionalism of the British Army. Conversely, the effectiveness of the German onslaught led him to remark: 'It is ... nothing short of phenomenal ... There is no doubt that they are most wonderful soldiers.'[35] Even before the Germans attacked on 10 May, Brooke expressed deep concern about the fighting effectiveness of the BEF and its commanders, reserving particular animus for Major-General Victor Fortune, commander of the 51st (Highland) Division. When Brooke returned to England he quickly began ridding the army of weak and unprofessional commanders.

In late July Brooke visited Major-General Kenneth A.N. Anderson's 46th Division. He condemned it for its 'lamentably backward state of training' and declared it 'barely fit to do platoon training.' Major-General Alan Cunningham's 9th Division was equally eviscerated; and of the commander of the 15th (Scottish) Division, Brooke recorded: 'doubtful whether he is good enough.' Indeed, between the time Brooke became C-in-C Home Forces in July 1940 and the end of the year, thirteen British division commanders were relieved, and it was he who cashiered most of them.[36]

Brooke's pursuit of Major-General Roger Evans's relief led to another direct clash with McNaughton soon after their falling out over the Brittany operation. Evans had commanded the 1st Armoured Division in France and was now in McNaughton's VII Corps. When Brooke paid a visit to Corps Headquarters he told McNaughton: 'You've got to give me an adverse report on that man, I don't want him [Evans] here in this thing.' McNaughton replied: 'I'll only give an adverse report when I've got an adverse opinion.' Evans had been with McNaughton at Camberley. McNaughton admitted that he had never had a high opinion of him, but added: 'You see, there's a lot behind it ... and we don't need to go into detail ... I wouldn't fire Evans on general principle because I don't fire people without reason, I don't fire people by order, I never have in all my life.'[37] Brooke could only have been angered by such a reply and relieved Evans on his own authority as C-in-C Home Forces. This verbal exchange, even if only partly accurate, demonstrated their inability to communicate effectively. It was perhaps because of the unwillingness to 'mark' Evans that Brooke first conceived the notion that McNaughton did not properly know his subordinate commanders or grasp their limitations.[38] Conversely, McNaughton felt some hesitation about forcing

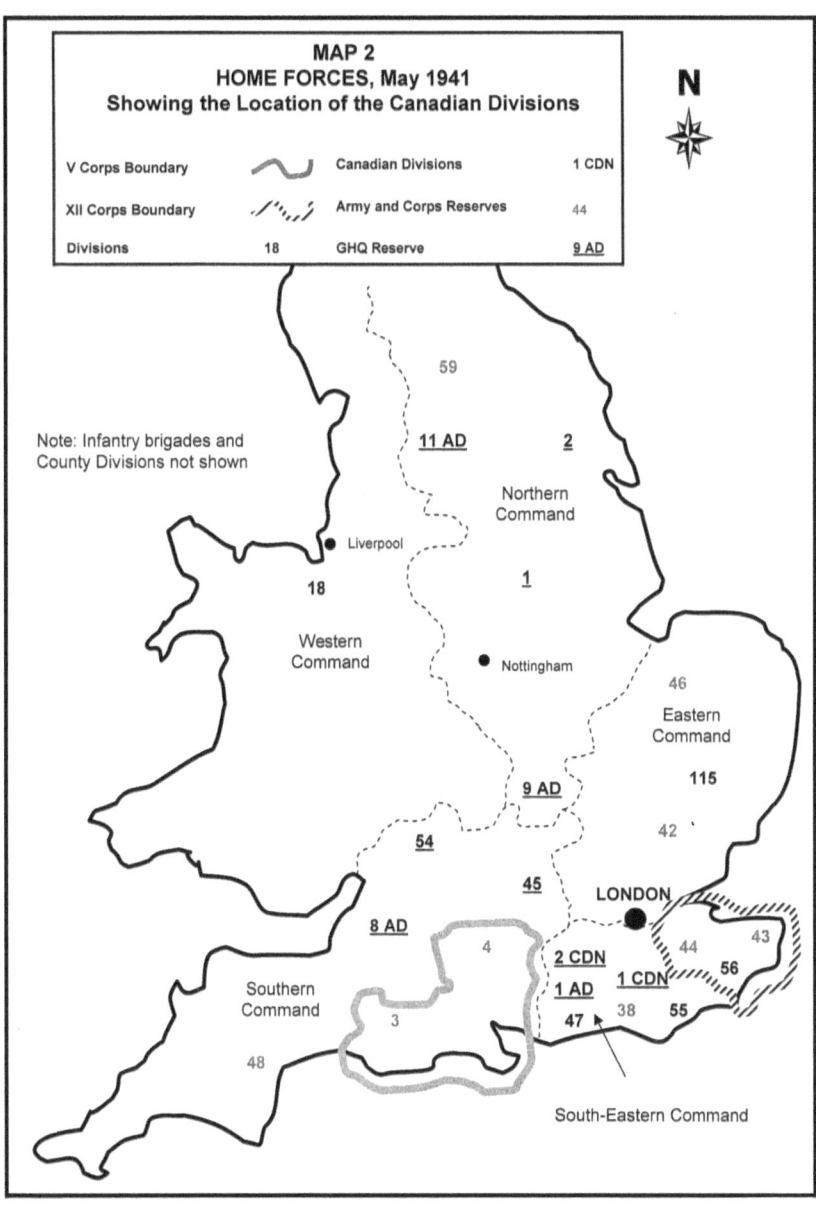

action against British generals (reflecting the way British generals felt about taking action against Canadians) and most certainly would have had reservations about relieving Evans on his own authority.

During FOX and DOG, McNaughton had been the exercise director. It was Exercise WATERLOO, run by Lieutenant-General Bernard Paget's recently established South-Eastern Command between 14 and 16 June 1941, that represented his first chance, after eighteen months in England, to actually command in the field. Moreover, it offered him a chance to exercise the complete Canadian Corps and to practise his neophyte corps commander skills. WATERLOO was conducted on the central Sussex coast and involved some 100,000 troops. Its purpose was to practise the corps working with (but not having under command) Major-General Richard L. McCreery's 8th Armoured Division in an anti-invasion role. Other training objectives involved moving formations through areas infested with enemy airborne forces and testing the effectiveness of infantry formations attacking armoured formations.[39]

The Canadian Corps got off to a slow start, but that seems to have been the result of miscommunication between Brigadier Eedson Burns, who had replaced a worn-out Turner as BGS in May, and Brigadier Thomas J.W. Winterton, Montgomery's BGS at XII Corps. Burns, who would go on to command I Canadian Corps in Italy, informed Stacey that McNaughton was to have been given the order to move at a scheduled conference; Winterton indicated that he had called Burns prior to the conference specifically to expedite the corps' movement. Burns, however, got no such impression from Winterton's phone call.[40] The situation was clarified only when McNaughton actually reached the conference.

Basic road movement continued to give the Canadians trouble during WATERLOO. Odlum maintained that it was 'typical of the confusion to be expected in war'; George G. Blackburn, a soldier in the 4th Field Regiment, described the exercise as 'a complete and utter shambles,' owing in part to heavy rains, to inexperience in map reading and the 'tortuous maze of English roads, from which all signs have been removed that might help an invader.'[41] Odlum had taken his division through a road-movement exercise, BENITO, between 9 and 11 April, so some improvement should have been noticeable on WATERLOO. As it happened, however, Winterton criticized Odlum's inability to get the division moving despite being at two hours Notice to Move (NTM). Mc-Creery was also heavily criticized by Major-General Giffard Le Q. Martel, Commander Royal Armoured Corps (CRAC), for his apparent lack of skill at forcing river lines. By failing to meet the timing for an important bridging operation by seven hours, McCreery had placed Pearkes's division in a bad way. McCreery seems to have drawn the sharpest criticism on the exercise.[42]

WATERLOO was Burns's first opportunity to see the Corps Headquarters function in the field, and he quickly identified 'certain deficiencies of organization and staff work.' The problem as he saw it was due in large part to the 'rapid changes in appointments which were then occurring, as a result of the expansion of the Canadian forces. An officer would hardly have had time to learn his job when he would be posted elsewhere, usually with promotion.'[43] This was an underappreciated impediment to higher training.

Brooke paid 'special attention' to the Canadians during WATERLOO and was 'not happy' with some of the senior officers.[44] He spoke to McNaughton on the way home about the command problem in the corps but unfortunately did not elaborate in his diary. This is a double loss, for McNaughton's reaction was not recorded, and neither was Brooke's method of raising with him the subject of training senior commanders. Indeed, the actual 'process' by which a superior corrects deficiencies in senior subordinate commanders remains a marginalized aspect of military history.[45] Historians often ignore the fact that Brooke seemed incapable of clearly expressing his concerns to McNaughton. A partial answer perhaps lies in Brooke's comment to Montgomery several months later that the Canadians were 'very touchy and childlike in many ways.'[46]

McNaughton's next opportunity to command the corps in the field came during BUMPER, a massive inter-army exercise held between 29 September and 3 October 1941. His corps (the 1st and 2nd Canadian Divisions) was under Lieutenant-General Harold Alexander's Southern Command. Lieutenant-General Lawrence Carr's Eastern Command played the German 'Sixth Army.'[47] Brooke ran BUMPER to give higher commanders the much-needed opportunity to handle large forces; to test the army's organization with a view to possible deployment on the Continent; and to practise the anti-invasion role.

McNaughton's BGS during BUMPER was Guy Simonds, who had assumed that post on an acting basis in August. He had been guiding most of the Canadian Corps' training and was deeply involved in preparations for BUMPER.[48] There is no indication that Brooke was displeased with McNaughton's performance on BUMPER, and Pearkes executed a successful counter-attack with his division. But Montgomery, who was Chief Umpire for the exercise, criticized Odlum (and the 6th Armoured Division) for missing opportunities to take offensive action.[49] It is possible that this subjective judgement of Odlum further implicated McNaughton in Brooke's eyes on the charge that he was failing to train his division commanders. Odlum had been slow during WATERLOO and had seemingly not improved. Brooke also contended that the laborious move of Odlum's division in the rear of the British 25th Army Tank Brigade had robbed Alexander of operational flexibility.[50]

It is clear that during BUMPER, Brooke focused his attention on the newly established British armoured divisions. McNaughton had no armoured divisions under his command during BUMPER, and after two years in Britain he still had little direct experience with them. What he did have was the benefit of attending the armoured conferences with exercise demonstrations conducted by Brooke at Camberley in early January and March 1941. These had been attended by all the C-in-Cs of Commands, by four corps commanders, and by all the armoured division commanders. Paget, Alexander, Montgomery, McCreery, John T. Crocker, and many others had all been there. Brooke had lectured on armoured division organization, signals, and administrative layout. His intent had been to instil 'a little more offensive spirit' in the army, and he had expressed concern regarding the stagnation of higher-level training for commanders. During the second conference he had told his audience: 'I think that these exercises ... have done good in educating formation commanders as to how to handle their commands in the event of an invasion.'[51] Brooke's pedagogy was now being tested on big exercises like BUMPER and was being found wanting in certain instances.

After watching Alexander and Carr manoeuvre their armoured divisions on BUMPER, Brooke lamented the 'sad mishandling ... by Higher Commanders' and concluded: 'It is lamentable how poor we are as regards Army and Corps Commanders, we ought to remove several of them, but heaven knows where we shall find anything very much better.' Alexander survived his weak performance, probably because of his disarming personality, but Brooke judged that Carr lacked the required qualities for a higher commander and 'started planning for his relief.'[52] Not surprisingly, the field exercise quickly became Brooke's principal method of evaluating commanders – albeit not a perfect one by any means.

The manoeuvres conducted in Britain during 1940 and 1941 faced several significant handicaps. Efforts to simulate actual battle were undermined by space limitations. George Kitching would remember Pearkes arguing after BUMPER that England was no place to train an armoured division because there were too many restrictions on cross-country movement.[53] The U.S. Army was able to conduct large-scale exercises at home across whole states; yet Crerar while CGS chose to forgo such training in Canada on the grounds that proper facilities did not exist. When the Canadian press pressured Crerar to conduct Canadian-based manoeuvres, he sought McNaughton's help in emphasizing the importance of individual training in Canada.[54] Ironically, only a few weeks before soliciting McNaughton's assistance, Crerar had informed Major-General Price Montague, the Senior Officer at CMHQ: 'The ideal which we are working to is to so equip and train our Army formations and units in Canada that, on arrival overseas, they will be ready to do a fairly good battle without further noticeable delay.'[55] This was wishful thinking and unachievable as long as opportunities to conduct manoeuvres in Canada were wasted.

McNaughton understood the problem of make-believe exercises. He told Stacey during Exercise ALBERT at the end of July 1941 that lack of realism was a real detriment to training.[56] In addition, supervising the conduct of these templated exercises was a cadre of umpires. Every commander – be it McNaughton, Alexander, Carr, or whoever – was to a considerable degree at the mercy of the umpires, many of whom had themselves been 'retired' by Brooke, Montgomery, Paget, or others for not measuring up to field command. Umpires relied on an arithmetical process to make their decisions in the context of a pre-programmed exercise, which meant there was little tolerance for unorthodox solutions. One recent scholar of the British Army has noted that outcomes of large-scale exercises were 'usually predetermined' because each 'took place within a perimeter established by timetabling, geographical and conceptual limitations'; thus it was 'not at all certain' that umpires accurately judged tactical performances.[57]

Brooke also used large-scale field exercises to test the army's doctrine. 'Doctrine' referred to guidance approved by the highest military authority for conducting battles; it bound together the functional aspects of training, equipment, and organization of military forces.[58] There is widespread agreement among historians that considerable doctrinal confusion existed throughout 1940 and 1941 and well into 1942. Anglo-Canadian doctrine evolved pragmatically but slowly.[59] The string of British defeats in North Africa following O'Connor's early victories at Sidi Barani and Beda Fomm against the feeble Italians can be traced to the Bartholomew Report of July 1940.[60] That report, which evaluated the BEF's performance during the Battle of France, recommended making the brigade group the primary manoeuvre formation instead of the division. Such decentralization was thought to be the answer to rapid German decision making.

The report's findings were quickly captured in the War Office's *Army Training Instruction No. 1,* subtitled *Notes on Tactics as Affected by the Reorganization of the Infantry Division* and published in January 1941. The pernicious effect was that in the desert, British armour quickly demonstrated a pronounced habit of piecemeal attack and defence. Thus during CRUSADER at the end of 1941 the 7th Armoured Regiment was left to defend itself against two panzer divisions while other British armoured brigades continued to defy the principle of concentration.[61]

British doctrine naturally exercised a powerful influence on the unblooded Canadians. Yet McNaughton correctly argued against what he saw as fundamental weaknesses in the Bartholomew Report's recommendations.[62] He strongly opposed Paget's suggestion that divisional artillery be reduced to forty-eight guns, with the remaining twenty-four permanently attached to brigade groups to achieve decentralization; and he was justifiably annoyed that he had not been consulted about artillery matters.[63] The fact that Montgomery quickly stopped the habit of parcelling out artillery before the battle of Alam El Halfa in August 1942 proved that McNaughton's instincts had been right in 1940 and 1941.[64]

Anglo-Canadian doctrinal confusion throughout 1940 and 1941 was a serious problem that negatively affected the training of formation commanders. As Dominick Graham and Shelford Bidwell have argued, tacticians do not have to be geniuses but they must have a 'clear view of the aim [and] an understanding of the tools.' Only when they understand the relationship between the types of weapons they possess and the organization established through doctrine can they develop 'a sense of what is practicable.' Because of changes in weaponry and constantly evolving doctrine, it was difficult to determine what to teach. Montgomery's comments on Exercise TIGER, conducted in May 1942, highlighted the problem. TIGER represented the first trial of the new model infantry and armoured divisions. Montgomery stated: 'All commanders must understand clearly that the new model division, whether armoured or infantry, is an entirely different instrument to the old type ... Obviously, much practice will be necessary before we reach a final doctrine; and it will probably require the practical experience of battle to prove the right use of the new equipment.'[65] In a broad sense this sensible appreciation was difficult for British and Canadian commanders to achieve. Brooke may have infused the army with a new spirit after Dunkirk, but he never succeeded in imposing a consistent battle doctrine on the army to mitigate the effects of decentralized training programs.[66]

BUMPER was McNaughton's last opportunity to command the Canadian Corps in the field and to test doctrinal problems associated with division and corps-level operations. He had had no opportunity to practise division command, and his two opportunities to exercise corps command were hardly an acceptable substitute. Moreover, his next opportunity to command in the field, during Exercise SPARTAN in March 1943, would not come for another sixteen months. In the meantime, important changes would be made to the command structures of the British and Canadian armies that would directly influence how the Canadian Army trained from that point on.

8

Enter Montgomery

> Those commanders who are known to be bad must be told they are bad, *and must be taught so that they will become good.*
>
> Montgomery, 14 April 1942

On 17 November 1941 Montgomery took over South-Eastern Command from Paget and immediately began to make his influence felt on the Canadian Corps. Paget moved up to take over Home Forces from Brooke, who in turn, on 1 December, assumed the duties of CIGS from John Dill. While Montgomery, Paget, and Brooke were all rising to greater positions of authority and influence over both the British Army and the Canadian Corps, McNaughton fell ill with a low-grade chest infection and went on sick leave. Crerar believed that McNaughton had actually suffered some form of breakdown.

When Guy Turner departed for a short rest in May 1941, Crerar wrote McNaughton: 'I only trust that his absence does not result in you consistently overworking yourself ... Mike Pearson ... told me that you had not been at all well and had been off duty for ten days or so.' A few months later, Crerar was getting reports that McNaughton was 'in fine fettle' and was not working himself as hard as in the past. Clearly, though, McNaughton was exhausted by the end of 1941. As far back as March 1940, Sansom had noted that McNaughton was 'tired out.'[1]

The War Office already had concerns about the ages of various commanders. When Mackenzie King visited Canadian troops in early September 1941, Montgomery spoke highly of McNaughton but quickly added that many Canadian officers were too old. In Britain the following month Crerar was informed 'very confidentially' by Major-General

John Noble Kennedy, the DMO&I, that the War Office was worried about the age of some senior Canadian commanders. Crerar replied that the matter had not escaped 'our own attention.' By the beginning of January 1942, McNaughton had instituted age limitations in the Canadian Army.[2]

Brooke and Montgomery had been strongly prejudiced against older senior commanders since the breakdown in France of forty-nine-year-old Major-General H.C. Loyd and fifty-six-year-old Lieutenant-General Michael Barker, GOCs of the 2nd Division and I Corps respectively. Eisenhower had held the same bias since watching the large-scale exercises in the United States. Corps, army, and even higher commanders, he declared, 'have to be relatively young men.'[3] Exceptions like Patton, already fifty-eight years old in 1942, were rare indeed. McNaughton, at fifty-four, was clearly in Brooke's cross hairs.

Perhaps the most important development at the end of 1941 that was to affect McNaughton's future was Crerar's arrival in London in late December. Crerar was desperately seeking a field command, telling McNaughton in mid-July 1940: 'I'm hot after the 3rd Division ... [I] count on you to help me get it.' Crerar lobbied hard on his own, with considerable support from Brooke, whose diary entry for 16 October 1941 noted: 'Busy trying to fix up Crerar to command 2nd Can Div ... Am getting Canadian Defence Minister [Ralston] to lunch with me on Monday to try and square matters up.'[4] Crerar replaced Odlum in command of the 2nd CID on 19 November (as a result of the new age limitations) but never actually commanded that division on any exercises.

When McNaughton temporarily stepped aside for health reasons, Pearkes, the senior divisional commander, became Acting Corps Commander. However, McNaughton did not feel that he was capable of commanding the corps on a permanent basis.[5] McNaughton considered James Roberts the best available officer in Britain for that task, though he also felt that Crerar – who had never commanded anything since 1917 – should be considered. McNaughton's reasoning seems to have been simply that Crerar was senior to Pearkes. Brooke concurred, even though in June he had specifically criticized Auchinleck's decision to give Eighth Army to Neil Ritchie, who had never commanded a division in action.[6] Incredibly, on the last day of 1941 Crerar assumed command of the Canadian Corps with no formation training – let alone battle experience – under his belt in the present war.

McNaughton's judgement in selecting certain personnel was open to criticism. Choosing Turner to be his GSO 1 was clearly a mistake. However, giving Crerar the corps made even less sense at the time. Major-General D.C. Spry, McNaughton's one-time PA and later commander of the 3rd CID, felt that Crerar should have 'stayed in Ottawa' to deal with the politicians.[7] Crerar was completely unqualified by recent experience for corps command, but he enjoyed Brooke's confidence, which was a powerful boost. '"Brookie" and I,' reflected Crerar, '[had] frequently tramped and reconnoitred together [during the First World War and] formed a firm friendship, happily refreshed by being brought together several times in our later service careers.'[8] Indeed, they often met socially for dinner whenever Crerar was in London, especially after he took over the corps. McNaughton might well have fended off Brooke's agenda on his own, but bringing Crerar into such close proximity with the CIGS was to create problems that he never anticipated.

With Crerar firmly installed in command, the final decision on the army's size and composition followed, with important repercussions for McNaughton's influence on training. When Ralston returned to Ottawa from Britain in November 1941 he told the CWC that as far as the War Office was concerned, the most important contribution Canada could now make was another armoured division.[9] The War Office never saw the need for a Canadian army headquarters, and from a command-and-control perspective five divisions should not have posed a problem for a corps commander. By this time, Brooke and 'other' officials at the War Office had suggested to Crerar and Ralston – during their October 1941 trip to London – that McNaughton's true abilities lay on the scientific side.[10] Brooke was not the first to suggest this. The idea of somehow displacing McNaughton – perhaps by elevating him to army level – seems to have taken root several months earlier.

On 18 August 1941, Colonel Eedson Burns, representing the General Staff on the Tank Production Committee in Camp Borden, Ontario, prepared a curious document titled 'Command of the Canadian Army in the United Kingdom' in which he suggested that McNaughton's dual role of corps commander and Senior Combatant Officer Overseas 'gives rise to certain anomalies, and difficulties.' The numerous administrative burdens, Burns noted, 'take up a great deal of the Corps Commander's time, and make it impossible for him to devote as much attention as he would like to training and operational questions. Due to the Corps Commander's special qualifications, a great many scientific and technical questions are referred to him, which makes further drafts on his time and energy.' Burns suggested elevating McNaughton to GOC-in-C,

Canadian Army in Britain. In that post, his main responsibility would be to advise the government on military operations. The Canadian Corps, most probably, would be placed 'in combination with' GHQ Home Forces.[11]

McNaughton had only recently sent Burns back to Canada after British censors discovered he had made embarrassing comments in a letter to his mistress in Montreal. It is highly unlikely that Burns – having been reduced in rank to colonel, removed from the prestigious position of BGS of the Canadian Corps, and thoroughly chastised by Ralston – would have made such a proposal on his own. It was Crerar who had saved him, sending him to Camp Borden to administer the new Canadian Armoured Corps. It is tantalizingly possible that Crerar influenced Burns's proposal, for the CGS always maintained that the reason for creating an army headquarters related to concerns over McNaughton's command and training ability.[12] Moreover, when Crerar returned to Canada to become CGS in July 1940 he had brought Burns with him, given him the title of Assistant Deputy CGS, and instructed him to analyse procedure and organization within NDHQ.[13] Perhaps Crerar was still giving Burns special assignments while the latter was in Borden.

When the Assistant Chief of the General Staff (ACGS), Brigadier-General Maurice Pope, perused Burns's draft document, he told him: 'I have no reason to believe that General McNaughton would wish to relinquish command of the Corps. His range of interest and activity has always been extremely wide.' Pope felt that 'so long as the position does not become too awkward, and I don't think it has, we should leave matters as they are.'[14] The issue lay dormant until 7 January 1942, when Crerar lunched with Brooke. According to Brooke, they discussed 'the organization of the Canadian Force now that it was expanding beyond the strength of a Corps. The question of the command of this Force was also giving me grave anxiety. I knew Andy McNaughton very well ... A brilliant brain best suited for matters connected with scientific development, but not gifted with the required qualifications of a Commander.' Brooke never told McNaughton face to face that he did not have the 'required qualifications.' Relieving him on his own authority would have been a 'drastic measure,' especially since he had been 'worked into the position of a national hero by the Canadian press.'[15]

Instead, Brooke plotted with Crerar, suggesting to McNaughton on the very same day he had lunched with Crerar that the Canadian force 'seems to me to be growing too big to be handled by one Corps Com-

Table 8.1 The expansion of the Canadian Army, 1939–1943

Formation	Date authorized	Date formation complete in UK
1 CID	September 1939	December 1939
2 CID	September 1939	September 1940
Canadian Corps	May 1940	December 1940
3 CID	May 1940	September 1941
4 CID*	May 1940	–
5 AD (formerly 1 AD)	January 1941	November 1941
1 ATB	January 1941	December 1941
First Canadian Army	January 1942	April 1942†
4 CAD	January 1942	November 1942
2 ATB	January 1942	June 1943
II Canadian Corps	January 1942	January 1943

*Reorganized as 4 CAD.
†Skeleton form only.

mander.' An army headquarters would 'free the Corps Commanders' hands for the job of commanding and training the fighting formations. That in itself is a full time job!'[16] There is no record of McNaughton's immediate reaction to Brooke's proposal. It is apparent, however, that Pope was suspicious. He later suggested: 'I always felt that the recommendations which eventually led to the formation of the First Canadian Army were not wholly objective. I shall say no more.' Nigel Hamilton has argued that establishing an army headquarters was 'in fact, Brooke's way of trying to kick McNaughton upstairs.'[17]

McNaughton, on sick leave since mid-November 1941, departed for Canada on 23 January 1942 for a further rest and only returned to Britain in late March. Immediately on his return to duty he became immersed in working out the details of establishing First Canadian Army. It officially stood up on 6 April. As it happened, the creation of a new, high-level headquarters placed a significant burden on the staff officers and only compounded the training problem. McNaughton recognized that the army headquarters would have to operate on a 'nucleus basis and would be developed gradually as suitably qualified officers became available.'[18] As a consequence the headquarters would, in the words of Brigadier Matthew H.S. Penhale, 'remain aloof from any operational control of Canadian troops placed under command SECO [South Eastern Command].' In late December, before Brooke discussed the subject with McNaughton, Paget had agreed with Crerar that an army headquarters should do as much administration as possible. McNaughton's strongest focus and effort was now on studying the difficulties and training for a cross-Channel invasion.[19]

Under this new arrangement McNaughton was effectively displaced from the training sphere. The I Canadian Corps would remain under Montgomery's command until First Canadian Army Headquarters became fully staffed and functional. As Montgomery told his one-time Canadian aide-de-camp, Major Trumball Warren, a few months later: 'The [Canadian] Army has no direct control over the Corps; it is under me for training, operations, etc, etc.'[20] McNaughton visited Montgomery on the last day of March and had a long talk about the latter's control of the corps, but specific details of their discussion are elusive. Considering McNaughton's strenuous efforts to maintain Canadian autonomy, it was remarkable that he was willing to grant Montgomery such control. McNaughton was left with a skeleton army headquarters and no formations to train. That responsibility had now passed to the inexperienced Crerar. This meant that after March 1942, two full years before D-Day, McNaughton had no responsibility for training divisions. Therefore, to be fair, he could hardly be held accountable for the apparently lamentable state of training below corps level in Normandy. On the contrary, it is Crerar who should be under the microscope.

Crerar was ill-equipped to conduct formation-level training, and he knew it.[21] Simonds, who was his BGS, considered him 'neither a tactician nor a trainer of troops.'[22] Because of Crerar's dearth of training knowledge, Montgomery bore a great deal of responsibility for the training of Canadian formation commanders prior to departing for the desert in August 1942 to fight Rommel. Montgomery had almost nine months to observe and train Canadian commanders from regiment to division, and it is important to understand his methodology.

Montgomery's reputation as a trainer was firmly established by the time the Canadian Corps came under his direct influence. Brooke was greatly impressed with his extensive and challenging exercises while commanding V and XII Corps, and he had seen first-hand Montgomery's mastery of military movement while commanding the 3rd 'Iron' Division in France. During the night of 27 and 28 May, Brooke had identified a large gap between his left flank and the nearest French division. The situation was critical, and he ordered Montgomery to move his division from the corps' right flank to the left flank to fill the gap. Montgomery promptly disengaged his division from active operations against

Table 8.2 Command and staff element, First Canadian Army/I Canadian Corps, April 1942

First Canadian Army:	Lieutenant-General A.G.L. McNaughton
BGS	Brigadier M.H.S. Penhale (also BGS CMHQ)
DA&QMG	Major-General Guy R. Turner
CCRA	Brigadier H.O.N. Brownfield, Howard Oswald Neville
CE	Brigadier C.S.L. Hertzberg
CSO	Brigadier John Ernest Genet
GSO 1 (Ops)	A/LCol. G.P. Henderson
GSO 1 (I)	Major Lord Tweedsmuir (16 November 1942)
I Canadian Corps	Lieutenant-General H.D.G. Crerar
BGS	Brigadier Guy G. Simonds
GSO 2 (Training)	Major George Kitching
DA&QMG	Brigadier Alfred Ernest Walford
CE	Brigadier J.C. Melville
CSO	Colonel W.L. Laurie

Source: Historical Report No. 78, 'Situation of the Canadian Military Forces in the United Kingdom, Summer, 1942: I, Recent Changes in Commands and Staffs, 31 July 1942.'

the Germans, loaded them onto transport, conducted a night move on minor roads a distance of twenty-five miles, and was firmly entrenched by dawn to receive any new attack. As Ronald Lewin has observed, few major-generals in the British Army at the time could have made such a complicated move look so easy. Montgomery made it look easy because he had already exercised the division in complex night moves over greater distances well before the battle of France commenced.[23]

One observer has noted: 'No one, either sympathizer or critic, will deny that Montgomery was an excellent trainer of men.' Another has suggested that his training skills reflected 'genius.'[24] Montgomery's training philosophy for higher commanders was, like Brooke's, the by-product of the BEF's performance in France. Montgomery declared with utter conviction: 'Commanders and staff officers at any level who couldn't stand the strain, or who got tired, were to be weeded out and replaced – ruthlessly.' Eisenhower expressed the same sentiment after Kasserine, telling Major-General Leonard Gerow that commanders who failed to properly train and lead their units 'must be ruthlessly weeded out.'[25] As soon as the Canadian Corps came under his authority in November 1941, Montgomery insisted that it adopt a more systematic approach to training than he believed had hitherto been evident under McNaughton. But it was Simonds, not Crerar, who made sure the new Montgomery philosophy was disseminated and accepted.[26]

Montgomery's principal observation about Canadian division commanders was that they did not understand what he called the 'stage-management' and 'technique' of battle. His concept of what an infantry division commander was supposed to do is best captured in 'Some Notes on the Conduct of War and the Infantry Division in Battle,' produced in Belgium by the 21st Army Group in November 1944. A division commander required certain personal qualities to succeed, including leadership, initiative, the drive to get things done, the ability to get the most out of tired troops, moral courage, resolution, determination when things hung in the balance, cheerfulness at all times, and confidence.

These unquantifiable traits spoke to the moral cohesion that good division commanders could generate. Montgomery translated that concept into a list of tasks for division commanders, including the following: (1) creating a good atmosphere at headquarters, (2) passing on clear intentions and directions, (3) influencing personally the morale of the division by instilling confidence, (4) being seen, (5) maintaining discipline, (6) maintaining the offensive spirit, (7) considering ways to rest and relieve troops, and (8) carefully watching subordinates for fatigue.

On the planning side, the division commander made the plan himself and then left the details to be worked out by his staff. He kicked off the battle procedure for the division, but then he had to give his subordinates enough time to consider the elements of the problem and make their own plans. In the set-piece attack the division commander was expected to organize his forces in depth, secure the start line, support the attack by fire all the way to the objective, ensure that assaulting infantry and tanks hugged the artillery fire closely, and ensure that supporting weapons moved forward quickly. The impetus of the attack could not be allowed to slacken, for it was critical to overrun mortar and gun positions.

Though division commanders had to allow their brigades to fight their own battles and had to 'avoid interfering with their detailed arrangements and cramping their action and initiative,' it was unquestionably the division commander, and the division commander alone, who could maximize the synergy of his brigades in order to ensure that momentum was not lost. Once the set-piece battle was in motion, Montgomery wanted division commanders to keep a tight grip on the tactical battle. To do so the commander had to know what was going on. That meant he 'must go forward and find out the situation for himself.' Montgomery's approach was based on the premise that success would come 'only after hard and prolonged fighting [and a] hard slogging match.'[27] The uncertainty of such attrition-based fighting called for men of considerable nerve and that unquantifiable attribute, guts.

This, then, was the standard to which Montgomery held formation commanders while watching them on exercise. Coupled with this standard was his concept of *how* to train formation commanders to achieve that standard. The best expression of his training 'method' for higher commanders was contained in his mid-April 1942 notes on Exercise CONQUEROR. It was brutally simple:

> Divisional Commanders should be spoken to very severely on the subject, and be ordered to see to it that Brigadiers and unit commanders are taught how to tee-up the various types of operations. Those commanders who are known to be bad must be told they are bad, *and must be taught so that they will become good* ... Conferences after exercises must be conducted in such a way that it is made *quite clear* what was good and what was bad ... Only in this way will commanders learn and will progress be made ... Any commander who cannot prepare himself, or his command, for the battle is obviously of no value to the Army. He must therefore be given advise [*sic*], guidance, and instruction by his superior commander. If, after having had instruction, he is unable to profit from it and makes no progress, then he must be removed.[28]

McNaughton never produced any sort of definitive training concept like this for senior commanders. Moreover, it is unlikely that he could have been as cold-blooded about it as Montgomery. It was not McNaughton's way.

Montgomery closely observed the Canadian division commanders during one exercise after another throughout the spring and was damning in his comments. His first impression of Major-General Basil Price, commander of the 3rd CID since March 1941, was that he 'knows very little' about training or commanding the division; he recommended that he be relieved. Price would recall that he once told Montgomery: 'You know, Sir, it's quite a thing for an honourary soldier to be pitchforked into a job like this and get no opportunity to train himself for it.'[29] Montgomery's reaction was to immediately see to it that Price attend a TEWT.

After watching Pearkes command the 1st Division in an offensive role on Exercise BEAVER III in April, Montgomery concluded that he was 'unable to formulate a sound plan'; more specifically, he criticized him for attacking without regard for his flanks. Montgomery had very clear views on how infantry divisions were to attack: they needed to penetrate deeply on narrow frontages. The Germans would quickly attack the flanks to cut the lines of communication; Montgomery's solution was to establish strong hinges on both flanks of the initial break-in. Beyond the hinges, flank security was best provided by having counter-attack units 'suitably positioned.'[30]

Attacking without regard to flanks was precisely what Montgomery emphasized prior to D-Day in order to meet objectives. Indeed, this is what Major-General Rod Keller's 3rd CID had to do in order to achieve its final objectives on D-Day. Both D-Day and the type of offensive action practised on BEAVER III were time-sensitive operations requiring that risks be taken. In each case a methodical advance would have proved fatal. Nevertheless Montgomery concluded that Pearkes was 'unfit to command a Division in the field.' Montgomery's criticisms must be considered in the context of his poor relations with Pearkes while the latter was Acting Corps Commander. Pearkes later reflected: 'Oh, it was a difficult period those six weeks.'[31] First impressions carried great weight with Montgomery.

The only division commander Montgomery actually considered 'teachable' at the time was Major-General James 'Ham' Roberts, Acting Commander of the 2nd CID since November 1941. Roberts had some problems at the beginning of BEAVER III, but Montgomery saw something in him and declared: 'He will do very well as he gains in experience ... [He was the] best Divisional Commander in the Corps.' Yet Montgomery added an important caveat. Roberts was 'not in any way brilliant' but had been well served by his excellent GSO 1, Lieutenant-Colonel Churchill C. Mann. Indeed, it was an article of faith with Montgomery that an average division commander 'would always do well' if he had a good GSO 1.[32]

The fundamental premise of Montgomery's training method was that without proper instruction from division commanders, brigade commanders were lost. Yet when Montgomery assessed all nine Canadian infantry brigadiers between February and March he determined that at least five were very good material.[33] Two of Roberts's brigadiers, G.V. Whitehead and William W. Southam, were judged very capable. Based on Montgomery's method, the logical deduction was that Roberts had trained them well. Using similar reasoning, the assumption was that since Pearkes was no good he had failed to properly train his brigadiers, Arthur E. Potts and Hardy Ganong. However, Montgomery's method fell apart when it came to the 3rd Division: Price was crucified for knowing noth-

Table 8.3 Montgomery's assessment of Canadian brigadiers, January–March 1942

Commander	Unit	Comments
R.L.F. Keller	1 Inf Bde	Ill at the time. No assessment.
Arthur E. Potts*	2 Inf Bde	Not fit to command and train an Inf Bde in 1942.
Hardy Ganong*	3 Inf Bde	No great trainer, but a grand fighter.
Charles B. Topp*	4 Inf Bde	Not fit to command and train an Infantry Brigade.
G.V. Whitehead	5 Inf Bde	Firm, decisive, inspires confidence.
W.W. Southam	6 Inf Bde	First class; will be a Div. Comd. before war is over.
H.L.N. Salmon	7 Inf Bde	Really high class Brigadier. Fit to command a Div.
Ken Blackader	8 Inf Bde	Will be first class; a future Div. Comd.
Eric W. Haldenby	9 Inf Bde	A good Brigadier. Runs a good show Keen, enthusiastic.

*Relieved of command.
Source: Montgomery, Notes on Inf. Bdes of Canadian Corps, January to March 1942, CP, Vol. 2, File 958C.009(D182).

ing, yet all three of his brigadiers – Harry Salmon, Ken Blackader, and Eric W. Haldenby – were favourably assessed. Exactly who 'taught' them, if not Price, remains unclear. Natural aptitude and personal drive to learn on the part of these particular brigadiers is one possible explanation.

Montgomery rated Salmon 'Really high class' but also felt that he was not properly supervising the training of the 7th CIB. Crerar took issue with this assessment – 'I am only "somewhat" inclined to agree with you' – and went on to explain that it was Montgomery's own training schedule that was taking Salmon away from hands-on supervision of the brigade.[34] The difficulty in focusing on training was an issue Crerar had raised with Montgomery shortly after becoming Corps Commander. On 6 January 1942 he informed him that the rapid expansion of the army (which Crerar had pushed so vigorously) was bleeding formations and units of good commanders and staff officers. As a result, there was 'little opportunity to develop that teamwork which is essential to the smooth operation of a Headquarters.' The present weakness in unit training, he added, was due to the lack of opportunity for uninterrupted training.[35]

After Exercise BEAVER IV in mid-May 1942, Montgomery once again stressed that the weakness in I Canadian Corps was the inability of division commanders to stage-manage a battle. It was 'quite impossible to make progress' in training brigadiers if division commanders were incompetent. He reiterated that Roberts was 'good,' then added that the performance put on by Price in a defensive role versus Roberts 'was lamentable.'[36] After BEAVER IV, McNaughton wrote to thank Montgomery

Table 8.4 Brigadier Salmon's extra-brigade responsibilities, December 1941–February 1942

12 Corps Study Week	15–20 December 1941	(5 days).
Senior Officer's Infantry Course	3–10 January 1942	(6 days).
Rehearsal, Canadian Corps Study Week	17 January 1942	(1 day).
Canadian Corps Study Week	19–23 January 1942	(5 days).
BEAVER II	3–5 February 1942	(3 days).
VICTOR II	15–19 February 1942	(5 days).

Source: Crerar to Montgomery, 25 February 1942, CP, Vol. 1, File 958C.009.

for his 'frank statements which are of a great help to me in assessing the situation and determining the steps to be taken with a view to improvement.' It is not clear how much McNaughton resented Montgomery's method, but it is apparent that some animosity was being generated. In late June, Montgomery wrote to Major Trumball Warren that his training of senior Canadian commanders was 'having very curious repercussions in that I have to proceed very carefully and to be very tactful – there is considerable jealousy in a certain quarter.' The reference is obviously to either McNaughton or Crerar, possibly both.[37]

The final major exercise Montgomery oversaw before heading to the desert was TIGER, designed not as an anti-invasion exercise but as an encounter battle. It took place between 19 and 30 May and pitted Crerar's I Canadian Corps against Lieutenant-General James Andrew H. Gammell's XII Corps. The I Canadian Corps consisted of Pearkes's 1st CID, Price's 3rd CID, and Major-General Eric C. Hayes's 3rd British Division, with the 5th CIB under command, and the Sussex Brigade. The 2nd CID Headquarters and the 4th and 6th CIBs were preparing for Dieppe. Gammell had two infantry divisions, the 43rd and 53rd, as well as the 11th Armoured Division.

Montgomery heaped praise on Crerar for his handling of I Canadian Corps during TIGER, but this seems to have been a motivational tool. Montgomery also made it quite clear that 'you have a 1st class B.G.S. in Simmonds [sic].'[38] Crerar, with Simonds as his BGS, apparently orchestrated a fairly good battle and had Gammell on the ropes. However, when Crerar executed a withdrawal during the night of 29 May, two brigades of the British 3rd Division and the entire Sussex Brigade and 5th CIB were judged by the umpires to have been encircled and written off by the 11th Armoured Division. The reason cited was traffic control. Moreover, the following day, owing to poor passage of information from

Crerar's headquarters to Price's division (or poor understanding by the division), the entire 3rd CID was isolated and destroyed between the 53rd Infantry Division and the 11th Armoured Division.[39]

McNaughton reported to Ottawa that TIGER reflected the 'satisfactory state of tactical training and endurance now reached by Canadian units and formations taking part.' Staff work, road discipline, and supply arrangements 'were on the whole excellent.' He was 'particularly pleased' with Crerar's handling of I Canadian Corps.[40] Somehow, despite losing 50 per cent of his force, Crerar had escaped TIGER with an enhanced reputation. That Montgomery and McNaughton judged him to have acquired the skills necessary to fight an entire corps without having commanded anything since 1917 – and having bypassed division command – again speaks to the problems inherent in assessing formation commanders during the training period in Britain. Conversely, Montgomery concluded that Pearkes and Price, commanding their divisions for an extended period of time, had failed to master the stage management of battle.

To summarize, there were inconsistencies in Montgomery's assessments, and even in his actual operations, despite his generally professional approach. He relieved Herbert Lumsden of X Corps and Alex Gatehouse of the 10th Armoured Division after Alamein. In Normandy he fired David C. Bullen-Smith of the 51st Highland Division, and Bobby Erskine and Gerald L. Verney, both commanders of the 7th Armoured Division, in rapid succession, as well as Gerald Bucknall of XXX Corps. Regarding Verney and Bucknall, Montgomery frankly admitted: 'I made a mistake.'[41] This admission leads one to ponder whether Montgomery also made mistakes when identifying senior Canadian commanders for disposal. Some subjectivity was inevitable during the training in Britain because there was no War Office publication titled 'How to Assess Formation Commanders' for guidance. Even with the outbreak of war the Army Council did not establish a specific course for training senior commanders, instead choosing to rely on the existing approach: assessment and exercises under the guidance of the CIGS.[42]

By 12 August 1942, Montgomery was in Cairo preparing to take over Eighth Army, but as a result of his assessments, Price and Pearkes were relieved in September. Only two days earlier Crerar had dropped the hammer on Price, telling McNaughton that as a division commander he was handicapped in 'two important respects': he had no sense of tactics, and he was 'unable quickly to appreciate a military situation and to dominate it with the requisite speed and decision.' The subjective nature of first impressions is borne out by the fact that when McNaughton decided to promote Price to command the 3rd Division (and Sansom to command the 5th CAD) in early 1941, Crerar declared that 'these are good appointments and should result in credit to the Commands as well as to their Commanders.'[43] Rod Keller, the youngest major-general in the army at the time and the only senior officer who had not served in the First World War, replaced Price. Salmon, commander of the 7th CIB, replaced the veteran Pearkes.[44]

Because of his illness and recovery, McNaughton was not directly involved in any higher-level training for the four-and-a-half months between starting sick leave in mid-November 1941 and his return to England at the end of March 1942. Faith in his judgement was undermined when he declared, prior to going on sick leave, that the Canadian Corps was 'thoroughly prepared for battle' because of the 'ceaseless training by day and night' in all the intricacies of combined-arms warfare.[45] Between 1940 and 1941 he had trained the Canadian formations to defend against an invasion. When McNaughton did return from sick leave he gave up control of I Canadian Corps, as per Brooke's intentions, to Crerar and Montgomery. From November 1941 to August 1942, a period of nine months, Montgomery imposed his training method on Crerar and the senior Canadian commanders. Montgomery seems to have succeeded in part in 'teaching' Crerar, but it is not at all apparent that Crerar succeeded in teaching his division commanders. His solution – fully backed by Montgomery – was to relieve commanders who did not measure up quickly enough. It is not inappropriate to suggest that the results achieved by Crerar and Montgomery in the offensive stages of training up to Montgomery's departure were not demonstrably superior to what McNaughton had achieved in 1940 and 1941 when the defensive mindset dominated.

As for Brooke's charge that McNaughton was a poor commander, it must be remembered that McNaughton only had two opportunities to command in the field, during WATERLOO and BUMPER. During the other exercises – FOX, DOG, BEAVER III, BEAVER IV, and TIGER – he acted as exercise director or as an observer. He would not get another opportunity to command in the field again until Exercise SPARTAN in March 1943, and then not as a corps commander improving on his first performance at corps level, but as an army commander. As it would turn out, it would be his only opportunity to command an army-level organization in the field. SPARTAN was to become the graveyard of Canadian reputations in Britain, including ultimately, McNaughton's.

9

Exercise SPARTAN

> Started by motoring to Godalming to see Andy McNaughton commanding Canadian Army, and proving my worst fears that he is quite incompetent to command an army! He does not know how to begin the job and was tying up his forces into the most awful muddle.
>
> Brooke Diary, 7 March 1943

SPARTAN, a large-scale army-versus-army exercise held in the south of England in early March 1943, proved to be a turning point in McNaughton's command of First Canadian Army. Involving ten-plus divisions, it was the largest field exercise since BUMPER. Brooke had designed it to test the Canadian Army in the dual tasks of breaking out of an established bridgehead and making the transition to open warfare. In late December 1942, McNaughton confidentially characterized SPARTAN to an Army public relations officer as 'a dress rehearsal for full-scale invasion on the continent.' Indeed, he believed it entirely conceivable that it 'might very well be the last dress-rehearsal before actual battle.' He had assured Paget the previous month that the army would be ready for action on the Continent by 1 August 1943.[1]

McNaughton also viewed SPARTAN as a 'strict test' of the ability of commanders and staffs to administer, handle, and fight their formations and units.[2] But as far as Montgomery was concerned, the only 'strict' test was battle experience. So he told Crerar in a letter of 9 January 1943 from Tunisia.

> Many of the [Canadian] generals have never seen a shot fired in this war – most of them in fact; they are not in touch with 'the feel of the battle'; they

cannot possibly teach their subordinates if they themselves do not know the game. We ought to make more use of our active front to train our generals; Corps and Divisional generals *must* know the game, and if they do not you can achieve little. Not only the theoretical side of it, but the stern practical side, and the repercussions that arise if you do NOT know what you are doing ... Unless we do something about it I fear there will be very rude awakenings when the Army now in England is launched into battle.[3]

First Canadian Army headquarters in particular was in need of a serious shake-out. The key positions were occupied by familiar faces. Simonds was appointed BGS in January. He had consistently proven himself as an exceptionally competent staff officer in all previous assignments and was the best possible choice for BGS. The selection of Guy Turner for DA&QMG, however, was questionable. During SPARTAN he would suffer from ongoing health problems. The important positions of GSO 1 (Operations) and GSO (I) were held by Lieutenant-Colonel Thomas G. Gibson and Lieutenant-Colonel G.P. Henderson, respectively. Simonds, Turner, Gibson, and Henderson were the key people in the headquarters that made the army function.

Most of southern England was divided into three areas for the exercise. 'Eastland,' with its capital at Huntingdon, was a 'German' stronghold. 'Southland,' with boundaries extending to the outer defences of London, was also under 'German' control; however, it had recently been 'invaded' by the Allies and served as the established bridgehead from which Canadian forces would have to break out. 'Westland' was neutral territory, and McNaughton was under strict orders not to violate it. For exercise purposes, the Germans received orders from 'Headquarters, Army Group West' while McNaughton received orders from 'GHQ BEF.'

McNaughton commanded the British Second Army and had three corps. Crerar's I Canadian Corps contained Major-General James Hamilton Roberts's 2nd CID, Major-General Rod Keller's 3rd CID, and Brigadier R.A. Wyman's 1st Canadian Army Tank Brigade. Lieutenant-General Ernest W. Sansom commanded II Canadian Corps, which functioned as an armoured corps for the exercise. It contained Major-General Allan H.S. Adair's Guards Armoured Division and Major-General Charles Ramsay S. 'Bud' Stein's 5th CAD. Lieutenant-General Montagu G.N. Stopford's British XII Corps, with the 43rd and 53rd Infantry Divisions and the 31st and 34th Tank Brigades, rounded out the order of battle. Major-General Harry L. Salmon's 1st CID did not participate because it was conducting combined training in Scotland.[4] McNaughton main-

Table 9.1 Staff element, First Canadian Army, during SPARTAN, March 1943

Commander	Lieutenant-General A.G.L. McNaughton
CE	Major-General C.S.L. Hertzberg
BGS	Brigadier Guy G. Simonds
DA&QMG	Major-General Guy R. Turner
GSO 1 (Ops)	Lieutenant-Colonel T.G. Gibson
GSO 1 (Int)	Lieutenant-Colonel G.P. Henderson

Source: Historical Report No. 91, 'Press Conference Concerning Organization of First Cdn Army and Arrangements for Press Representatives, 25 February 1943,' 16 March 1943.

tained the Canadian organization for the infantry divisions of three infantry brigades but had adopted the British model of one armoured and one infantry brigade for the armoured divisions.[5]

McNaughton's opponent was Lieutenant-General James Gammell, commanding the 'German Sixth Army.'[6] Gammell had only two corps to McNaughton's three. The VIII Corps consisted of two armoured divisions, the 9th and 42nd, while XI Corps contained two infantry divisions, the 49th and 61st, both based on the new British organization of two infantry brigades and one armoured brigade. The air situation replicated what existed at the time between the RAF and the Luftwaffe in France. McNaughton had established airfields in the base area, from which 'Z' Composite Group RAF supported him with twelve fighter squadrons, one night fighter squadron, four fighter-bomber squadrons (Army Support), three light bomber squadrons, and six reconnaissance squadrons. The 'X' Mobile Composite Group of the RAF was supporting Gammell with seven fighter squadrons, one night fighter squadron, two fighter-bomber squadrons (Army Support), four light bomber squadrons, and five reconnaissance squadrons.

The best tank country was along the 'Westland' border; however, it contained barriers to manoeuvre such as the Grand Union and Oxford canals and the Evenlode and Windrush rivers. In the east there were additional significant water obstacles. The Thames ran west out of London, then forked at Reading, continuing northward while the westward extension turned into the Kennet River. The southeast portion of 'Eastland' close on London was characterized by additional canals as well as by built-up areas known as the Chiltern Hundreds.[7]

McNaughton's mission was to launch an invasion of Eastland from the bridgehead already notionally established by Lieutenant-General Gerald

W.R. Templer's British First Army and seize the capital, Huntingdon, as rapidly as possible. A secondary task was to secure and develop airfields southwest of Oxford. The best way to assess how he intended to achieve his mission is by looking at his 'Appreciation of the Situation' prepared before the exercise started. An Appreciation is a military review of the situation based on all available information; it culminates in a statement of the measures the commander will take to meet the situation. McNaughton's Appreciation is reproduced here as paraphrased in the Narrative of Events:

(a) The available German forces would probably be superior to his in infantry but inferior in armour and more especially in artillery. They were disposed in two groups divided by THE WASH and the FEN country. The northern group was the stronger and included the bulk of the armour. The GERMANS might be able to concentrate in the BEDFORD area one armoured division, one division and one infantry division on D, and a further armoured and two infantry divisions on D+1.
(b) The fortress of LONDON was garrisoned by static troops and possessed heavy guns which could command the roads within a radius of approximately nine miles. The initial delay in the concentration of 12 Corps provided a protection against any major sortie from LONDON.
(c) If he could position his force between these two groups, he could defeat each in detail. The most favourable area for such action was BANBURY-OUNDLE-SAFFRON WALDEN-OXFORD.
(d) In order to give maximum effective fighter cover up to the line IPSWICH-WISBECH-NOTTINGHAM, forward airfields in the area CHIPPING NORTON-TOWCESTER-AYLESBURY-WALLINGFORD-OXFORD must be gained and protected.
(e) A thrust to HUNTINGDON on D with I Canadian Corps would entail big risks, but would prevent the concentration of enemy forces and would cover the airfield area. When the SECOND ARMY was concentrated in the forward area, it could then attack the two German groups in turn. The need to cover the airfields and the position of his armour on his left flank favoured an attack in the first place on the northern German group.
(f) Available transport was just sufficient to maintain the Army over the estimated required distance of 200 miles provided no abnormal demands were made.[8]

The maintenance plan was very important because it spoke to the fundamental ability of the army staff to keep the formations moving.

132 McNaughton as Military Commander and Trainer

Figure 9.1 McNaughton's logistical plan for SPARTAN

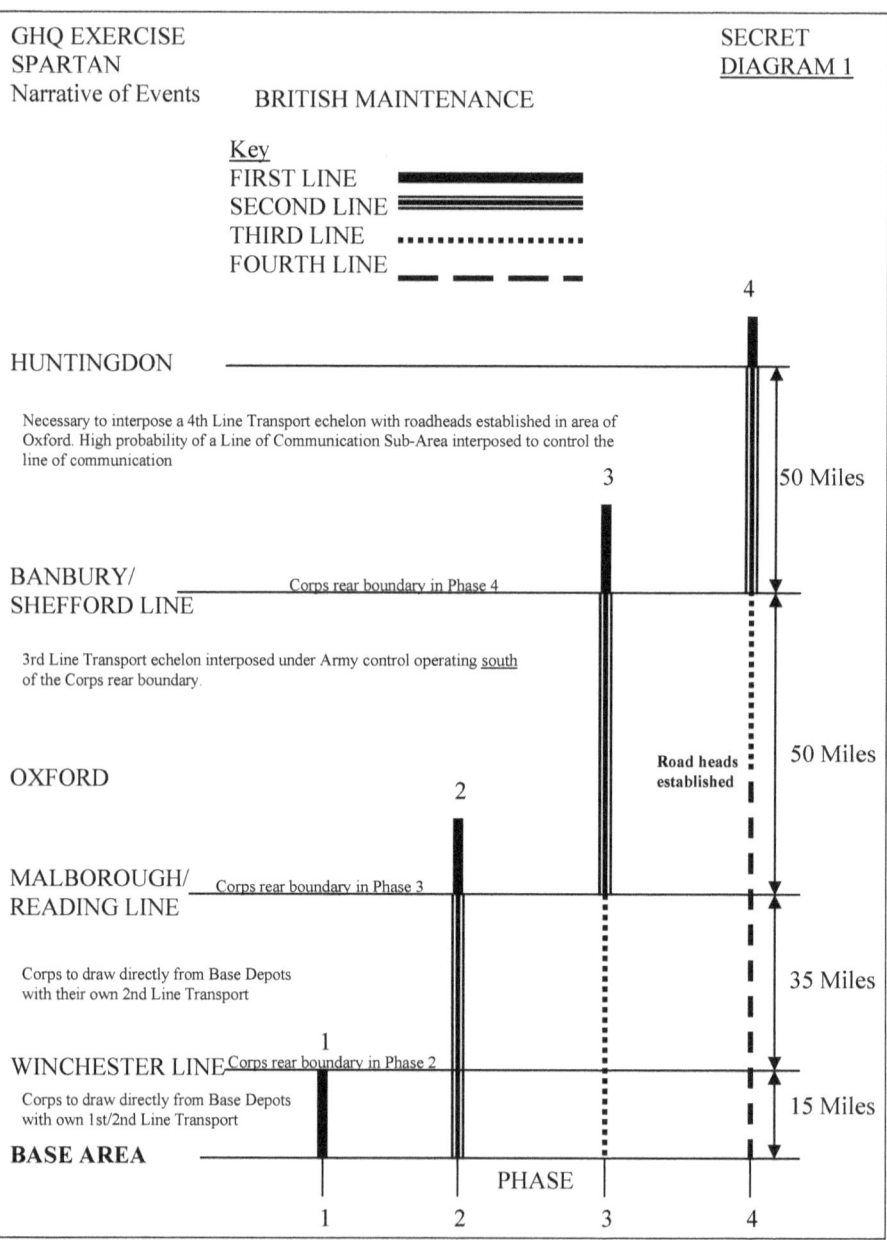

Source: GHQ Exercise SPARTAN Narrative of Events

McNaughton had to maintain Second Army from a port and base depot area without the use of railways for a distance of 150 miles. To do so he had designated the normal first- and second-line units as troop-carrying companies; he also had nine third-line general transport companies (GTCs) with a total lift of 3,700 tons as well as seven GTCs with a total lift of 2,500 tons. The first- and second-line transport worked well up to the Newbury–Reading line, but beyond that point, complications arose as Second Army pushed forward large amounts of airfield construction materials and engineering stores.

From this appreciation, McNaughton made his plan, which had two phases. In Phase One, Crerar was to push I Canadian Corps as rapidly as possible to the area Wellingsborough–Thrapson–Huntingdon–Bedford. Sansom was to advance as rapidly as possible to the area Banbury–Towcester–Bicester in order to destroy any German armour attempting to move south in the neighbourhood of the Westland frontier or attempting to operate against Crerar's left flank. Stopford was to advance when available to secure the line Saffron Walden–Cambridge and to protect Crerar's right flank.

In Phase Two, Crerar and Sansom were to advance to destroy German forces in the area west and north of the line Peterborough–Banbury, protected on the right by Stopford less one of his divisions, which was to move back to become Second Army's reserve. 'Z' Composite Group was to discover the movements of the German forces in order to attack unarmoured columns and provide air cover for the army's forward movement.

McNaughton's original plan called for Sansom to advance west of the Thames and for Crerar to advance astride it. Paget suggested in his comments issued soon after the exercise that the better course for McNaughton would have been to push both corps through the more open country and across smaller obstacles west of the Thames to threaten Gammell's lines of communication and pry him out of successive positions. Such action would have mitigated the need for a direct assault across a serious water obstacle.[9] Gammell had anticipated the movement of the British armour on the west in his own appreciation.

Gammell's mission was to resist a British invasion of Eastland. Specifically, he was to prevent any enemy crossing the Gloucester–Swindon–Newbury–Reading line from penetrating a ten-mile perimeter around Huntingdon for eight days. From an appreciation of the ground, Gammell believed that McNaughton would commit no more than one infantry division in the Chilterns, concentrate the bulk of his infantry for a crossing of the Thames near Oxford, and employ his armoured divisions

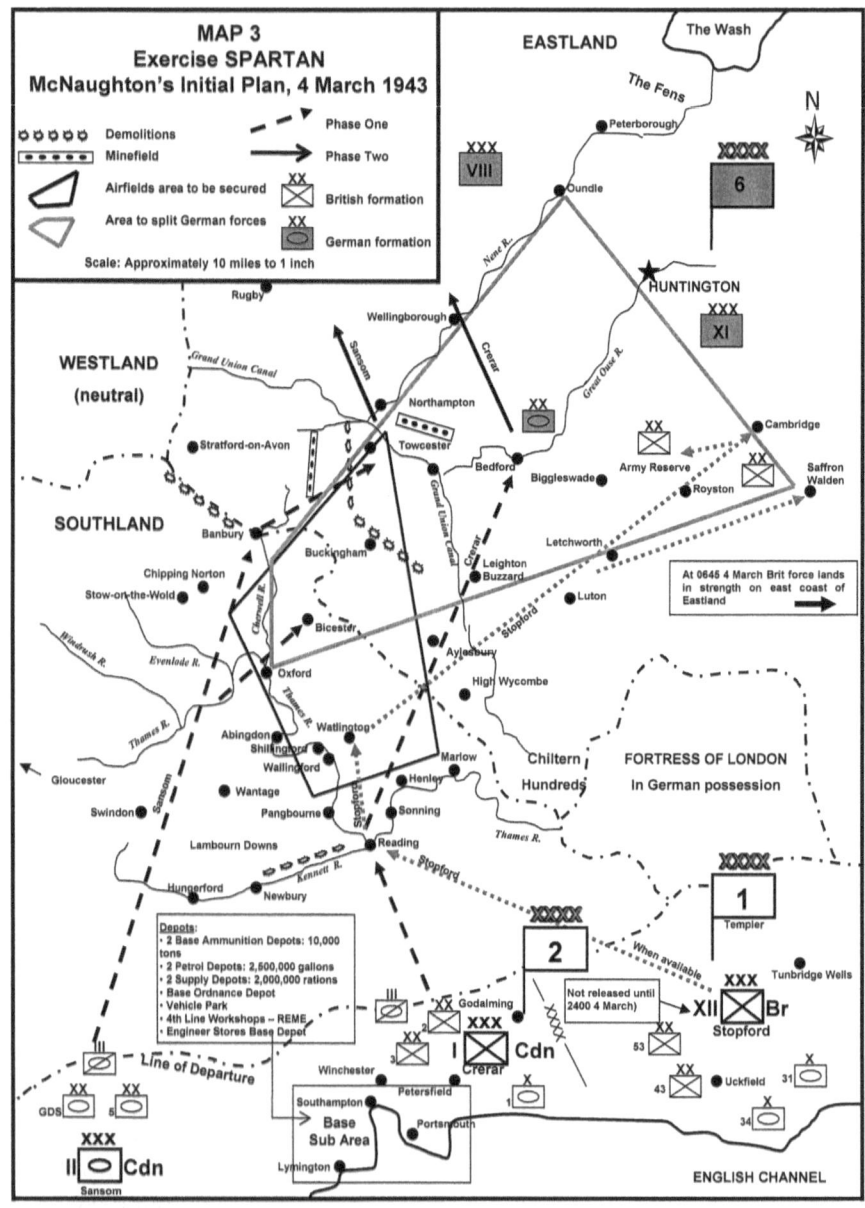

along with the infantry divisions of XII Corps west of the Thames for a push north through the Banbury–Stratford-on-Avon gap.

Based on this appreciation, Gammell devised a multiphase plan. In Phase One he would advance to the Gloucester–Reading line, with VIII Corps (less the 9th Armoured Division) deployed between Gloucester and Swindon. In the centre, XI Corps would deploy astride the Thames near Oxford between Swindon and Newbury. On his east flank he intended to establish a strong hinge in the Chilterns to allow VIII and XI Corps to swing back to the Thames and Cherwell rivers through Banbury. The Buckinghamshire (Bucks) Brigade was to establish the hinge from Newbury to Marlow. The withdrawal under pressure of VIII and XI Corps would be aided by a series of significant demolitions. Gammell designated the entire 9th Armoured Division as Army Reserve for the initial phase.

For Phase Two, Gammell had considered concentrating and counter-attacking McNaughton north of Banbury on prepared ground (covered by minefields and extensive demolitions) in the area of Towcester–Banbury–Stratford-on-Avon, but he decided against it. He had little desire to seek decisive armoured action in the west until the situation was in his favour.[10] In any event Sansom's axis of advance would only skirt the eastern edge of Gammell's Towcester–Banbury–Stratford-on-Avon 'kill zone.' Instead, Phase Two would be a withdrawal under pressure to the east, falling back on prepared positions between Northampton and the Chilterns. Phase Three called for a further withdrawal to prepared defended areas at Bedford, Biggleswade, and Letchworth. Gammell's plan did not mirror German doctrine, which called for immediate and heavy tactical counter-attacks as well as massed armoured counter-attacks at the operational level.

At the very start of the exercise, GHQ presented McNaughton with an unexpected situation to test his plan's flexibility. He had not considered advancing before first light on 5 March, but HQ Army Group West allowed Gammell to move first. He quickly invaded Southland at 2400 on the night of 3 and 4 March to gain much-needed depth and effected considerable demolitions – especially of bridges – well forward of his main position while McNaughton was still tied to his bridgehead on instructions from GHQ. In Paget's view, McNaughton did not fully appreciate the effects of Gammell's move. Note that GHQ had passed information on to him as an 'unconfirmed report' received in the Operations 'I' (Information) Filter Room at 0415 on 4 March.[11]

At 0600, McNaughton requested confirmation from GHQ of when he could move and was given no definitive answer. The umpire assigned to

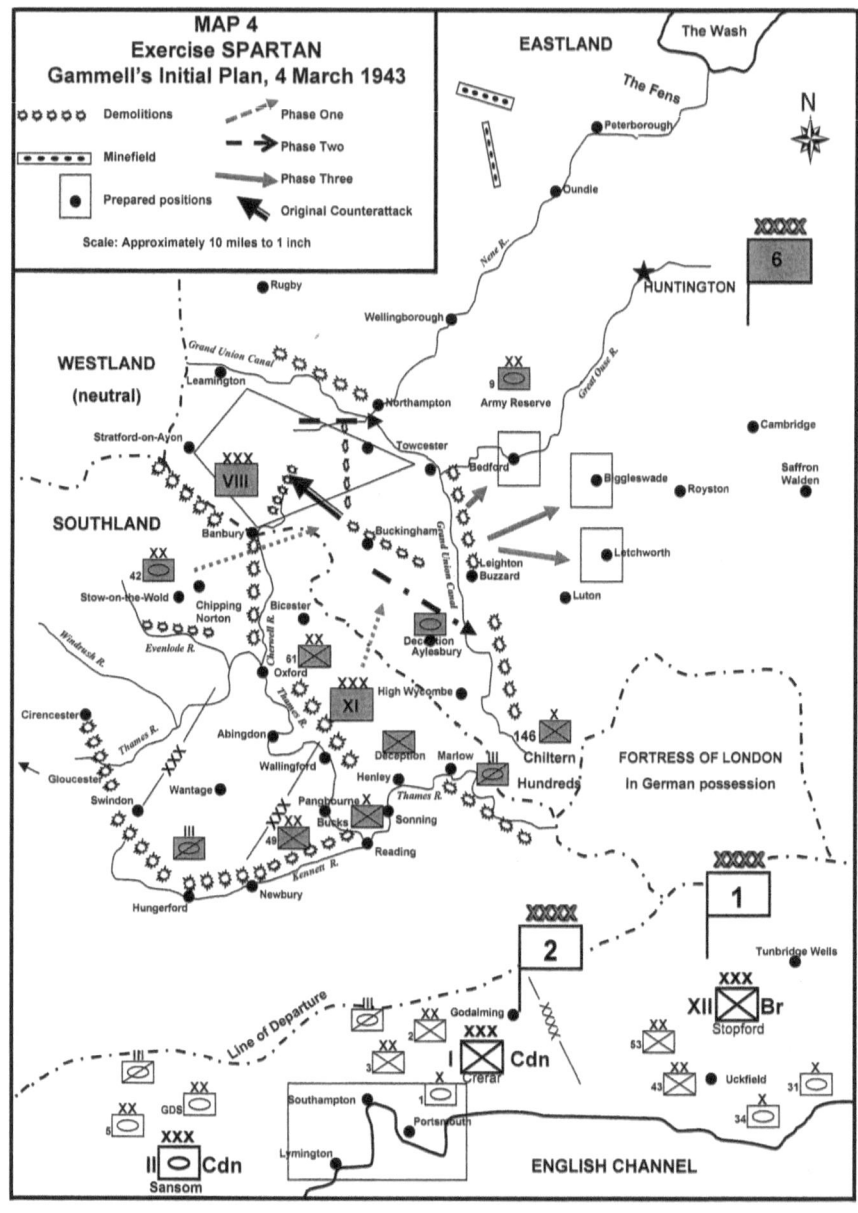

Second Army refused him permission to bump up his zero hour. He did not receive confirmation of Gammell's movement until 0844, at which point GHQ ordered him to begin his advance as soon as possible. At 0944 and 0948 Crerar and Sansom respectively were ordered by phone to pass through the bridgehead at 1200 that day. Paget identified some confusion in the early stages of this accelerated movement – some divisions were not grouped in their concentration areas – but concluded that 'in view of the difficulties the speed with which the advance began was a most credible effort.'[12]

In less than four hours McNaughton had Second Army on the move. Sixth Army was no longer in its original areas divided into two parts, as McNaughton had first anticipated; even so, he stuck to his initial plan out of fear that Stopford's XII British Corps – which was scheduled to complete its concentration by 2400 hrs and to be released to Second Army – would not be able to get into line on a northeast axis if Crerar did not push north.[13] The exercise rules stated that XII British Corps could not be concentrated until forty-eight hours after jump-off owing to lack of port facilities. Moreover, McNaughton had received word that additional Allied forces had successfully landed in Eastland at 0645 that day. Thus he deserves some credit for the swiftness of his initial advance and for taking the changed circumstances in stride.

McNaughton's forces made contact with Gammell's reconnaissance elements soon after advancing beyond the bridgehead. First serious contact, however, was made by McNaughton's reconnaissance units on the Newbury–Hungerford line in the early afternoon. Farther east, the 2nd CID reached the Thames at 1830 and by 2400 had established a small bridgehead. At first light on 5 March, Roberts had a full brigade across at Sonning (the bridge was captured intact) and was pushing back the Bucks Brigade holding thirty miles of river. According to Paget, this local success had two 'major repercussions' on the operations. First, it had turned the German defences at a critical point and enabled successive bridgeheads to be formed across the major water obstacle of the Thames. Second, it cracked the main pivot and forced Gammell very quickly to start withdrawing forces from the west to bolster it.

The leading troops of the 5th CAD had reached the Kennet between Kintbury and Mildenhall by 2000 and were still fighting for crossings three hours later. At 2130 hrs on 4 March, McNaughton issued new orders: Crerar was to seize bridgeheads over the Thames and the Kennet from Reading to Newbury during the night and advance the following morning north to the Wallingford–Abingdon area. Sansom was directed to occupy the high ground northeast of Swindon in the Vale of White Horse.

At 0001 hrs on 5 March, Gammell began withdrawing XI Corps in the centre back to the Abingdon–Wallingford–Schrivenham–Goring line. By sunrise (roughly 0730), four battalions of the 3rd CID were across the Kennett at Newbury and the 2nd CID had upgraded the Sonning bridge for wheeled traffic. An hour later Stein's 5th CAD had established a bridgehead at Hungerford; at 0930 both of Sansom's divisions were still under orders to advance on their original lines. Sansom's reconnaissance elements had lost contact with the patrols from the 42nd Armoured Division about 0800 but regained contact along the Abingdon–Lechlade line around 1000. Unfortunately, at 1030 McNaughton ordered Sansom to halt and occupy the Wantage–Hungerford–Swindon area with reconnaissance elements well forward.

On Crerar's front, heavy fighting erupted at Reading between the 2nd CID and the Bucks Brigade. By 1100, Roberts had two full brigades across at Sonning and had expanded the bridgehead six miles. By 1700, Keller had the entire 3rd CID on the north bank of the Kennett moving north into the bend of the Thames to regain contact with the 49th Infantry Division. Gammell had managed to destroy most of the bridges, but McNaughton was able to retain one bridge at Sonning on the Thames as well as the Marlborough and Hungerford bridges on the Kennett.

By the evening of 5 March, Gammell was worried about his eastern pivot. Reconnaissance was telling him that II Canadian Corps was no longer pushing north and west of Banbury, so he decided to begin shortening his line through phased withdrawals. By 1000 on 6 March, VIII corps had been ordered to fall back to the Leamington–Banbury–Cherwell River line. Gammell now planned to withdraw the Bucks and 146th Infantry Brigades and to close the widening gap between the Chilterns and the Thames with the 61st Infantry Division.

At 0335 on 6 March, McNaughton concluded that he was being held up by light forces and had yet to come to grips with the enemy's main body. He believed he had identified the 9th Armoured Division near Abingdon, but he was uncertain about the 42nd Armoured Division, placing it astride the Cherwell River at Oxford. He felt that Crerar was susceptible to an armoured attack and decided to bring Stopford's XII Corps into play to continue the attack in relief of I Canadian Corps so that Crerar could deal with the armoured threat. Crerar's intelligence estimate predicted an 'all-out attack by German armour on the left flank of 1 Cdn Corps' the following day.[14] The relief was to take place once

Crerar had secured the line Henley–Wallingford, hopefully during the night of 6 and 7 March. Crerar was to extend his left flank to Wantage to give Stopford room to deploy; Sansom was to stand fast and protect I Canadian Corps' left flank.

Historical opinion has it that at this juncture of the exercise, 6 March, McNaughton demonstrated an acute inability to determine a course of action. He had a host of visitors at his headquarters that morning, which probably did not help. Air Marshals Sir Charles Portal and Sir Trafford Leigh-Mallory appeared at 1030 in McNaughton's Battle Room. Sir Archibald Sinclair, Secretary of State for Air, showed up just before noon. Then Paget and Sir James Grigg, the Secretary of State for War, arrived at 1245. Grigg would recall being appalled by McNaughton's indecision: 'Intelligence was coming in and McNaughton stood in front of his situation map hesitating as to what to do and what orders to issue.'[15]

Grigg's assessment may or may not have reflected the reality of the situation. Montgomery knew Grigg quite well and felt that he was prone to snap judgements.[16] What Grigg viewed as indecision on McNaughton's part was more likely the ongoing appreciation that all commanders conduct. An army commander does not make split-second decisions about moving formations over great distances based on raw intelligence 'coming in.' First it has to be authenticated; then it has to be evaluated. It must not be forgotten that Gammell was employing infantry and armoured deception groups.

Intelligence first went through the 'I' Filter Room. Each message was assigned a serial number, summarized in an Operations Log, and put in its proper place on the 'battle board.' Once confirmed, the changes reflected by the new intelligence were shown by the movement of symbols on the situation map. There was always a time lag in this procedure, and McNaughton had told Stacey during Exercise ALBERT in July 1941 that he wanted to reduce the lag as much as possible.[17] Before any decisions were made based on new intelligence, McNaughton would also have to confirm his own positions. The GSO 1 (Operations), Gibson, would send messages out asking for the LOCSTATS of the formations and units. It all took time.

Moreover, at some point before 1630 on 6 March (perhaps at 1130, when he was briefed by Henderson on the latest intelligence reports), McNaughton received a captured operations order from the 49th Infantry Division outlining its intention to withdraw from the area Wallingford–Abingdon to the left bank of the Thames. The same order provided valuable intelligence on the future moves of the 61st Infantry

Division.[18] McNaughton's subsequent decision based on these captured orders lay at the heart of Brooke's belief that he should never command First Canadian Army against the Germans.

It is important at this time to understand McNaughton's thought process. The Exercise Narrative stated that he 'appreciated that, whatever he might do, the Germans, temporarily inferior in numbers, might withdraw further if threatened with any major attack ... He considered that the time and space factor ruled out any successful action against the German right flank. Further, it would not now provide any element of surprise, and surprise he was determined to obtain.'

McNaughton knew that Gammell was trying to canalize the British armour along the Westland border, and through ground reconnaissance he had a rough idea of the demolitions and minefields in that area. He was being cautious with his armoured divisions in the early stages, and Paget was justified in questioning his decision to guard the army's left flank throughout 6 and 7 March with II Canadian Corps when the two opposing armoured forces were fifty miles apart.[19] Yet the probability of achieving the large armour-on-armour action that Paget, Brooke, and others were looking for was low, for the minefields and demolitions would only slow and disrupt Sansom's armour and, in addition, would confound Gammell's own efforts to counter-attack with armour. The maze of tactical defences virtually guaranteed that the opposing armoured forces would remain at arm's length.

In an effort to maintain the initiative, McNaughton cancelled the relief of I Canadian Corps by XII Corps at 1600 on 6 March, issuing new orders by telephone to Crerar and Stopford and in person to Sansom, who was visiting Army Headquarters at the time. Stopford, who had begun to move forward from his concentration area at 2000 on 5 March, was to send the 43rd Division – less its tank brigade but including one infantry brigade from the 53rd Division – across the Thames at Sonning and Reading and advance through the Chilterns in the direction of Watlington–Luton. The I Canadian Corps was to advance on the night of 6 March and form a bridgehead in the area of Watlington–Wheatley. On 7 March, Crerar was to continue this movement north to the general line from Wheatley west to the Thames and around the loop of the river to North Moor. Sansom was directed to move his armoured corps to an assembly area *east* of the Thames and south of Oxford and prepare to move northeast to cut XI Corps' lines of communication. This was to be followed by a push to dominate the high ground north of Buckingham. To facilitate this new movement McNaughton directed that two major routes be cleared for the corps and that Class 40 bridges be completed at Wallingford, Shillington, and Abingdon by the afternoon of 7 March at the latest – less than twenty-four hours away.[20]

The decision to move Sansom's II Canadian Corps east across the Thames meant that all three of McNaughton's corps would be jammed between Oxford and Reading on a frontage of only thirty miles. Constricting the army's frontage to such a degree was not optimal but would have appeared less questionable had it not required movement across Crerar's lines of communication. At the time he issued the orders McNaughton deferred the decision as to whether to do it by day or by night.

By 2130 on 6 March, Second Army had issued a SITREP that helps situate McNaughton's thought process at the time: 'Contact continues with Bucks Bde Gp [in the] Henley area. Captured doc indicates enemy intention to hold line THAMES – OXFORD CANAL – CHERWELL between BANBURY and WALLINGFORD with 61 Inf Div right and 49 Inf Div left. Brs [bridges] over THAMES on 2 Corps front reported blown. Contact lost with enemy armour on 2 Cdn [Corps] front. Very widespread demolition of minor brs behind enemy front.'[21]

At 0045 the following morning GHQ sent word to McNaughton that Gammell was retreating behind the Ouse River heading for the Grand Union Canal. As the Exercise Narrative stated, GHQ deemed it 'essential [that McNaughton] should bring the enemy to battle with the maximum British force before he reached the canal position.'[22] The quickest way to do this would have been by using Sansom's armoured divisions if they were in position, but they were not. Wyman's 1st Canadian Army Tank Brigade was well located a few miles southwest of Pangbourne in reserve in Crerar's corps, but neither Crerar nor McNaughton considered using it for pursuit – most probably because army tank brigades were not designed for that role. However, McNaughton did just the opposite. At 1100 he ordered Crerar to establish a firm base south of Oxford, anticipating a counter-attack by the 9th Armoured Division against Crerar from the direction of Towcester.

At 0730 on 7 March, McNaughton was analysing the bridging problems confronting the movement of II Canadian Corps across the Thames. At midday he considered the situation unsatisfactory at Wallingford and Abingdon, though the Shillingford bridge was nearly complete. He was concerned, but he did not rush off to the bridging sites, as Swettenham has asserted. McNaughton never left headquarters the entire day, and his War Diary shows this clearly.[23] And even if he *had* gone to the bridges, he cannot be criticized for doing so. Those bridges were critical to his

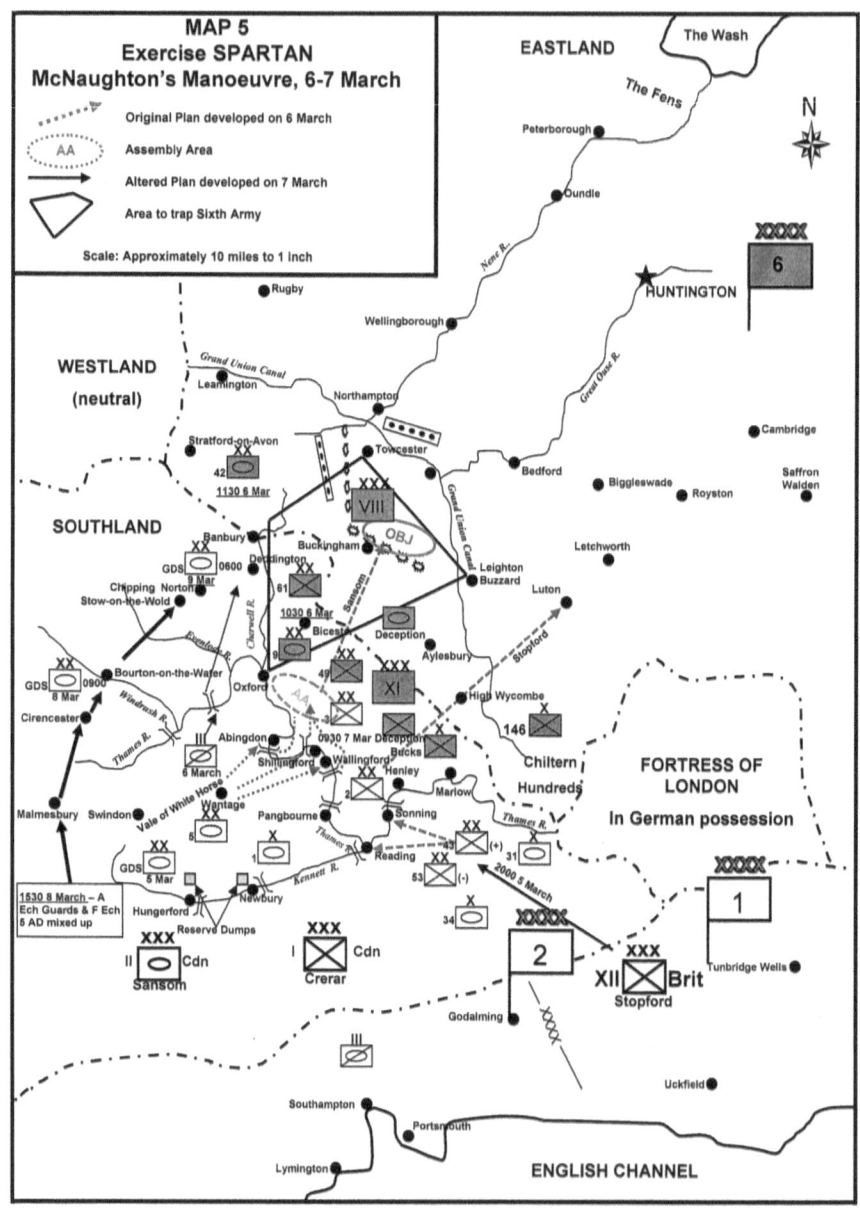

selected Line of Operations, and at *that* moment they represented his main effort. A student of military history has only to reflect on where the Emperor Napoleon was during his second attempt to cross the Danube in his 1809 campaign.

McNaughton was not guilty of micromanagement at this stage (as has often been suggested). Even so, his visits to the front clearly bothered Paget, who felt that the tendency of commanders to go forward 'must be curbed ... Commanders on the higher levels can fight their battles only from their HQ where they are fully in the picture and have full signals facilities.'[24] This was nothing more than Paget's personal view and was far too rigid.

In practice, Montgomery established his TAC HQ close to the headquarters of the main effort. Rommel and Guderian were constantly at the front to good effect. During the battle for France, Guderian had moved forward from XIX Panzer Corps HQ and personally intervened to hurry tanks across the bridge at Gaulier on the Meuse River. Guderian had actually watched the crossing of his infantry elements for two hours before crossing himself to roam around the forward lines. The disaster that befell French arms had been partly due to the fact that many French generals remained tied to their command posts and did not go forward to get a feel for actual conditions.

At 0950 on 7 March, Brooke and General Sir Ronald Adam, the Adjutant General, visited McNaughton's HQ, where they were joined by Lieutenant-General Jack Swayne a little later. Swayne's function was to examine the headquarters layout, including the Operations and 'I' Filter Rooms and the Battle Room. They stayed for less than an hour, but in that short span of time, Brooke's long-standing anxieties about McNaughton's command ability were reinforced. He noted in his diary at the time that McNaughton was 'quite incompetent to command an army! He does not know how to begin to cope with the job and was tying up his forces into the most awful muddle.' How McNaughton explained his plan to Brooke is not known, but the CIGS was obviously upset at the idea of pushing Sansom's corps east across the Thames. His diary comment has been enough ever since to convict McNaughton of incompetence in the eyes of military historians. But Brooke, like Grigg, was prone to snap judgements. For example, Brooke would note sarcastically during the Battle of the Bulge that Patton's scheduled attack from the south would be nothing more than 'a half-baked affair.'[25]

Brooke and Adam left by 1040. Within an hour, McNaughton had rethought the complicated corps move and reconsidered the left swing through Banbury. He called Sansom thirty minutes later and told him that a final decision would be made by 1600. According to Stacey the 'circumstances in which McNaughton changed his mind are not recorded,' but there are several possible reasons.[26] The SPARTAN narrative stated that McNaughton's thinking was influenced by the apparent difficulty of moving II Canadian Corps across the rear of I Canadian Corps as well as by the risk of concentrating both corps between Oxford and Reading in a bottleneck with only four bridges.[27] Another possible explanation is Brooke's presence. Brooke's disapproving tone or body language could have been the catalyst; or perhaps he suggested outright how difficult it would be to pass one corps through another.

McNaughton was probably rethinking his decision based on the reasons cited above. It is probable, though, that the final decision was based on intelligence received from GHQ at 1525 and on tactical reconnaissance indicating that Gammell had ceased his withdrawal to the canal. Additional intelligence revealed that Gammell was soon to be reinforced from 'Russia' (i.e., McNaughton's right flank). Whatever the true circumstances, by 1605 on 7 March McNaughton had conferred with Simonds and decided to push Stopford and Crerar north and northeast and Sansom west. Paget considered this decision 'undoubtedly the correct one,' given that the best way to exploit superior forces is to stretch the enemy's front.[28] Warning orders went out at 1615 to Sansom and at 1630 to Crerar and Stopford, with detailed written orders following at 1740. The main points of McNaughton's orders were as follows: Second Army was to envelop and destroy the German Army in the Banbury–Towcester–Leighton Buzzard–Oxford area on 8 March. The XII Corps was to advance towards Luton and I Canadian Corps to Stony Stratford–Kimbolton. The II Canadian Corps was to move via Cirencester–Southam to the Towcester area, sever Sixth Army's communications south of Northhampton, and bring VIII Corps to battle and destruction.

McNaughton's new plan had merit. Stopford's divisions were fresh, Crerar had made steady progress in the centre, and the movement of Sansom's armour west was guaranteed to pose greater problems for Gammell. But however wonderful pincer operations look on a situation map with its large arrows, success depends on sound staff work, good communications, and aggressive commanders. It is not apparent that McNaughton's plan could depend on any of these. Stopford, for instance, was reluctant to make so sudden a move because he was not fully concentrated. This is the excuse of an infantry-trained corps commander – in practice, one concentrates during the move. McNaughton telephoned him at 2110 on 7 March and 'impressed upon him the necessity for determined advance.'[29]

Only by moving quickly could McNaughton gain positional advantage over Gammell, but his change of orders on 6 and 7 March made that advantage difficult to achieve. For twenty-four hours II Canadian Corps had been realigning itself for a move east across the Thames; now, as a result of McNaughton's new orders, most units and formations began moving west up to three hours late. As the narrative would correctly state: 'It was problematical whether II Canadian Corps could advance with sufficient speed to gain the advantage.'[30]

At 1930 on 7 March, less than three hours after Sansom received his new warning order, the Guards Armoured Division began a forty-eight-mile move to its first concentration area. By 0900 the following morning it had arrived at Bourton-on-the-Water, where it waited almost ten hours for the 5th CAD to concentrate. The average rate of advance of Sansom's armoured divisions was a paltry five miles an hour. Stacey felt that this lack of speed – most apparent in Stein's division – was not surprising, considering the lack of signals equipment and the ill-trained signals personnel.[31] McNaughton's situational awareness was degraded because Sansom was out of touch with Army Headquarters for fifteen hours due to imposed wireless silence. Better use could have been made of aircraft and liaison officers, and Signals Dispatch Riders (SDRs) could have been sent out.

The real problem with the advance of II Canadian Corps, however, was traffic control – the same deficiency so evident throughout 1940 and 1941 in other Canadian formations. Between 1530 and 1730 on 8 March the administrative echelons of the Guards Armoured Division were mixed up with the fighting elements of the 5th CAD at Malmesbury. Throughout this period McNaughton was out visiting (he left his HQ at 1330 and had returned by 2040). He spent little more than an hour inspecting the bridges at Pangbourne, Wallingford, and Shillingford. The rest of 8 March was taken up with visits to the airfield at Lambourne Downs across the Kennett (construction had just begun) and the rear headquarters of the 3rd CID.[32] Preparing airfields was essential to his mission, so visiting them was hardly inappropriate. Perhaps he should have attempted to sort out II Corps personally, but it was unlikely that he could have done anything positive. It was a pure staffing issue from army down to division.

At 1900 on 8 March the Guards Armoured Division moved to its forward concentration area southwest of Banbury. It arrived at 0600 the following morning. Movement remained sluggish because of ongoing traffic problems and the demolitions that Gammell had prepared at the start of the exercise. Moreover, VIII Corps' Armoured Car Regiment had picked up the move and during 9 March, Sansom's armour was subjected to enemy air attack. A major jam developed throughout 9 March in II Canadian Corps' rear areas from Tetbury to Cirencester, where double lines of vehicles were immobilized for such a length of time that fighter cover was requested.[33]

Because of the lack of 'coming to grips' evident during the exercise – a result of Gammell's withdrawals and McNaughton's new orders – HQ Army Group West changed Gammell's mission on 8 March. He was no longer responsible for defending Huntingdon; rather, he was to inflict maximum damage on McNaughton. As a result, Gammell stayed in his current positions and did not withdraw rapidly behind the Grand Union Canal. However, he did not think that his army was capable of counter-attacking, and HQ Army Group West had to repeat its desire for offensive action, ordering that no further withdrawals were to be executed without permission.

From 9 to 11 March, McNaughton urged his corps commanders forward. Crerar and Stopford were making steady progress, but Sansom was slow. McNaughton spent much of 9 March at the front. At 1515 he was at Reading. At 1700 he arrived at XII Corps HQ at High Wycombe and found that Stopford was out. He then moved on to visit the 53rd Division HQ. Sansom's divisions had reached the canal northeast of Banbury by 1600, and light reconnaissance troops had established a bridgehead at Brackley. The narrative characterized II Canadian Corps' advance as 'extremely slow' but added that the pressure exerted by Crerar and Stopford prevented Gammell from achieving any major regrouping: 'An opportunity seemed to exist for a British "pincer movement," but only speed could make it possible.'

McNaughton arrived at his new Advanced HQ at Mongewell Park south of Wallingford at 2215. He and Simonds actually drove two miles down the road to a No. 9 wireless set to try to communicate with Sansom. Second Army Intelligence Summary no. 10, issued on the morning of 9 March, indicated that communications with II Corps had broken down 'fairly early on' and that the locations of Sansom's armoured divisions were unconfirmed. At 2330, McNaughton sent off a message by SDR to Sansom to move forward 'as fast as humanly possible.'[34]

By 1600 the following day, 10 March, Sansom's armour was finally across the canal in force and in a position to threaten Gammell's position. The 5th CAD twice reported fuel shortages. Both divisions were replenished by second-line transport two hours later, but it seems that they could have been on the move much sooner. Third line had replenished second line as early as 1130, but then second line halted for three hours. Compounding the delay, the petrol trucks had difficulty finding the forward units. Orders for a further advance were cancelled at 1830, shortly after the refuelling.[35] The narrative does not state who ordered the cancellation, but the logical suggestion is Sansom, for McNaughton was trying to urge him forward instead of halting him.

McNaughton realized late on 10 March that his pincer operation had failed. Consequently, at 2130 he ordered Sansom to destroy enemy forces in the Towcester–Buckingham area and then advance northeast towards the final objective, Huntingdon, the following day. Yet by 2300 Sansom had still not established contact with Gammell's armoured divisions.[36] In carrying out this new assignment, Sansom had taken the incredible step of stripping the 5th CAD of its armoured brigade, exchanging it for the Guards Armoured Division's infantry brigade. In other words, the 5th CAD (with infantry only) was tasked to hunt tanks, and the Guards Armoured (with tanks only) was sent off in an advance. This sort of radical regrouping had been sharply criticized by Montgomery in past exercises. When McNaughton heard about it the following morning he immediately did the right thing – he sent off this Emergency Operations message to Sansom at 0950: 'Army Comd directs you to re-establish normal organization armd divs forthwith. Reliable information indicates there is no enemy force on your front which you need fear and it is imperative you push on with utmost speed on axis TOWCESTER – HUNTINGDON establishing contact with new sup route being built fwd at BUCKINGHAM. Repeat utmost importance you push on vigorously and immediately. Ack personally to Army Comd.'[37] This was impossible before 1800. The result was the near destruction of the Guards Armoured Brigade when, without infantry support, it attacked carefully selected infantry positions reinforced by substantial anti-tank assets.[38]

McNaughton remained active throughout 11 March. At 1510 he arrived at Sansom's HQ, having travelled by road from Wallingford–Oxford–Banbury, and again urged him to press on with the utmost vigour. He also spent some time reconnoitering the front in an Air OP and grew concerned that Stopford was not moving quickly enough. Stopford felt that he had to secure his flank before moving on. Indeed, the day before, the 43rd Infantry Division had been attacked by the newly arrived

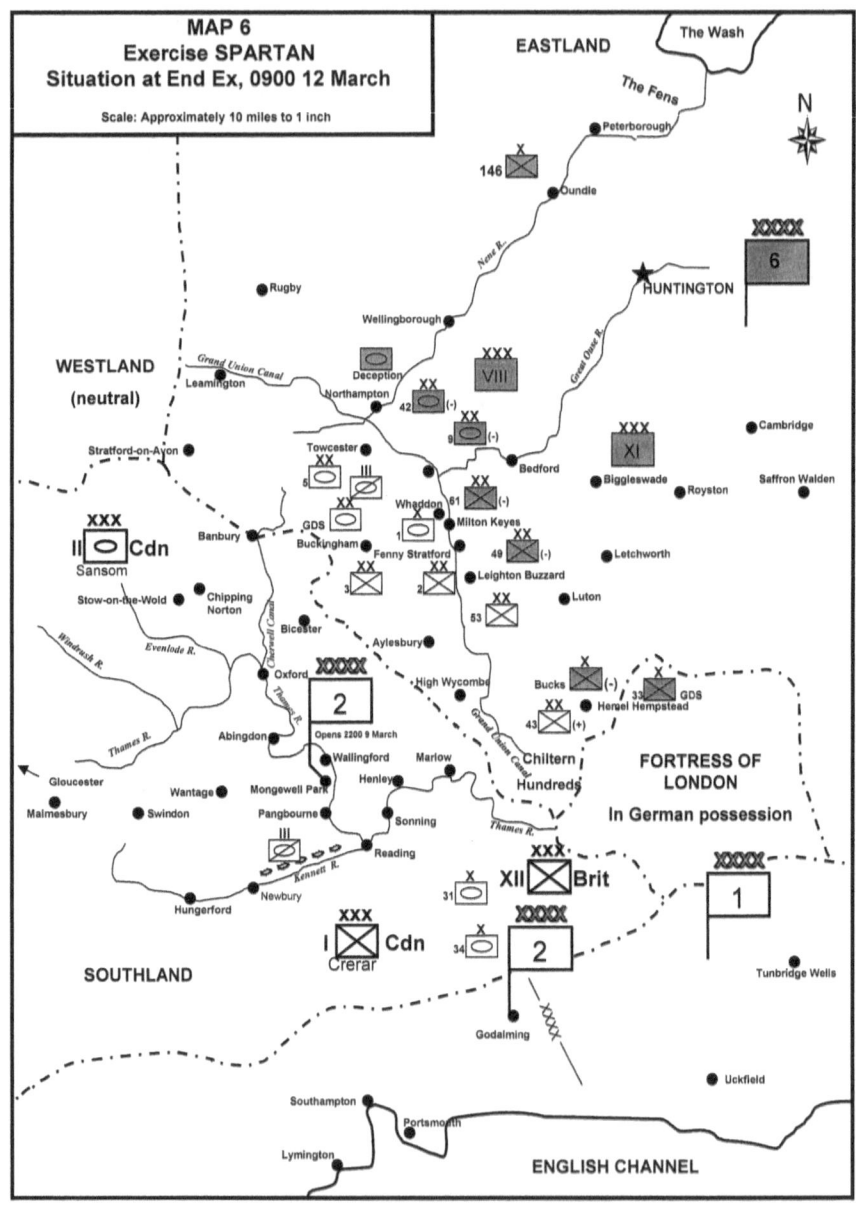

33rd Independent Guards Brigade supported by the 75th Medium Field Regiment from Hemel Hempstead.

At 2255, McNaughton issued his orders for the next day, essentially urging his corps commanders to press ahead on their assigned axis. However, communications with II Canadian Corps remained inadequate – so much so that McNaughton sent off this message by SDR to Sansom at 0830 on 12 March: 'Please give your personal attention to insuring the prompt passage to me concerning enemy dispositions & progress your attacks. This has been most unsatisfactory to date. ACK.'[39] Thirty minutes later GHQ announced the ceasefire.

With the conclusion of the week-long exercise the evaluation of commanders and staffs began in earnest. SPARTAN had involved large forces moving over considerable distances within the confines of the exercise rules. McNaughton did not have to exhibit Napoleonic brilliance, but he did have to pull off a solid, workmanlike performance. Achieving a satisfactory grade from his superiors would have done much to solidify his position as commander of First Canadian Army. It also would have undermined Brooke's fundamental premise that he was incapable of commanding troops in actual operations. The question, then, is whether McNaughton had done enough on SPARTAN to demonstrate sufficient professionalism to parry Brooke's agenda. That is the subject of the next chapter.

10

The Long Shadow of SPARTAN

As the commander of the whole army, McNaughton naturally tended to get the blame. That was natural, but in my view mistaken – a view based on subsequent examination as well as on observation at the time.

Basil Liddell Hart

McNaughton's performance on SPARTAN has not received the attention it deserves. His lone biographer, John Swettenham, found little fault with it, declaring that he displayed 'superior generalship.' John English has taken the complete opposite view. Outside the official history, little serious analysis has been offered over the years even though SPARTAN was an important factor in McNaughton's relief and in the relief of several other senior Canadian officers at the time.[1] One possible explanation for the lack of interest in this critical training event is that Canadian military historians have accepted the judgements of Paget and Brooke without conducting their own independent analyses.

To fully appreciate McNaughton's performance on SPARTAN, it is important to understand certain decisions he made before the exercise started. He was originally to have commanded I Canadian Corps and two British corps, one of which was to be armoured.[2] This changed when he substituted II Canadian Corps Headquarters, newly created on 15 January 1943, for the British armoured corps headquarters only four days later. With Paget's agreement McNaughton promoted Sansom to command it.[3] The II Canadian Corps, however, was severely deficient in signals equipment and had never even conducted a staff exercise or a TEWT. Sansom's BGS, Hugh A. Young, told Stacey after the war that 'a number of the staff officers only reported to Corps Headquarters a few days before the move to the assembly area took place. This precluded the possibility of holding even a staff exercise. Furthermore, a considerable amount of transport and practically all of the signals equipment did not arrive until just before the exercise commenced. As we were about to embark on the exercise I remember remarking that we have a conglomeration of officers, men and material now but it cannot be called a Corps Headquarters.'[4]

English argued that without such fundamental preparation the training value of II Corps' participation 'was clearly questionable.' But one should not overlook Paget's observation that getting two large armies into the field required 'large scale improvisation, particularly with regard to signals ... and the events of the exercise as a whole must be judged against this background before any deductions or criticisms can be made.'[5]

McNaughton originally intended to use Crerar's corps in the mobile role and to have Sansom's headquarters assume responsibility for the administration of those units and establishments which were not directly involved in the exercise.[6] Though there appears to have been no compelling need to test II Canadian Corps HQ so soon after its establishment, it is apparent that McNaughton wished to take advantage of the opportunity provided by SPARTAN in case it did in fact represent the final dress rehearsal before invasion. Brooke had already made it clear that no invasion was possible in 1943; even so, by early January McNaughton was working on the assumption (based on discussions with Paget) that the army's components needed to be ready for battle in the order of priority in which they might be engaged. That priority was as follows: (1) divisions; (2) basic corps troops; (3) division increments; and (4) army, GHQ, line-of-communication, and base units.[7]

McNaughton and Crerar already had great confidence in the level of training in I Canadian Corps.[8] Moreover, Stacey was essentially correct when he stated that the rarity of such large-scale exercises made it 'an opportunity not to be missed.'[9] He later reversed himself, declaring that McNaughton's greatest mistake on SPARTAN was to employ Sansom's raw corps headquarters: 'Andy, I think, was looking at the affair from the point of view of its training value [and because he was a] very secure person ... I doubt whether he ever considered the possibility that a poor performance by the army might reflect on himself.'[10] The double-edged sword of SPARTAN was apparent in Stacey's observation.

McNaughton certainly recognized the tremendous difficulties a green staff would encounter in trying to execute mobile operations with thousands of vehicles for the first time with insufficient signals facilities. Furthermore, he understood that he could only expect so much from Sansom's corps during the exercise. SPARTAN would reveal that Sansom was clearly out of his depth as the commander of an armoured corps; unfortunately, a suitable replacement was not at hand. His deficiencies could have been offset somewhat by a strong BGS, but Young was a First World War veteran and was by no means an expert in armour. He had been the BGS at CMHQ before taking over the 6th CIB following Southam's capture at Dieppe.

The 5th CAD had significant limitations as well. Sansom had commanded it from 1941 to early January 1943, but during that time opportunities for formation-level training were almost non-existent owing to equipment shortages. When the division arrived in Britain in late 1941 it was placed under CMHQ until its individual training had progressed to the point where it could assume an operational role in the Canadian order of battle. It did not receive any tanks until 7 February 1942, when forty M3 General Lee tanks arrived. Only in late June of 1942 did the division come under First Canadian Army and attempt formation training. The equipment situation remained so bad that in mid-November McNaughton informed Lieutenant-General Frederick E. Morgan, Chief-of-Staff Home Forces, that even by the end of the year the division would have only 75 per cent of its allotted tanks.[11] The training situation in the 5th CAD had not demonstrably improved by the time Major-General C.R.S. 'Bud' Stein, commander of the 2nd Canadian Armoured Brigade for seven months during 1942, took command in mid-January 1943.[12]

McNaughton recognized early on that the Canadian armoured divisions needed assistance in training. In late 1940 the CIGS, General Sir John Dill, established HQ, Royal Armoured Corps (RAC), and appointed Major-General Giffard le Quesne Martel as Commander, Royal Armoured Corps (CRAC), under Home Forces with the mandate to guide armoured training. He took up his duties on 8 December 1940 with his headquarters near St Paul's School in Hammersmith. On 22 June 1942 McNaughton suggested to Martel that Major-General Montague Brocas Burrows's 2nd Armoured Group Headquarters, originally intended to stand down, be retained to liaise with First Canadian Army and help train its armoured formations. Martel agreed, and by 8 July so had Paget.[13] Despite the goodwill and assistance of Burrows's headquarters, progress in training Stein's division continued to lag far behind that of other divisions.

Stacey cautioned his readers in *Six Years of War* that SPARTAN was the 'first occasion on which the whole of the 5th Division was actually exercised together as a formation.' One historian has suggested that even as it prepared to go to Italy, the division was only beginning to grasp the complexities of divisional operations.[14] Major-General Frank F. Worthington's 4th Armoured Division had only arrived complete in Britain the previous October. It had not yet exercised as a complete division, and it would not do so until Exercise GRIZZLY II in October 1943. It did, though, have the benefit of a few unit-level exercises. McNaughton could have substituted Worthington's division for Stein's, but the advantages would have been minimal.

McNaughton has faced further postwar criticism for concentrating his two armoured divisions in the green II Canadian Corps under Sansom, the most inexperienced of his three corps commanders.[15] In June 1942, Montgomery drew attention to the lessons of Exercise TIGER, declaring that though a well-balanced corps had two infantry divisions and one armoured division, an army commander could, as the battle progressed, 're-group his divisions, forming a Corps of two, or even three, armoured divisions.'[16] Before El Alamein, Montgomery created his own *corps de chasse*, X Corps, composed of the 1st and 10th Armoured Divisions. But he was dissatisfied with the results, announcing afterward: 'I do not agree with the policy of keeping Armd. Divs in separate Corps.'[17]

McNaughton could have made I and II Canadian Corps more balanced by giving them each one infantry and one armoured division, but Crerar was facing poor tank country in the centre. In practice, he failed to effectively employ even the 1st Canadian Army Tank Brigade to maintain contact with enemy forces during the exercise.[18] Stopford's zone near London offered no better prospects for armoured manoeuvres. McNaughton's options regarding the placement of the armoured divisions were limited by the terrain, and it made sense to concentrate them on the left.

Brooke had given considerable attention to what he considered the proper employment of armour and clearly would have looked disparagingly on McNaughton's inability to bring Gammell's armour to battle. There were only three armour-on-armour actions during the entire exercise, and two were initiated by Gammell. However, what exactly constituted 'proper' employment of armour was still open to debate even by the time of SPARTAN.

As far as Paget was concerned, the handling of armoured forces during the exercise 'did not conform to the doctrine I laid down and issued last October.' He pointed to three factors that were necessary to maximize effectiveness: surprise, suitable tank-going unimpeded by frequent obstacles; and suitable objectives not covered by heavy anti-tank defences. He did not think that armoured divisions should attack the head of the enemy; instead, the tail should be the true objective. Military Training Pamphlet (MTP) no. 41, *The Tactical Handling of the Armoured Division and Its Components,* Part 2, *The Armoured Regiment,* had been distributed in February 1943. It stated clearly that 'the necessity for pressing on unceasingly and relentlessly to carry out the task of cutting the enemy's lines of retreat must be constantly in the forefront of the mind of every commander of an armoured unit.'[19] This was different from Brooke's concept of locating enemy armour and bringing it to battle; it was also completely different from Montgomery's concept of seizing vital terrain to force enemy armour to counter-attack. One historian has argued that the armoured divisions in England were 'trained according to the War Office concept of their role but ultimately had to function in battle according to Montgomery's different policy.'[20]

MTP no. 41 also indicated clearly that an opponent's voluntary withdrawal of armoured forces – such as the one executed by Gammell – was probable evidence of his intention to draw one's own armour onto an anti-tank screen. Rommel had repeatedly employed this tactic to great effect in the desert. The doctrinal countermeasure advocated by MTP no. 41 was '*being extremely wary before committing our armour* [emphasis added]' in the first place.[21] This in itself partly disarmed Paget's criticism that McNaughton had failed to close with Gammell's armour. McNaughton had quite properly withheld his armour until he felt the time was right to fully commit it. Indeed, if one applies Montgomery's template, McNaughton had to 'break in' against Gammell somewhere *before* committing his armour. Paget's concept was not entirely clear, in that he also wanted armour to be 'held far enough away to threaten several alternative objectives' in order to cause the enemy to disperse his anti-tank defences.[22]

McNaughton would have been able to identify much more quickly where he needed to break in and thereby commit his armour had Second Army not suffered from persistent failures of ground reconnaissance. Lieutenant-General Henry C. Loyd, the Chief Umpire for SPARTAN, was highly critical of both Gammell and McNaughton regarding their use of reconnaissance assets to locate the enemy: 'The dissemination of information on both sides was not at all good. This can be partly attributed to wireless silence (when enforced) and to machinery of staff work not working smoothly.' The absence of information about the location of Gammell's armour during the early stages of the battle 'was largely responsible for the slowness of the advance of the British Army.'[23] Loyd had been judged unfit to command a division after his performance in France; even so, his observation was valid. It must be said that McNaughton did understand that there was a problem with the passage of information backward – something that was critical to his decision making.

Another factor that heavily influenced how McNaughton shaped the battle was bridging. Problems with this were evident early on. McNaughton's Chief Engineer (CE) was fifty-seven-year-old Major-General Charles S.L. Hertzberg, former CE of the 1st Division, VII Corps, and I Canadian Corps. For reasons of age, McNaughton had had to request special authority from Ralston to promote him to brigadier and major-general. On 23 June 1943, Hertzberg was finally retired and was replaced by Brigadier James C. Melville. Hertzberg's initial engineer appreciation was based on the assumption that no bridging would be required up to the Eastland border and that few demolitions would be carried out in Eastland. Paget quite rightly questioned this assumption in that 'no hint of this nature was given in any of the available intelligence.' It probably came from McNaughton, but Hertzberg does not appear to have updated his appreciation when Gammell moved pre-emptively.[24]

Second Army constructed sixty bridges of varying classification during SPARTAN, but only half were used. More disturbing is that no instructions on bridging policy were issued during the early planning. According to Paget, bridging plans 'never caught up with the operations and were never more than one day ahead of the advance.' A CE had to be intimately involved in any army-level planning process. Had this been the case, Hertzberg would have completed his own appreciation based on McNaughton's intentions and then would have briefed him on what he could do to achieve the army's plan. In other words, Hertzberg would have forecasted bridging requirements. But without a clear expression of the army's scheme of manoeuvre several days in advance, his planning efforts could have no real focus. As the CE of Home Forces noted: 'By far the greater part of the delay imposed on the British advance was accounted for by the time taken to deploy resources. In an advance a CRE (Commander Royal Engineers) must be planning for tomorrow; he must be able to look ahead 48 hours. The CE (Chief Engineer) of a Corps must be able to look ahead 3 days. The CE of an Army must be able to look ahead 4 days at least. There must be a plan.'[25]

Table 10.1 McNaughton's compressed battle procedure during SPARTAN

4 March	2130 hours for Army operations on 5 March.
6 March	1630 hours cancels relief of I Cdn Corps by XII Brit Corps and orders II Cdn Corps to move east across Thames (confirmed in writing 2335 hours).
7 March	1615 hours (warning order) for II Cdn Corps to move west.
8 March	2025 hours for Army operations on 9 March.
9 March	1800 hours for Army operations on 10 March.
10 March	2130 hours for II Cdn Corps operations on 11 March.
11 March	2255 hours for Army operations on 12 March.

Source: 'GHQ Exericise SPARTAN: Narrative of Events, March 1943,' DHH, File 545.033(D1).

McNaughton did not demonstrate the ability to forecast army operations four days in advance. He visualized what he wanted to do operationally in very short time frames, and the result was excessively compressed Battle Procedure. In his Operations Order (Op O) for 6 March he had stated that 'speed is essential ... Formations will be handled boldly and widely, full advantage being taken of possibilities of accelerating movement and action against the enemy, by proceeding across country.'[26] However, his timings for next-day operations were far too optimistic to ensure orderly execution. His appreciation of the need for speed was good, but at times there was a clear weakness in his planning, the result of an imprecise understanding of what was possible in time and space for large formations.

The negative effects of compressed Battle Procedure were at times compounded by weak traffic control, especially in Sansom's corps. Paget felt that the repeated traffic jams throughout McNaughton's army (and Gammell's) had been avoidable, since there were 'ample roads in both army areas and the density of movement was comparatively light'; but 'no army traffic plan was formulated' in Second Army.[27] If so, then the blame ought to be placed on Simonds, as well as Turner, the Deputy Assistant and Quartermaster General (DA&QMG), and Gibson, the GSO 1 (Ops). At corps level, Young and Brigadier Horace V.D. Laing, the DA&QMG, ought to have assigned practical division routes within the corps' boundaries and 'de-conflicted' route problems between Stein and Adair. Loyd concluded that the selection by Sansom's staff of narrow, second-class roads (McNaughton had directed that major routes be used) for the night move on 7 and 8 March without proper route reconnaissance had made traffic control very difficult; it had also led to considerable congestion. Moreover, these routes had inflicted severe mechanical strain on the tanks.[28] In the 5th CAD the GSO 1 (Ops), Lieutenant-Colonel James D.B. Smith, had been responsible for disseminating the division traffic plan.[29] That there were breakdowns in fundamental staff work at all levels is beyond doubt. Worthington, perhaps the only true 'armour-minded' general in the entire Canadian Army at the time, also observed the manoeuvre and felt that the traffic jams were not McNaughton's fault.[30]

A sound army traffic plan was essential because it facilitated the army's logistical support. Paget criticized McNaughton's initial appreciation, which assumed that the advance to the frontier would be unopposed, which in turn 'naturally coloured the maintenance plan.' The crux of the administrative problem facing McNaughton was the economical use of transport. Paget concluded that despite the problems, after 8 March the situation had 'swung steadily' in favour of Second Army, which succeeded in overcoming the handicap of its lines of communication across the Thames. Apart from the disastrous operations of Sansom's corps on 11 March, 'it became only a matter of time' before Second Army achieved its objective.[31] Indeed, McNaughton had managed to exert fairly consistent pressure on Gammell and for the most part had managed to retain the initiative throughout the exercise.

As far as Brooke was concerned, McNaughton's greatest mistake on SPARTAN was his decision to pass II Canadian Corps through I Canadian Corps on 7 March at night. Brooke told Ralston that he doubted McNaughton's command ability after he 'in his presence ordered Sansom to move across Crerar's lines of communications.'[32] As planned, the Guards and the 5th CAD would have crossed Keller's lines of communication back to the 2 CID divisional dump at Hungerford. Crerar's corps lines of communication from the forward divisional dumps back to the base sub-area and ports would not have been crossed.

McNaughton's decision to commit his armour was not in itself a bad one, but he misjudged the operation's complexity. He allowed little more than twenty-four hours to stop II Canadian Corps, reorient it, move it into new assembly areas, move it across the Thames bridges, and concentrate it in new assembly areas. Considerable time would have been required after all of these preliminary moves before the divisions could start their attack in a new direction. As Paget correctly stated in his comments: 'Corps cannot be swung about on the battlefield like battalions and should be given at least 24 hours' warning.'[33] Moving at night would have definitely slowed the movement down. Loyd concluded: 'It is very questionable if night moves for an armd formation are ever justifiable except to secure tactical surprise or when the enemy has complete air superiority. However carefully they are arranged the additional length of time involved in their execution as compared with that required for a day move for a similar distance usually results in less thorough maintenance than is desirable.'[34]

McNaughton's manoeuvre does not look so bad when compared to some other operations actually executed during the campaign in Northwest Europe. Patton pushed two corps through the narrow Avranches corridor in August 1944 to begin Third Army's breakout. Major-General Troy H. Middleton, commander of VIII Corps, reflected: 'You couldn't work it out on paper; you could never have gotten by with saying that it could be done at any of the service schools – an automatic F would have resulted.'[35] Patton himself admitted that losses would have been terrific had the Luftwaffe appeared in strength.

An operation of far greater complexity was GOODWOOD. The three armoured divisions involved – the 7th, the Guards, and the 11th – were under command of O'Connor's VIII Corps. All had to cross the lines of communication of either II Canadian Corps or British I Corps to reach their start lines. They then had to cross the Orne River into a small salient held by the 6th Airborne Division. The three divisions attacked in echelon on a very narrow front and had to negotiate a gap in a minefield and funnel through small villages. The 7th Armoured was caught up in heavy congestion in the rear areas and had difficulty even making it to the start line.[36]

Montgomery and Dempsey took a huge risk in executing GOODWOOD. The difference between it and McNaughton's manoeuvre on SPARTAN was that O'Connor had several days to plan and execute the move of the three armoured divisions compared to the twenty hours McNaughton allowed Sansom and his inexperienced staff. McNaughton does not seem to have considered simply placing one or both of Sansom's armoured divisions under Crerar's command to achieve the kind of concentrated breakthrough he was looking for. Doing so would have hugely simplified the move. Moving II Corps complete across the Thames meant that all the corps troops had to cross as well, which only magnified the complexity.

Not everyone present during SPARTAN condemned McNaughton for his decision. Brigadier Harry Latham, who later would work on the British official history, watched the preliminary moves of the intended manoeuvre unfold. After the war he told Stacey (who had been present in

Table 10.2 Staff element, I and II Canadian Corps, during SPARTAN

I Canadian Corps	Lieutenant General H.D.G. Crerar
BGS	Brigadier Churchill C. Mann
2 Canadian Infantry Division	Major-General James H. Roberts
3 Canadian Infantry Division	Major-General R.L.F. Keller
1 Canadian Army Tank Brigade	Brigadier R. Wyman
II Canadian Corps	Lieutenant-General Ernest W. Sansom
BGS	Brigadier H.A. Young
DA&QMG	Brigadier H.V.D. Laing
GSO 1 (Ops)	Lieutenant-Colonel M.P. Bogert
AQMG	Lieutenant-Colonel D.K. Tow
5 Armoured Division	Major-General C.R.S. Stein
GSO 1 (Ops)	LCol J.D.B. Smith
AA&QMG	LCol J.R.R. Gough

the 'I' Filter Room): 'McNaughton's order led of course to chaos but as I watched it the whole [thing?] was sorting itself out well ... McNaughton had the sympathy of all us "observers."'[37]

Basil Liddell Hart was also on hand for SPARTAN and saw nothing wrong with McNaughton's manoeuvre. 'Unfortunately things went wrong in the early stages,' he maintained after the war, 'and a bad traffic jam occurred in getting 2nd Canadian Corps through the bottleneck area around Oxford.' However, Sansom never executed the manoeuvre: McNaughton cancelled it, so the fighting elements of Sansom's armoured divisions never moved across Crerar's lines of communication. Liddell Hart's recollection, therefore, was faulty. As he saw it, the principal fault was not with McNaughton and the decision to move Sansom across the Thames but with 'some of his chief subordinate commanders.'[38] In a post-war interview with the CBC, Liddell Hart added: 'As the commander of the whole army, McNaughton naturally tended to get the blame. That was natural, but in my view mistaken – a view based on subsequent examination as well as on observation at the time.'[39]

Lost in the complete focus on McNaughton is the role played by Simonds. After evaluating the various SPARTAN reports, Simonds concluded that Paget's main conclusions regarding weaknesses in organization and training were 'substantiated by events during the exercise.'[40] After the war he reflected on McNaughton's performance to Trumbull Warren, maintaining that 'the fiasco which resulted ... was clearly *all* his own doing.'[41] As subsequent events during the war would demonstrate, Simonds was very good at blaming others and deflecting blame. There is a serious issue at hand here. When confronted with Paget's observation that there was no army traffic plan, no army bridging plan, and a faulty maintenance plan, the first question any military historian should ask is this: Where was the great Guy Simonds, *the* principal staff officer in Second Army and according to many the greatest staff officer in the entire Canadian Army?

Simonds has left us a glimpse of McNaughton's command technique at the time. After the exercise he spoke to McNaughton privately: 'I told him I would serve him in any capacity he wished, except that I would not remain as Chief of Staff at Army HQ unless he would accept my advice on the matter of organizing Army headquarters to fight a battle and concentrating on training and operational aspects ... He kicked me out of his office and within 48 hours I left on attachment to 8th Army.'[42]

This suggests that McNaughton did not listen to Simonds during the exercise. McNaughton's War Diary reveals that they were together constantly, and it is hard to imagine Simonds not saying something when McNaughton proposed to give Sansom a day to cross the Thames. There are really only two possibilities: either Simonds said nothing, or he raised concerns and was told to be quiet. McNaughton could be dismissive, but it is hard to imagine him failing to appreciate the time and space limitations had they been presented with the clarity Simonds was known for. Indeed, it is difficult to accept the idea that McNaughton, who possessed a sharp, analytical mind, would not have appreciated the complexity on his own. There exists the very real possibility that he did in fact understand the problem but was prepared to accept a great deal of risk in carrying out his manoeuvre.

The day after the exercise ended McNaughton reported optimistically to Stuart: 'I feel we have learnt most valuable lessons for the future ... As you know, our Army staff was new and partly set up ad hoc. Nevertheless by the conclusion of the exercise it was working smoothly and efficiently and our officers have proved their capacity.' He was pleased with the overall performance of the engineers, signals, supply and transport, medical, ordnance, and other administrative services, 'particularly as this was the first occasion in which we have ever had an opportunity to give them actual practice full scale.'[43]

On 15 March, McNaughton held a press conference at First Canadian Army Headquarters, where he indicated that he was completely satisfied with all commanders, staffs, and units. He said much the same thing to Stacey while the latter was drafting *Six Years of War*. After reading McNaughton's comments on the draft, Crerar told Stacey that he was doubtful of the desirability of including them in the history:

As a matter of fact the Commander-in-Chief, General Paget issued a memorandum on Spartan Exercise (distributed down to battalion commanders) very shortly after the Exercise terminated which was definitely critical of General McNaughton's conduct of the manoeuvre operation and also of the part played by 2nd Canadian Corps. General McNaughton, I remember, was naturally upset at this public criticism, aspects of which also appeared in the Daily Press at that time. I suggest that the circulated comments of the C.-in-C. Home Forces (of which there should be a copy on file somewhere) should be read before it is decided to publish General McNaughton's own comments as quoted in the draft. The two viewpoints were not in agreement – that is certain.[44]

McNaughton was not debriefed on his apparent shortcomings on SPARTAN when he met with Nye on 16 March or during any subsequent meeting with Paget or Brooke.[45] No direct criticism reached McNaughton except for what was contained in Paget's post-exercise comments. But the British press, which had so often sung McNaughton's praises, offered significant criticism. According to Richard Malone, the press criticism was 'extremely sharp and damaging' immediately after the exercise, and he and Jim Bone, the London editor of the *Manchester Guardian*, took steps to water down the coverage. As a result, the following morning 'not a single word of criticism against ... McNaughton was carried in any paper.'[46]

Within three days of the end of the exercise, Paget asked McNaughton to immediately relieve Sansom of command of II Canadian Corps.[47] Loyd had criticized Sansom for the excessive number of vehicles, 143, making up his tactical headquarters, noting that it 'proved quite unmanageable since it was not possible to get forward rapidly in the prevailing traffic conditions.' The concentration areas chosen by corps headquarters had been 'very small and poorly supplied with roads,' and this had made it difficult to marshal columns for moving off. In his draft comments Paget had been much harsher than in his printed version, stating that II Canadian Corps' movement had been 'slow and deliberate and hampered by tactical and administrative mistakes.'[48]

On 17 March, McNaughton had a private conversation with Paget regarding Sansom's performance. Paget pressed for his immediate relief; McNaughton, shocked and surprised at the criticism, refused outright. He defended Sansom by taking the blame for including the inexperienced II Canadian Corps Headquarters, and he argued that Sansom simply needed more time to improve the efficiency of the corps, having been in command only two months.⁴⁹ Paget backed down for the moment, but it seems that he had pressed for Sansom's relief on his own authority without backing from Brooke. According to McNaughton, Brooke stated on 5 April that he had no personal knowledge of Sansom and that he 'fully agreed' with the decision to give him another chance.⁵⁰ That Paget and Brooke would not have compared notes on various commanders, however, seems unreasonable.

Shortly afterwards Crerar summarily dealt with Roberts, Commander of the 2nd CID. Crerar felt that Roberts needed to be assessed as to his real capacities in the field quite apart from his role in the Dieppe Raid; to this end, he paid close attention to his performances on MAPLE I, MAPLE II, and SPARTAN. Regarding MAPLE I, Crerar considered Roberts's solution to a minefield 'unsound' and criticized his fire plan. Regarding MAPLE II, Roberts was castigated for his faulty disposition of units, which proved incapable of covering one another by fire, and for his weak understanding of dominating heights.⁵¹ Major-General D. Charles Bullen-Smith, the umpire assigned to Crerar's corps during SPARTAN, had observed that there was 'no coordination between infantry and artillery' in the 2nd CID.⁵² On 9 April 1943, Crerar wrote to McNaughton that he had examined all the umpire reports and concluded that

> there was, in fact, very considerable confusion, loss of time and inadequate co-ordination of effort in respect to the 2 Cdn Div during this period, and in the following days of SPARTAN. It is also to be remarked that 2 Cdn Div suffered very heavy losses by umpire decision, resulting in theoretical [*sic*] destruction of the 4 Cdn Inf Bde, and the reduction of fighting strength of 5 and 6 Cdn Inf Bdes to 40% and 80% respectively. The above is submitted for record purposes only, in case the real *reasons* for the transfer of Maj.-Gen. Roberts should be questioned, on some future occasion.⁵³

Roberts had not taken part in BEAVER IV or TIGER but *had* taken part in the inter-divisional exercise BEAVER III in April 1942. Crerar did not agree with Montgomery's positive assessment of Roberts on BEAVER III, citing what he viewed as important errors of judgement – specifically, compressed Battle Procedure. Brigade attacks had been prepared 'very hurriedly,' Crerar declared, 'and in my opinion could not possibly have been delivered in time.' The failure to appreciate the time required between the ordering of a brigade or division attack and the launching of the battalions supported by coordinated fire power was 'shown on more than one occasion by HQ of 2 Cdn Div.'[54] Crerar told McNaughton: 'While Maj.-Gen. Roberts has shown himself to be a definite and forceful Commander and a good administrator, recent Exercises and Manoeuvres, in which he has participated, have now convinced me that he lacks the high tactical abilities in a Divisional Commander when engaged in mobile operations.'[55] McNaughton accepted Crerar's assessment, and Roberts was relieved of command in April. His replacement was Major-General Eedson L.M. 'Tommy' Burns, a favourite of Crerar.

By 26 July, Paget was again pressing for Sansom's relief, this time by trying to impose the age limit. McNaughton again said no, arguing that Sansom had been administering and building up the corps in a satisfactory manner and adding that his tactical efficiency 'was yet to be tested but that the exercises now in progress ... would give me some opportunity to form a conclusion.'[56] Indeed, McNaughton had visited Sansom's headquarters on 19 April to observe the II Canadian Corps Study Period. The topic was the advance guard of an armoured division, and McNaughton was 'somewhat concerned to find some disregard for basic principles which he had always stressed to formation commanders and he also noted a considerable misuse of military terms.'[57]

After observing Sansom on Exercises SNAFFLE (9 to 11 August) and LINK (13 September), Brooke accepted Paget's judgement. Together they pushed McNaughton once more to relieve him. McNaughton stated that he 'was not prepared to accept General Paget's views,' which had been 'formed as result of Exercise SPARTAN when 2 Cdn Corps had been made up ad hoc from untrained personnel. This had not been an adequate or a fair test. Exercise LINK was now in progress. I had seen something of it myself earlier in the day, and the work of 2 Cdn Corps, particularly as regards traffic, had been very good. I would make up my mind after the Exercise, on the results.'[58]

McNaughton and Paget had gotten along remarkably well except for a few minor disagreements over organizational matters. They were both straightforward, no-nonsense individuals, Paget even demonstrating the strength to challenge Churchill on occasion.[59] The argument over Sansom ate away at some of this mutual respect. Paget no doubt questioned McNaughton's judgement in retaining an ineffective senior commander. Liddell Hart, who knew Sansom well, blamed him for the problems on SPARTAN: 'Much as I liked him, I should never have conceived that he would become commander of an armoured corps in mobile warfare.' Perhaps the harshest criticism came from Lieutenant-Colonel M. Pat Bogert, who considered Sansom the most incompetent general in the

Table 10.3 Canadian senior officers relieved during training, 1940–1944

Officer	Unit	Date	Prompted By	Reason
LGen A.G.L. McNaughton	First Cdn Army	Nov/43	Brooke/Paget	SPARTAN
LGen E.W. Sansom	II Cdn Corps	Jan/44	Paget/Brooke	SPARTAN
MGen J.A.H. Roberts	2 CID	Apr/43	Cerar	SPARTAN
MGen C.R.S. Stein	5 CAD	Oct/43	Sansom	SPARTAN
MGen V. Odlum	2 CID	Nov/41	Brooke	Age limit/ Ineffective
MGen F.F. Worthington	4 CAD	Feb/44	Simonds/Crerar	Age limit/ unsound
MGen G.R. Pearkes	1 CID	Aug/42	Montgomery	Ineffective
MGen C.B. Price	3 CID	Aug/42	Montgomery	Ineffective
MGen L.F. Page	4 CID/AD	Dec/41	McNaughton	Age limit
BGen P.E. LeClerc	5 CIB	Mar/41	McNaughton	Medical
BGen A.A. Smith	1 CIB	Nov/40	McNaughton	Medical

whole army.[60] McNaughton waited so long to drop Sansom partly out of the loyalty he felt he owed friends and associates. Perhaps more important, there was no one to replace him in mid-1943. Charles Foulkes told Stacey that McNaughton actually did intend to replace Sansom and was only waiting for Simonds to gain the necessary experience.[61]

Another divisional commander who did not survive SPARTAN was Bud Stein of the 5th CAD. Ironically, Sansom had passed harsh sentence on Stein well after SPARTAN, concluding in October that he lacked the 'tactical ability and strength of character' to command an armoured division in combat.[62] Stein was medically diagnosed with 'progressive anxiety neurosis' and was relieved of command by McNaughton on 14 October. He was replaced in November 1943 by Simonds. Sansom gave up his command on 29 January 1944 after McNaughton had been replaced.[63] Once again, it was Simonds who filled the vacuum by assuming command of II Canadian Corps. All of this highlighted the shortage of skilled senior commanders. SPARTAN thus served as a clearing house, adding dramatically to the list of senior Canadian officers relieved for command deficiencies during the training period in Britain.

The fact that three of McNaughton's senior commanders were judged to have performed poorly on SPARTAN could only have reinforced Brooke's view that McNaughton himself was an inadequate trainer. The average time in command for Canadian division commanders to the end of 1943 was ten months, but it did not seem to matter whether one was

Table 10.4 Time in command for Canadian Division commanders to the end of 1943

Commander	Division	Months in command*
LGen A.G.L. McNaughton	1 ID	6½
MGen G.R. Pearkes	1 ID	24
MGen H.L. Salmon	1 ID	8
MGen G.G. Simonds	1 ID	6
MGen V. Odlum	2 ID	17
MGen H.D.G. Crerar†	2 ID	1½
MGen J.H. Roberts	2 ID	12
MGen G.G. Simonds	2 ID	½
MGen E.L.M. Burns	2 ID	8
MGen B. Price	3 ID	18
MGen R.L.F. Keller	3 ID	15
MGen L.F. Page	4 ID	7
MGen F.F. Worthington	4 AD	22
MGen E.W. Sansom	5 AD	33
MGen C.R.S. Stein	5 AD	9
MGen G.G. Simonds	5 AD	2

*Average time of divisional command = 10 months
†Crerar never actually commanded the division

in command for a long time or a short time. Stein had commanded the 5th CAD for nine months but had had little real opportunity to practise manoeuvring his formation in the field. Sansom had commanded the same division for *thirty-three* months before taking over II Canadian Corps and had suffered from the same deficiency as Stein. Roberts had commanded the 2nd CID for twelve months and in Montgomery's eyes had done a decent job. By condemning Roberts, Crerar was inadvertently underlining his own failure to correct Roberts's supposed deficiencies. It is not at all evident that Burns represented an upgrade. Roberts's relief highlighted the subjective nature of assessing formation commanders during this period of training.

SPARTAN had been a valuable learning experience for the entire Canadian Army, and many of Paget's criticisms of McNaughton's performance as Army Commander were justified. Montgomery considered Paget 'completely ignorant of practical battle knowledge'; even so, several of the C-in-C Home Forces' observations – especially regarding administrative issues and staff planning – were completely justified.[64] That said, Paget harboured confusing ideas about the employment of armour, the proper location of higher commanders, and – more specifically – what

McNaughton's proper line of operations should have been. Despite his overt criticisms of McNaughton, Paget had tempered his comments by adding that more practice was required to build on the lessons learned. Moreover, it is quite possible that Paget did not interpret McNaughton's performance as negatively as Brooke did, for shortly after the exercise Brooke recorded that Paget 'has no idea how bad Andy McNaughton really is.'[65]

McNaughton had not performed as poorly as has often been suggested. English was entirely justified in observing that McNaughton would have performed better had he acquired more basic professional knowledge through self-study.[66] Yet he still would have had to translate theory into the reality of the ground on the exercise. That transition from book to battlefield was, for McNaughton, not seamless. Furthermore, he most definitely had experienced considerable skill fade in the seventeen months since his last opportunity to command in the field. Conversely, Crerar had taken part in BEAVER III, BEAVER IV, and TIGER before taking part in SPARTAN.

McNaughton did some good things on SPARTAN. He overcame Gammell's pre-emptive move and got the army moving quickly. Quite properly, he went forward in order to see his corps and divisional commanders as well as to be seen and to get information. He pushed his commanders hard when necessary, and he quickly corrected his mistake of trying to pass Sansom's corps across the Thames. Indeed, he essentially admitted his error (a good quality for a commander in its own right) and quickly made the correction. He has not received enough credit for these things. The counter-orders certainly created problems, but they still reflected the right course of action.

In lieu of any official War Office grading system, a simple one presented here will help place McNaughton's performance in context. An 'A' performance means that a commander achieved his objectives and produced maximum effectiveness against the opponent. He was able to synchronize his combat power (i.e., concentrate his forces in place and unify them in time) to a high degree against the enemy with great effect. A 'B' performance means that a commander failed to achieve his objectives but still positioned his forces for ultimate success. He retained the initiative and did not sustain debilitating casualties. Perhaps, though, he demonstrated some inability to synchronize his combat power. A 'C' performance means that a commander achieved few if any of his objectives, possibly sustained prohibitive casualties, and more than likely lost the initiative. When *all* factors are considered, McNaughton probably deserved a 'B–' on SPARTAN. In academic terms, it was not enough to get him a scholarship but sufficient to 'keep him in school.' It is entirely conceivable that with his sharp mind, he would have performed much better given another opportunity to command the army in the field. One does not simply take an army to the field for the first time and perform like Napoleon.

Ultimately, however, there was a finality to SPARTAN – in Brooke's mind – that McNaughton never appreciated. Brooke had mentally compared McNaughton's performance with Crerar's. During the exercise Brooke had visited I Canadian Corps Headquarters and noted that Crerar had put on 'a real good show' and had 'improved that Corps out of all recognition.'[67] Interwoven with Brooke's assessment of McNaughton on SPARTAN was his gut reaction to McNaughton as an individual. That the one influenced the other seems beyond doubt. That long personal 'history' was destined not to improve: events soon conspired to bring the two men into direct conflict once again – not over SPARTAN, however, but over what seemed to be the relatively minor issue of visiting Canadian troops in Sicily.

PART FOUR
The End of an Idea

11

The Sicily Incident

You cannot have visitors when you are fighting hard.

Montgomery to Crerar, 23 July 1943

At the beginning of the invasion of Sicily in July 1943, McNaughton attempted to visit Canadian forces fighting under Montgomery's command in Eighth Army. It was perfectly understandable that McNaughton would want to see his men in action after their long training in England.[1] Any other senior commander of any other national force would have wanted the same. He had missed that opportunity during the ill-fated Dieppe Raid. Yet Montgomery refused outright to grant him access to his troops. This cold rebuke was one of the most confusing and frustrating episodes confronted by McNaughton during his tenure as commander of the Canadian Army. It was symptomatic of his long struggle to maintain a reasonable degree of Canadian autonomy within the larger British Army. It also highlighted the impact of personalities.

On 27 May 1943, McNaughton informed Lieutenant-General Sir Archibald Nye, the VCIGS, that he intended to visit the Mediterranean with one staff officer around mid-June. He wanted to see 'at first hand, the arrangements which were being made for the administration of the Cdn force in Operation HUSKY.' McNaughton added that he and Montgomery had already discussed such a visit and that the latter 'had been entirely in sympathy with the proposal.' Montgomery had suggested that he visit the Middle East 'during Operation HUSKY, rather than before,' but McNaughton thought this inappropriate, making it clear to Nye that he had no intention of interfering with operational matters and that he would rely 'entirely upon the good sense' of Alexander, commander of 15th Army Group, and Montgomery.² There is no indication in McNaughton's War Diary of a meeting with Montgomery for several weeks prior to 27 May, except for 19 May, at which meeting no trip to Sicily seems to have been discussed. This does not mean that such a conversation never took place, but it is difficult to verify McNaughton's version. Considering subsequent events, McNaughton may have given Montgomery and Alexander far too much discretionary power by relying 'entirely' on their good sense. Nye assured McNaughton that he would furnish him with every assistance in the way of transportation.³

McNaughton met Montgomery at the War Office in London on the last day of May and reiterated his desire to visit, perhaps in late June or early July. Montgomery replied that he would be in the Red Sea area on exercises during June and that from 1 July onwards he would be at his HQ in Malta. He hoped to see McNaughton during his visit, but there is no indication from this discussion that Montgomery anticipated McNaughton visiting Sicily – only Malta.⁴ Any chance of McNaughton visiting the Mediterranean in late June was dashed when Nye virtually ordered him to attend RATTLE, a combined-arms exercise for senior commanders and staff officers at Largs, Scotland, to discuss OVERLORD. As the leader of one of the possible invasion armies, McNaughton had no choice but to attend that conference, held from 28 June to 1 July.⁵

A great deal of uncertainty exists regarding McNaughton's intentions in visiting Sicily. Without question, he wanted to see Canadian soldiers. However, Canadian official historian Gerald W.L. Nicholson concluded that 'no authority appears to have been obtained beforehand for any visit to subordinate formations in the Mediterranean.'⁶ It may well be that McNaughton did not feel it necessary to get permission to see his own men. Nye had clearly demonstrated a willingness to provide transport, but the situation became even more clouded on 19 June when the War Office sent a bizarre message to General Dwight Eisenhower at Allied Forces Headquarters (AFHQ) in Algiers. The message indicated that the Canadian CGS General Kenneth Stuart and one staff officer would 'visit you on 7 July for about one week ... particularly in connection with organization and preparation of Canadian forces in Canada for later use in the war against Japan.' As a virtual afterthought, the cable added that McNaughton had asked to tag along with three staff officers.⁷ Yet on the same day, Major-General John Noble Kennedy, the DMO, telephoned McNaughton indicating that Brooke had 'approved the arrangements for the North African visit' and had cabled Eisenhower 'requesting facilities and special attention.' On 23 June, Brooke received confirmation of Eisenhower's willingness to see McNaughton.[8]

Nye, Montgomery, Brooke, Kennedy, and Eisenhower all knew of McNaughton's intention to visit North Africa. However, Brooke's approval said nothing about going to Sicily during operations. Like Nicholson, Stacey concluded that the clearances gained for visiting Sicily were 'inadequate ... It is fairly clear that the War Office was unwilling to put pressure on the British commanders in the theatre in connection with it.' There is, however, no direct evidence for Stacey's assertion.[9] Armed with vague authorization, McNaughton, Stuart, and four other officers headed to North Africa on 7 July.

Even before his confrontation with Montgomery, Brooke, and Alexander, McNaughton had clashed with his own service chief, Stuart. In early May, Stuart had sent Montgomery a cable in North Africa that finished with the words 'Hope to see you soon' – an indication that he was contemplating a visit to the Mediterranean in the near future. McNaughton jumped on this as a clear breach of security that 'might well endanger the lives of Cdn soldiers' and told Stuart so via transatlantic telephone. This outright rebuke could only have been received with anger, for it implied gross incompetence on Stuart's part. According to Swettenham, Stuart 'resented the interference and refused to change his message.'[10] In an instant, McNaughton had probably alienated someone who had up to that point had been an ally.

On 10 July the 1st CID waded ashore in Sicily as part of Eighth Army. The same day, McNaughton's party reached Tunisia after stops at Gibraltar and Algiers.[11] They visited Alexander's Main Headquarters 15th Army Group at La Marsa and there were admitted to the War Room, where they saw some maps though they could not locate the reinforcement depots for the Canadian division. Alexander was no longer there, having relocated to a former British dungeon/tunnel complex on Malta for the actual invasion. When McNaughton and his party attempted to travel to Malta afterwards they were denied entry by an unknown source. Eisenhower, himself now co-located with Alexander in Malta, was informed of this development by his Chief of Staff, Major-General Walter Bedell Smith, soon after the invasion. On 12 July, Eisenhower, who had spent the day on the Sicilian beaches, replied: 'There was no repeat no intention to refuse McNaughton permission to visit Malta and he may certainly come [to Malta] if he wishes.' Eisenhower then instructed

Bedell Smith to inform Montgomery's headquarters of McNaughton's intentions.[12]

On 13 July, McNaughton, still waiting in Tunisia, was greeted warmly by Eisenhower, who instructed General Carl A. 'Tooey' Spaatz, commander of the Northwest African Air Force, to fly him to Malta to stay with the governor, Field Marshal Lord Gort. At Spaatz's request, McNaughton spoke to American airmen at the headquarters of the Mediterranean Air Command. The following day he was flown to Malta in Spaatz's private aircraft. That evening McNaughton met Alexander at his headquarters. Alexander was heading for Sicily the following day and told McNaughton he would ask Montgomery if there was any objection to McNaughton coming over. Clearly, if Alexander or Montgomery had known of McNaughton's intentions at the time there would have been no need for Alexander to clear it with Montgomery.

On leaving Alexander's headquarters McNaughton ran into four Canadian staff officers attached to the 21st Army Group Mission.[13] They reported that they had received little information on the progress of the battle in Sicily and had so far been denied access to the battlefield. McNaughton's temper was slowly rising after five days in the Mediterranean with little to show for it. 'It should be noted,' he recorded, 'that no arrangements appeared to have been made' for the reception or quartering of the Canadians and that 'no effort was made by any member of the staff at Tac HQ 15th Army Group to assist these officers in any way.' One of the accompanying staff officers, Charles Foulkes, BGS of First Canadian Army, observed: 'It was obvious from the start that we were not wanted ... The only time any enthusiasm was shown by the British was when McNaughton asked for transportation to go back to England.'[14]

While conversing with Lord Gort at his palace in Valleta on the morning of 15 July, McNaughton was informed by one of Alexander's staff officers that Montgomery had refused his request to visit Sicily because of a vehicle transport shortage in Eighth Army. When Eisenhower's headquarters notified Alexander of the impending visit and requested that Montgomery be advised, the Eighth Army Commander had erupted, apparently going so far as to tell Alexander he would have McNaughton arrested if he set foot on the island.[15] It is not clear whether Montgomery made this threat on 12 July, when Eisenhower's headquarters informed Alexander of McNaughton's intentions, or on 14 July, when Alexander actually went to Sicily.

Incensed, McNaughton requested an interview with Alexander, who had returned to Malta. According to various contemporary observers, Alexander was a gentle man with a light laugh who somehow managed to pacify even those who had a grievance with him. He had successfully dealt with General Joseph 'Vinegar Joe' Stilwell in Burma. Alexander's biographer went so far as to suggest that he 'never made a personal enemy [but was] instinctively repelled by neurosis of any kind' – something that he probably sensed in McNaughton in Malta. The trouble, Alexander explained to McNaughton, 'lay with the War Office in that he had not heard of the visit in time to make arrangements.'[16] McNaughton pressed his case, indicating that he had not only a constitutional right but also a duty to visit his troops. Alexander told him to wait or come back in a few weeks. Stuart was apparently prepared to do so and stayed until 25 July, but he never did get to Sicily.[17] McNaughton would have none of this and decided to return to Britain immediately. Alexander offered him the use of his own aircraft for the trip back to Algiers; in a feeble attempt to further placate him, he also suggested they tour the Tunisian battlefields. McNaughton politely declined the latter offer.

Based on his character and on his notorious dislike of visitors, it should have been no great surprise that Montgomery tried to keep McNaughton out of Sicily. He simply did not want anyone interfering in his show. The same arrogance that demanded that those in attendance at his conferences get all their sneezing and coughing out of their systems before he started to speak also demanded that 'inferior' minds keep their distance while he was teaching others how to win battles. McNaughton was not the only victim of Montgomery's policy. He apparently threatened to have Churchill's son, Randolph, arrested if he stepped anywhere near the front in Sicily; and when Brigadier Desmond Inglis, soon to be the Chief Engineer (CE) of 21st Army Group, tried to visit Sicily, he was bluntly told that he would be arrested if he did so. Montgomery also refused entry to observers from GHQ, Middle East Forces, in Cairo. He even attempted to prevent Churchill from visiting Normandy a year later.[18]

Yet not everyone was denied access to Sicily while the fighting raged. The Chief of Combined Operations, Admiral Lord Louis Mountbatten, was escorted around the island by Montgomery on 11 July. This was a curious turn of events in that Montgomery generally held an unfavourable impression of the young admiral. Mountbatten actually felt honoured that Montgomery had permitted him entry: 'It was only when I saw Andy McNaughton & the Canadian C.G.S. kicking their heels here [Malta] waiting for your permission to go to Sicily that I fully appreciated the great honour you had done me in letting me … come over.'[19] The following day Eisenhower made his way to Sicily and sought out the Canadians, but they were already inland fighting, so he left his sincere message of goodwill with a Canadian junior officer.

According to Guy Simonds, commander of the 1st CID, Brooke had warned McNaughton shortly before HUSKY that Montgomery would not permit visitors during operations.[20] Montgomery would justify his actions in his memoirs: 'I had a difficult decision to make soon after we landed ... Guy Simonds, the Divisional Commander, was young and inexperienced; it was the first time he had commanded a division in battle. I was determined that the Canadians must be left alone and I wasn't going to have Simonds bothered with visitors when he was heavily engaged with his division in all-out operations against first-rate German troops.'[21]

In fact, Simonds – at thirty-nine the youngest of the Canadian divisional commanders – did not encounter elements of the Herman Goering Panzer Division until 15 July at Grammichele. Denying McNaughton entry was hardly a 'difficult decision,' and the fact that Simonds had apparently begged 'For God's sakes keep him away' only reinforced Montgomery's position. Nigel Hamilton believed that Montgomery asked Simonds if he wanted the visit 'but in such a way that Simonds was encouraged to say no.' McNaughton was never told any of this at the time.[22]

Colonel Richard Malone, appointed by McNaughton as a liaison officer with Eighth Army HQ soon after the incident, openly sided with Montgomery in *Missing from the Record* (1946).[23] Malone felt that unfortunately, McNaughton had gone about the visit improperly and had not considered Montgomery's views. However, his discussions with Montgomery and Alexander after the war led him to modify his position somewhat. In *A Portrait of War, 1939–1943* (1983), he concluded that Montgomery had been chiefly to blame for the incident.[24]

Through thoughtlessness or as a result of bad advice, Montgomery was not warned in advance of the visit, nor was he asked whether McNaughton could visit his theatre. From a military standpoint this would have been but simple courtesy, as the 1st Division was completely under Montgomery's command.

Some Canadian historians have suggested that Montgomery may have done the right thing for Simonds in keeping McNaughton away during those first few days of action, but that was not the real issue. Montgomery may have been the operational commander in Sicily, but McNaughton was the commander of the Canadian Army, to which the 1st Division unalterably belonged. He was ultimately responsible for it regardless of who had operational command. Swettenham was essentially correct in concluding that Montgomery's action 'abrogated every principle he [McNaughton] had fought for, and seen established, over the years and amounted to a denial of Canada's status as a separate nation.'[25]

Major-General Francis de Guingand, Eighth Army's Chief of Staff, felt that Montgomery was absolutely right to keep away what he referred to as 'rubber-neckers' – a derogatory Eighth Army term for those poor saps not involved in the fight. He admitted that on the face of it, Montgomery's action seemed hard to justify; however, the Eighth Army Commander was merely protecting Simonds from undue distraction at the time of his first battle. 'I have every sympathy with McNaughton's feelings,' de Guingand added, 'and I blame to some extent the machinery at home which faced the Army Commander with a *fait accompli.*'[26] Montgomery's biographer, Alun Chalfont, agreed with de Guingand: 'One can only suppose that there had been some fatal breakdown in staff procedures, since the need to inform an army commander of a projected visit to troops under his command should have been obvious to the most junior staff officer.' To achieve this objective, Montgomery need only have sent a gentle signal that some other time might be more appropriate. This would 'undoubtedly have been received with understanding if not with great pleasure.' Nigel Hamilton, however, took a different tack: 'General McNaughton, anxious not only to see his men in battle, but also to gather lessons that would be of value in planning further Canadian operations in Europe, was understandably livid. It seemed an act of callous chauvinism by Montgomery towards a Canadian general he had despised since his command of South Eastern Army in ... 1941.'[27]

McNaughton failed to garner the same degree of support from Stacey, who played both sides of the fence while clearly leaning towards Montgomery:

> Looking at this painful Sicilian episode as a whole, one is tempted to say that every major actor in it was wrong, except perhaps Eisenhower. It can hardly be doubted that, constitutionally speaking, General McNaughton was right in claiming that the senior Canadian officer overseas was entitled to access to formations of the Canadian Army at any time; any other interpretation would make nonsense of Canadian sovereignty. And yet McNaughton surely chose a bad moment to assert this constitutional right. With the battle for Sicily at its height, the presence in the theatre of a senior Canadian general with no place in the chain of command for the operations could hardly have failed to be in some degree a disturbing element.[28]

If Stacey believed that McNaughton had the right to see his men 'at any time,' then the second half of his statement in no way satisfies that belief. It is difficult to accept his speculation that he would have been a 'disturbing element.' Chalfont's view of what should have occurred seems more reasonable.

On 16 July, Alexander informed Montgomery that McNaughton was 'very disappointed and rather snotty at not being allowed to go to Sicily but admitted the impossibility rather reluctantly ... He is most anxious to have two liaison officers with the Canadians ... I take it you have no objection and I advise acceptance. These liaison officers will be sent forward in the normal way as reinforcement personnel and no special arrangement need be made.'[29]

The next evening McNaughton returned to Tunisia and spoke with Eisenhower at his headquarters. When Eisenhower heard of Montgomery's reason (the transportation shortage), he, according to McNaughton, described it as 'silly.' Eisenhower's naval aide, Captain Harry C. Butcher, suggested that 'we get McNaughton a C-47, load a jeep in it, and send him to the American sector and let him make his way to the Canadians, telling Alexander and Montgomery what we are doing.' Eisenhower, however, was extremely reluctant to intervene in what he considered 'an issue between the British Empire and one of its Commonwealths' and felt 'disinclined to intrude in such a family matter.'[30]

Eisenhower seemed to have considerable respect for McNaughton despite feeling that he and the Canadians had a slight chip on their shoulder. In his personal diary for 9 March 1942, he recorded: 'General McNaughton (commanding Canadians in Britain) came to see me. He believes in attacking in Europe (thank God). He's over here in an effort to speed up landing craft production and cargo ships. Has some d— good ideas. Sent him over to see Somervell and Admiral Land. How I hope he can do something on landing craft.'[31] Eisenhower went to bat for McNaughton with Alexander by drawing attention to Canada's importance in the war and to the fact that tensions were already strained because of the failure to include Canadian forces in the communiqué announcing the invasion of Sicily.[32] These arguments failed to move Alexander, who refused outright to press Montgomery for acceptance. He even went so far as to suggest that McNaughton had been 'so persistent that if he were a junior officer I would put him under arrest.'[33]

When Eisenhower informed McNaughton of Alexander's final refusal, McNaughton dropped the matter for the moment and determined to 'minimize the unfortunate effects' of the refusal back in Britain even though it was his sole prerogative to press the matter into a serious dispute with the British. Eisenhower clearly disliked Alexander's methods, and Butcher questioned how the British 'had ever succeeded in holding together an empire when they treat the respected military representative of its most important Commonwealth so rudely.' Obviously, Alexander 'was not giving sufficient weight to the problem of a democracy conducting a war ... The Canadian public needs inspiration and some means should be worked out for accommodating McNaughton.' Major-General John F.M. Whiteley, the Deputy Chief of Staff at AFHQ, however, sided with Alexander, stating that McNaughton would have been a nuisance in Sicily and that if he were the military figure he was reputed to be he would have grasped the situation and accepted it like a soldier.[34]

Alexander informed Brooke of the episode with McNaughton on 18 July: 'McNaughton and five officers arrived suddenly in Malta and are angry and disgruntled at not being allowed to go to Sicily. I told him there was no transport available except bare minimum cars for key staff officers ... [the] Canadians were 40 miles inland and already on the march fighting ... Seventh and Eighth Army Commanders had urgently requested NO visitors at all for the moment. This was still only D plus 5 day and suggested he return later when he would be welcome. These incidents are very upsetting when we are trying to win battles.'[35]

On D plus 5 the Canadians were just bumping into the Herman Goering Panzer Division near the village of Grammichele. Brooke, having dealt with McNaughton for more than three years, replied to Alexander: 'You have my heartfelt sympathy but I am sure you will realize the need to do all we can to meet Canadian susceptibilities.'[36]

McNaughton returned from the Mediterranean early in the morning of 21 July. In the afternoon he and his wife, Mabel, attended the Army Troops Sports and Field Day in Leatherhead. After supper, McNaughton confronted Brooke at the War Office, having already cabled Stuart the day before: 'In the result while we had every courtesy and consideration from Eisenhower, Alexander at instance of Montgomery denied us permission to land in Sicily to see Cdn troops.'[37] Brooke, who had requested the meeting, asked him how his trip to North Africa had gone. McNaughton replied that for a 'frivolous' reason he had been denied access to his troops and that while it did not bother him personally (which seems hardly possible to believe), there was an important principle involved 'and that on this account the results might be tragic.' In his diary Brooke described McNaughton as 'infuriated [and] livid with rage! I spent 1½ hours pacifying him!'[38] The professional frustration of being effectively snubbed by Montgomery was certainly enough to have set McNaughton off, but it would be unreasonable to believe that the death of his son, Ian, in a bombing raid over Germany in June 1942, had not shortened his fuse and gave a sharper edge to his arguments. Unfortunately, little is known of the impact of Ian's death on McNaughton's ability to perform his duties. Swettenham referred to the tragedy only in passing. It is known that at the time, McNaughton had told his friend, Dr C.J. Mackenzie: 'We have had no news whatever of him since he took off on 22 June ... so we are still hopeful.'[39] Hope, however, had quickly turned to mourning.

McNaughton did not have to bear Ian's loss alone, for his wife, Mabel – whom he had written every day during the First World War – was with him in England. From Stacey's first-hand perspective, he was 'extraordinarily dependent' on her. She, along with Guy Turner and Major-General Robert R.M. Luton, the Canadian Army's Senior Medical Officer, provided McNaughton with solid emotional support; perhaps they were even a sounding board for policy decisions and for personal opinions of peers and colleagues. Charles Foulkes described this inner circle to Stacey as the 'kitchen-cabinet.'[40] McNaughton no doubt vented his frustration about his treatment in Malta to Mabel and Guy Turner, but there is no direct evidence that they influenced McNaughton in his relations with, for example, Brooke.

During his 21 July meeting with Brooke, McNaughton suggested that it was the CIGS who had first gone on the offensive – that he had expressed the view 'forcibly and with some passion, that I had no right to visit Cdn troops.' Brooke pointed out that Field Marshal Jan Christian Smuts had not pressed for visiting rights in Egypt when he was told that the South African Division was assembling for an operation because it would have entailed loss of time and inconvenience. McNaughton replied that 'the cases were not parallel, as I had proposed at the time only to visit troops assembled in reserve in the rear areas near the coast – I mentioned that at the time 1 Cdn Army Tank Bde was concentrated near SYRACUSE and that I could have seen them without inconvenience to anyone. There would have been no need to go near Eighth Army H.Q. or 30 Corps H.Q., or to interfere with operations in any way.'[41]

This was not consistent with what McNaughton had told Nye before leaving for North Africa or with his report of the visit prepared the following day after the meeting with Brooke. On 22 July he wrote that he had gone to the Mediterranean with two objectives in mind: he wanted to witness a combined-arms operation for lessons to take back to Britain;

and he wanted to ensure that the Canadian contribution to the operation was completely in accordance with Alexander's requirements.[42] In fact, it was Stuart who specifically wanted to see the Canadians in active operations. On 17 June he had written McNaughton: 'It is important for reasons of which you are aware that I should obtain first hand information in the conduct of combined operations. Consequently I am anxious to proceed to North Africa as an observer for about one week or ten days.'[43] It appears that McNaughton amalgamated Stuart's reasoning with his own and put forward two different versions of why he wanted to go to Sicily.

As their 21 July meeting continued, McNaughton bluntly told Brooke that 'there should be no doubt that representatives of the Cdn Army would have access to our troops at all times in their discretion.' McNaughton assured him that he need be under no anxiety that this right would be used to hamper the appropriate conduct of military operations, 'as we were just as much concerned as anyone else with the effective prosecution of the war.'[44] As far as Swettenham was concerned, the CIGS 'showed neither the slightest comprehension of Canada's constitutional position nor sympathy for McNaughton's championship of it.'[45] Brooke, however, realized that Montgomery had not handled the situation properly, and in a letter of 29 September 1943 he would tear a strip off him: 'You fairly infuriated him for forbidding him access to his troops in Sicily!!' Yet Brooke never really had any time for McNaughton's argument, adding that he 'wasted an hour of my time on his return explaining to me how strained imperial relations must in future be owing to such treatment, the serious outlook of his Government, etc. etc.'[46]

McNaughton believed that he had won a small concession from Brooke after indicating that the Canadian people expected him to 'at least see something of our troops' in their first battles: 'I think General Brooke realized my position, for he said he apologized for any lack of courtesy and consideration which might have been shown to me in North Africa.'[47] Brooke's true feelings were hard to mask, however, for at some point he told McNaughton that he felt 'inclined to tell him that he and his Government had already made more fuss than the whole of the rest of the Commonwealth concerning the employment of Dominion forces!' McNaughton noted: 'At times the atmosphere of the conversation became somewhat tense, and General Brooke was obviously disturbed and anxious.'[48]

Additional tensions between McNaughton and Brooke, going far beyond visiting rights in Sicily, soon arose during the meeting. McNaughton related that Brooke 'referred ... with some heat, to the letter I had written him when, in the course of an exercise [VICTOR] when he was C-in-C Home Forces, he had detached a couple of our battalions from 1 Canadian Division. He said these views [that the army was not to be split up] had influenced him against proposing operations involving only a part of our force.'

Once they had rehashed VICTOR, Brooke took the lively conversation even further back in time to the problems that had arisen for the British in the First World War because of the independent-mindedness of Arthur Currie. Brooke referred to Passchendaele, where Currie had insisted on bringing in the Canadian Heavy Artillery at great inconvenience so that the Canadian Corps could fight as a unified force. McNaughton challenged Brooke's assertions but did not press the point, noting later that 'very little sympathy was expressed for the Cdn Army and their difficulties.' McNaughton pointed to TONIC (the planned invasion of the Canary Islands by I Canadian Corps) and HUSKY as counterpoints to Brooke's assertion that the Canadian Army could not be used in pieces. In the tense atmosphere of one-upmanship, sound reasoning gave way to emotion. At the conclusion of the meeting the two men, no doubt taxed by the encounter, shook hands and agreed that the effective prosecution of the war 'was the important matter, and that all other considerations must be bent to this.'[49]

After the war, Brooke reassessed the incident:

> It was an excellent example of unnecessary clashes caused by failings in various personalities. In the first place it was typical of Monty to try and stop McNaughton *for no valid reason* [author's emphasis], and to fail to realize, from the Commonwealth point of view, [the] need for McNaughton to visit the Canadians under his orders the first time they had been committed to action. Secondly it was typical of Alex not to have the strength of character to sit on Monty and stop him being foolish. Thirdly it was typical of McNaughton's ultra political outlook to always look for some slight to his position as a servant of the Canadian Government.[50]

Pent-up frustrations on both sides had erupted during the encounter. Though Brooke and McNaughton both vowed to put aside their differences, there is little doubt that the 21 July meeting had only aggravated their already poor personal relations.

The following day Brooke wrote to Alexander that McNaughton had returned and was 'sorely indignant at refusal to allow him to visit Canadian troops. He is reporting the fact to his government and probably in terms which will suggest his constitutional rights have been ignored. I am doing my best to smooth things over but there is the prospect of a Canadian protest on a ministerial level being made.'[51] Brooke wanted further information in case this happened, and Alexander sent off a cipher the same day. After whining somewhat that 'I only have an operational HQ and cannot feed and ... put up guests,' he went on to outline McNaughton's sudden arrival. Alexander included the fact (never told McNaughton originally) that Simonds had urgently requested that the visit be postponed. Moreover, 'Both Army Commanders [Montgomery and Patton] had requested that NO repeat NO visitors at all be allowed for present.' Alexander finished his message by stating that 'McNaughton remained hurt but General Stewart [*sic*] very amiable and understood perfectly. I am sorry about this but I know I am right.'[52]

On 23 July, Montgomery offered Crerar his perspective on the incident. After explaining that the 1st CID had started slowly, was 'a bit soft,' and 'had too many fat officers,' he quickly let Crerar know that less progress would have been made had McNaughton shown up: 'I have allowed no visitors and as a result we have been able to get on with the battle ... You cannot have visitors when you are fighting hard.'[53] Montgomery did not tell Crerar that Simonds had specifically requested no visitors or that he had escorted Mountbatten around the island. A few days later, on 26 July, Montgomery even saw fit to write to McNaughton, but in typical 'Monty' fashion his attempt to smooth things over only aggravated the situation. After explaining his handling of a dispute between Simonds and Brigadier Howard Graham, commander of the 1st CIB, Montgomery added that 'Simmonds [*sic*] has got to learn the art of command just as his Division has got to learn the art of battle fighting. I shall teach him; and he is learning well. He will be a 1st Class DIV. Comd in due course ... *I cannot imagine why it is not announced that the Canadians are in the Eighth Army. The Germans know this quite well and must be highly amused at the efforts to make them think the Canadians are a separate Army*' [emphasis added].

In this letter Montgomery was mocking McNaughton's pretensions to giving the Canadian Army some unified identity. Yet Eighth Army truly was a Commonwealth army, containing forces from Britain, New Zealand, South Africa, India, and now Canada. As Montgomery informed Alexander on 6 August: 'It is definitely the will of every officer and man in Canadian formations here that they should be a part of the Eighth Army and be known as such.'[54] The irony in Montgomery's statement regarding the feud between Simonds and Graham – 'I imagine there were faults on both sides' – was that he was hardly inclined to admit any fault in how he handled McNaughton's desire to visit Sicily.

On 28 July, Ralston arrived in Britain. On 29 July, he and Stuart visited McNaughton for the first of many tough meetings over the next few days regarding the future employment of the army. During the first of these meetings McNaughton raised the incident in Sicily. Ralston had shown King McNaughton's two cables of 20 and 23 July, but the CWC was not privy to them. Ralston told McNaughton that King had been annoyed at the refusal to permit the senior Canadian officer to visit Sicily. However, the incident had hardly elicited the resentment that might have been expected, especially considering King's reaction to the Sicily communiqué. King felt that Canada had been discriminated against by the British: 'If the day ever comes that the people of Canada should incline in some political affiliation to the United States rather than to Britain, it will be that the British Government itself and their officials will be responsible for it.'[55]

McNaughton could not hide his feelings and told Ralston that Montgomery possessed an unfriendly personality and great egocentricity. Montgomery's habit of using the first-person possessive when referring to the 1st CID especially upset him. It would be unwise, McNaughton argued, 'to encourage this to develop still further.'[56] The division had gone to Sicily for the express purpose of gaining battle experience for the eventual cross-Channel invasion, so McNaughton had every right to view it as part of his army and to ward off Montgomery's attempts to claim it. Indeed, Montgomery did consider the 1st Canadian Division his own; as far as he was concerned, he had 'developed' it by his own close supervision. On 25 August he wrote to Crerar: 'Your Canadians have done magnificently. We must not draw false lessons; they have been operated by commanders, staffs, and headquarters who are real experts in this business of fighting battles and so your 1st Division was carefully handled and well looked after; it made the most of it.'[57] In the end McNaughton did not propose any further action, 'as he thought the principle at issue had been maintained and *perhaps* accepted. On the next appropriate occasion, he would take steps to ensure that specific provision for visits by Senior Cdn Officers, at their discretion, to Cdn troops would be included in the directive to the Cdn Comd, and for the formal acceptance of this by the War Office, in advance.'[58]

Unfortunately, Stuart's reflections on the incident are not known. It is apparent, though, that he never felt rebuked to the same degree as Mc-

Naughton. McNaughton, in turn, must have felt somewhat betrayed by King, Ralston, and Stuart in that they had not backed him more strongly in what he considered – with justification – to be a major incident.

McNaughton's pride was not sufficiently hurt by Montgomery's initial refusal that he refused to accept an official invitation from Alexander on 12 August to visit Sicily now that the campaign was over. He arrived in Algiers on 20 August. In between, he wrote to Montgomery: 'Many thanks for your letter 26 July just received. I do appreciate the great personal interest you are taking in the 1 Cdn Division and 1 Cdn Army Tank Bde.' This was pure sugar-coating, for McNaughton could only have been incensed at the condescending tone of Montgomery's letter. On 22 August, Montgomery replied: 'My dear Andy, I am glad to hear you have arrived and are with your Canadians. Welcome to the Eighth Army.'[59] McNaughton visited Canadian installations, including No. 1 Base Reinforcement Depot, No. 1 Canadian Convalescent Depot at Philippeville, and No. 15 General Hospital at nearby El Arrouch. On 21 August he arrived at Alexander's headquarters at Cassibile in Sicily. He toured the Canadian battlefields and had several conferences with Simonds.

On 24 August, McNaughton finally met Montgomery at his impressive new villa at Taormina. Both Richard Malone and Alun Chalfont insisted that on seeing the villa's rich surroundings, McNaughton stated: 'Not going soft, are you, Monty?' According to Chalfont this sent Montgomery into a 'condition of speechless rage.'[60] Yet there is no direct evidence for this remark, and it certainly did not sound like McNaughton, despite his antipathy for Montgomery. The meeting between the two apparently did not start off to McNaughton's liking:

> I was invited to admire his canaries and their very elaborate cages which had been made by the C.E. Eighth Army. I was particularly invited to note that the cages bore the Eighth Army sign, the Crusader Shield ... He had also acquired a peacock which had been trained to sit on the top of his caravan and not to stray away ... He had got the peacock a peahen, etc, etc. ... His BGS had secured a remarkable Italian car for his own use but he, Montgomery, at first sight had pinched it for himself. Did I not agree that he was right to take it for himself? After more of this personal and Eighth Army egotistical nonsense to which I can claim to have listened with all courtesy and attention we passed on to a discussion of matters related to the affairs of 1 Cdn Div, etc.[61]

This exchange no doubt solidified many of McNaughton's perceptions of Montgomery as a prima donna.

McNaughton and Montgomery agreed on a few things, such as the impropriety of breaking up divisions or brigades for regrouping, the value of concentrated artillery, and the faults in the present set-up of the 21st Army Group in England. Then, from out of the blue, Montgomery suggested that McNaughton had too much to do as both the Senior Canadian Officer Overseas and the Commander of First Canadian Army. He should concentrate on the former and become a minister. The look on McNaughton's face would be worth a thousand words to a historian, but he merely recorded: 'I thanked him for this (gratuitous) advice but made no comment.'[62] Montgomery was no shrinking violet when it came to saying precisely what was on his mind at any given moment, and he most certainly understood – and indeed shared – Brooke's concerns about McNaughton's command abilities.

After McNaughton left Sicily in late August, Montgomery recorded in his diary: 'He was most agreeable in every respect, and it is hard to realise that he has made so much trouble about not being allowed to come to SICILY when we were fighting, and that he has intrigued on other matters such as trying to get the Canadians recognized as nominally independent. He is I fancy very ambitious for himself, and very jealous. *But I like him very much as a chap* [emphasis added].'[63] This indicates the masterly job McNaughton did in hiding his animosity at the time, though Montgomery would reflect in his memoirs: 'It seemed to me that he had never forgiven me for denying him entry into Sicily in July 1943.'[64]

The Sicily incident amounted to a microcosm of the difficulties McNaughton faced during the war. First, it strongly underscored the fact that personality clashes were a real obstacle to cooperation. The incident should never have happened: McNaughton had every right to see his troops. Historians have been far too charitable to Montgomery and have accepted his argument about 'protecting' Simonds because it is more palatable than defending McNaughton. Second, it proved beyond a shadow of a doubt that Brooke resented McNaughton's entire position regarding control of Canadian forces – a position fully supported by King. Had the Sicily incident occurred in isolation it would have been nothing more than a minor hiccup. But it involved McNaughton with the three top British commanders of the war at precisely the moment that intense discussions were under way regarding sending an entire Canadian corps to Italy. Moreover, a conspiracy involving Brooke, Paget, Stuart, Ralston, and Vincent Massey was simultaneously under way to remove McNaughton from command. McNaughton remained in the dark as he clung to the mirage of control over his 'dagger pointed at the heart of Berlin.'

12

Broken Dagger: A Corps in Italy

I do not want another Corps Headquarters at this stage.
                                                                Alexander to Brooke, 16 October 1943

If the strategic plans developed as a major thrust in N.W. Europe it would be evident that the movement of 1 Cdn Corps to the Mediterranean had been a major waste of effort.
                                          McNaughton to Morgan, 12 December 1943

The circumstances that led to the establishment of I Canadian Corps in Italy in late 1943 raise many questions. At stake were the very idea of First Canadian Army, McNaughton's position as its commander, and the autonomy of Canadian forces. Stuart and Ralston consistently stressed to McNaughton that more of the army had to get into action to gain battle experience in anticipation of the cross-Channel invasion. As a result, they purposely steered a course towards sustained dispersion of the army between Britain and Italy. Simultaneously, they plotted with Brooke to get rid of McNaughton. The solution that Ralston, Stuart, and Brooke consistently advocated was to divide First Canadian Army to make McNaughton's position as commander redundant. This was not necessary, for even if the army had stayed intact, Brooke at the last hour would have intervened directly to prevent him from commanding it. What is not at all clear is why Ralston and Stuart pushed so hard for the army's dispersion. Perhaps they believed that the salient event of the war in the West – a major amphibious landing in France – could never be led by Canadian and American armies at the expense of British national honour.

Very soon after authorizing Canadian participation in HUSKY, McNaughton learned informally that the War Office was inquiring as to the readiness of the 3rd Canadian Division for BRIMSTONE, the invasion of Sardinia – an operation that had first been contemplated in December 1942.[1] On hearing of this he cabled Stuart the same day:

> In so far as Cdn Army is concerned the departure of 1 Cdn Div etc leaves our army out of balance and so considerable replanning will be necessary unless we can count on return of these troops before serious operations commence or on their replacement from British sources ... Alleged reason given for request to substitute 1 Cdn Div and 1 Cdn Army Tank Bde and ancillaries for British 3 Div and other units was to satisfy request for activity said to have been made by Canada. *My own view is that real reason is related to desire to maintain number of British divisions in expeditionary force vis a vis U.S. and ourselves* [emphasis added]. My reasons for recommending this operation [HUSKY] are of course not dependent on either of these considerations.[2]

There is direct evidence that King pressured Churchill to include Canadians in HUSKY – evidence that cannot be ignored. Even so, McNaughton's observation looks tantalizing. At the time, the proposed British–Canadian army group slated for the invasion of France contained eighteen first-line divisions, but six division equivalents were Canadian and another was a Polish armoured division.[3] The British could hardly match the number of American divisions, but the ratio of British to Canadian divisions was something that could be adjusted to British advantage by moving some around. Though the evidence strongly suggests that the British did not deliberately pursue such a policy, the request from Ottawa to send more formations to the Mediterranean inadvertently helped them in this regard.

Sensing that events were starting to slip beyond his control, McNaughton told Stuart in early May 1943: 'I must point out and warn you that if this request is made and accepted it will end the conception on which we have been proceeding namely that Canada's contribution to the war effort could best be through her own army.'[4] Two days later Stuart tipped his hand: he strongly implied that McNaughton should be exploring the possibility of grouping the 1st Canadian Division and the possible BRIMSTONE division 'in 1 operation under Canadian Corps HQ. I appreciate that the C-in-C might prefer to place each of our divisions in experienced British corps with prospect of grouping later. On the other hand our corps HQ need battle experience as much as our divisions and while dispersion if necessary is accepted on the principle we have all along expressed there are obvious reasons for minimizing dispersion if feasible by keeping together the components selected for Field Force.'

McNaughton argued: 'I think you will agree that if a corps of two divs plus ancillaries were positioned in Africa during 1943 it is very unlikely that they could be brought back to North West Europe for 1944. Time, space and shipping would be determining factors.'[5] McNaughton, of course, would be proven correct on this point.

Stuart's agenda suffered a temporary setback at the end of the month when Archibald Nye, the VCIGS, told McNaughton that as a result of the decisions reached at the recently concluded Washington (TRIDENT) Conference, the Canadian Army should prepare to cross the Channel with British and American forces (for a total of twenty-nine divisions) by 1 April 1944. The army should also prepare for a 'hasty return' in the event of a sudden German collapse. As far as Nye knew, 'there was no intention to propose sending abroad any more Cdn formations.' Indeed, Paget had told McNaughton only a week earlier that the sending of Canadian formations to Sicily 'very much upset the balance of First Cdn Army.'[6] There the matter stood until McNaughton returned from his ill-fated Sicily trip in late July.

On 27 July, McNaughton met with Stuart and Lieutenant-General Sir Frederick E. Morgan, Chief of Staff to the Supreme Allied Commander (COSSAC). By this stage McNaughton knew that Paget had been selected to command the newly designated 21st Army Group, that the target date for invasion was now 1 May 1944, not 1 April, and that the War Office (according to Nye) fully supported the idea of a unified Canadian Army.[7] Morgan told McNaughton and Stuart that the United States was committed to OVERLORD, a cross-Channel invasion, whereas the War Office favoured further operations in the Mediterranean. When McNaughton asked what impact the conflicting strategic emphasis would have on First Canadian Army, Morgan spoke of the 'adverse effect' on Anglo-American cooperation if further Canadian formations went to the Mediterranean.[8] In response Stuart raised the prospect of exchanging British divisions with Canadian divisions. The genesis of this idea is not known, but he may have got it from Brooke. Morgan admitted that it 'might not be harmful,' so Stuart pressed his position, declaring that since Canadians needed battle experience, an additional division and a corps HQ should be sent to the Mediterranean if further operations were seriously contemplated there. He mentioned adverse public opinion if more Canadians did not fight soon, and he implied that if the target date for the cross-Channel invasion was really as late as discussed, the army might be reassembled beforehand in Britain.[9] This last caveat was nothing more than an attempt to deflect McNaughton's concerns, for Stuart had ulterior motives.

Three weeks earlier Stuart had visited Brooke and made it clear to the CIGS that McNaughton was 'unsuitable to command the Canadian Army!' Brooke agreed, noting that 'we discussed possible ways of eliminating him, not an easy job!' They agreed that the best solution was to split First Canadian Army between the Mediterranean and European theatres. Before meeting with McNaughton and Morgan on 27 July, Stuart again saw Brooke earlier that day and suggested splitting the army 'so as to dispose of McNaughton.' Brooke noted: 'He is right. It is the only way to save the outfit.'[10] This was only six days after McNaughton had confronted Brooke about the Sicily incident in one of their stormiest meetings, and there can be no doubt that tensions between them were still running high.

Even after Brooke had time to reflect on his diary entry for 27 July he condemned McNaughton for being 'devoid of any kind of strategic outlook and would sooner have risked losing the war than agreed to splitting the Dominion forces.' This was a ridiculous statement in hindsight for it implied a degree of interdependence between the war's outcome and the employment of the Canadian Army – especially in Italy – that simply did not exist. If McNaughton was devoid of strategic perspective it may have had something to do with the fact that he was being forced to generate his perspective from scraps of oftentimes conflicting information. Moreover, he was not the only one considered short-sighted by Brooke, who similarly condemned George Marshall because he 'absolutely fails to realize what strategic treasures lie at our feet in the Mediterranean.'[11]

The palpable tensions among McNaughton, Stuart, and Ralston (who had just arrived by air) continued to build on 29 July. Stuart pointed out that the invasion might not be possible before 1 September 1944, even though Morgan had just indicated that it would be 1 May. The only other option for getting Canadians into action before September was RANKIN, a contingency operation designed to exploit a sudden German collapse.[12] In the absence of any compelling evidence of this possibility in the near future, Stuart 'favoured building up Cdn forces in AFRICA to a Corps, by *exchanging* Cdn divs for British, *even if there was no certainty that they could be brought back* [emphasis added]. He said that if this were done, H.Q. First Cdn Army as an operational command would be redundant and that it might therefore be advisable to combine H.Q. First Cdn Army and C.M.H.Q. to administer all Cdn Forces in EUROPEAN and AFRICAN theatres.'[13]

In effect, Stuart had adopted an extreme position with no room for compromise. In this he was strongly backed by Ralston. Both men emphasized – erroneously as it turned out – the important economies in manpower that would be realized by such a move. There is no doubt that Ralston had been concerned about manpower for some time. As far back as May 1941 he had informed the CWC that recruiting was being hindered because the army was not in action.[14]

In an obvious attempt to undermine the position of Ralston and Stuart, McNaughton pointed out that every step forward in the development of the army had been taken on War Office advice – advice that at the time seemed eminently logical.[15] Neither Ralston nor Stuart was in a strong position to argue this point, for they had fully endorsed Crerar's 'big army' agenda during 1940 and 1941. Despite the tremendous time and resources that had gone into building the army, however, Ralston concluded: 'I made it very clear that I wanted it to be clear that sending a Corps to the Mediterranean might result in having to give up the idea of Canadian Army.'[16]

Though presented with an extreme position, McNaughton did not respond in kind. Instead he tried to compromise. When the idea of dispatching a further division was raised back in early May he pointed out that a lack of shipping might not allow its return. Now that the shipping situation had greatly improved, he argued that by the early part of 1944 it 'need not be considered as a factor which would prevent the return of additional Cdn formations which might be sent to the Mediterranean theatre for a *short* tour of duty.'[17] In this way McNaughton was contradicting the argument that First Canadian Army HQ would be rendered redundant and was offering Ralston and Stuart another option.

McNaughton clearly favoured expanding to a corps in Italy if major operations were to continue there after Sicily. He presented Ralston and Stuart with the following proposal:

> (i) If operations are to continue in the Mediterranean, and if additional Cdn troops are required, or in replacement of British divs, Canada would be glad to raise our contingent in NORTH AFRICA to a Corps of appropriate composition.
> (ii) As regards planning and development in the U.K., we would continue

to go on the basis that the Cdn contingent in NORTH AFRICA would return to the Cdn Army in the U.K. before the date set for a major offensive on the Continent.

This compromise, recorded McNaughton, 'brought forth further argument' and it became 'quite clear' that neither Ralston nor Stuart accepted the second paragraph.[18] McNaughton's proposal represented the optimal solution to the problem of gaining battle experience, ultimately fighting as a unified force, and satisfying Canadian public opinion. Public opinion was a factor not to be taken lightly, though it seems that McNaughton was less sensitive to it than were Ralston and Stuart, who were entrenched in Ottawa.

King had been relatively free of political turmoil during the first months of 1943. Then, in early August, the domestic political scene in Canada erupted. On 4 August the Liberal government of Henry Nixon in Ontario was soundly defeated by George Drew's Conservatives. King worried that the collapse of Liberal power in Ontario 'may be the beginning of the end of the Liberal party federally.' Five days later, on 9 August, the Liberals lost four seats in federal by-elections. J.L. Granatstein has described those defeats – mainly the result of a sharp increase in support for the Co-operative Commonwealth Federation (CCF) – as 'decisive' and as 'hammer blows' to the fortunes of the Liberal Party.[19] Because of these domestic setbacks, King was in a frame of mind to acquiesce to greater use of the army, even though such a step had its own perils. Increased use would garner headlines and would help silence the harshest domestic criticisms about how the army was twiddling its thumbs in Britain; but it would also guarantee heavier casualties. More casualties would in turn raise the spectre of conscription. King was approaching the point where he was prepared to accept greater risk, but he felt far more comfortable doing so by sending more Canadians to Italy instead of having them spearhead a cross-Channel invasion.

In the event, Ralston and Stuart could not agree to McNaughton's proposal – not, however, because they were concerned about the effect on public opinion of another period of inactivity when the corps returned to Britain for the invasion, but because they were *already* fully prepared to sacrifice the idea of an army. Dividing the army would certainly have facilitated McNaughton's removal, but there seems to have been more to it than that. On 29 July (before the Ontario and federal elections) Ralston's aide, Colonel H.A. Dyde, recorded a critical entry in the Minister's 'War Notes.'

Minister arrived in U.K. (on arrival General Stuart proposed to the Minister:
(a) send one or more divisions to Sicily additional to 1 Div.
(b) move Army Commander to C.M.H.Q.
(c) Break up Army. Canadians now on two fronts and *never will be together* [emphasis added]. Anyway Brooke and Paget would not put Canadians under McNaughton who and whose staff lack battle experience).[20]

This suggests that Ralston and Stuart did not necessarily believe they had to break up the army in order to remove McNaughton. It seems they knew that Brooke would remove McNaughton for them at the appropriate time.

As the 29 July meeting continued, McNaughton told Ralston and Stuart that he wished to make it clear that his position was strictly impersonal: 'He would advocate and support the course which he considered right for the most effective contribution to winning the war, now and for Canada thereafter, *even if it meant the disappearance of the Army and the operation of a number of separate Cdn contingents scattered over the globe* [emphasis added]. In consequence, what he would say on this matter must not be taken as any special pleading.'[21]

This extraordinary statement appeared to contradict what he had told Stuart on 6 May. Moreover, it may have been the source of Ralston's belief that McNaughton 'showed his acceptance of the possibility that a move to the Mediterranean might mean that he would have to accept some other post.'[22] Yet Ralston seems to have inferred too much from McNaughton's statement.

After a few days to ponder these important matters McNaughton met again with Ralston and Stuart on 2 August. He now viewed a corps in Italy 'with more coolness,' sensing that the entire issue was driven by Brooke's irritation with Canadian constitutional restrictions. When McNaughton suggested that 'some people want to break us up,' the straight-faced Ralston asked whom he had in mind. McNaughton declined to come right out and blame Brooke (though he strongly implied it); instead he cautioned Ralston about the Mediterranean strategy, fearing it 'may turn out only to be garrison duty.'[23]

The following day Ralston and Stuart visited Brooke to discuss 'how we are to get rid of Andy.' Brooke declared it 'no easy problem.' Then, backpedalling on his previously strong support for a unified Canadian army, he advised that if a corps went to Italy it would be 'advisable to abandon' the idea of such an army. His original support for the army had been based on what he thought had been the preferences of the Canadian government. Ralston was surprised by this statement: 'This hardly explains his advice as reported by McNaughton. Brooke knew – he ought to have known – in November 1942 when McNaughton went to him re our wires, that the Canadians were to be used *in whole or in part.*'[24] It is unclear precisely why Brooke offered such advice: Was it to aid in McNaughton's removal or because of military practicality? This exchange proved, however, the incompatibility of the creation of First Canadian Army in the first place with the 'in whole or in part' policy. Vincent Massey had observed as early as December 1942 that the large army was already giving the government 'great anxiety.'[25]

When Ralston and Stuart presented Brooke's argument in favour of disbanding the army to McNaughton on 5 August, McNaughton – long accustomed to the CIGS's support on this issue – reacted strongly and blamed the minister:

> I told the Minister that I thought that as a result of the statements he had made, and the suggestive suppositions he had advanced, it could now be taken as a foregone conclusion that Cdn Army would be dissipated. Mr. Ralston denied this emphatically, but I said that his statements, however conditionally he had intended them, would play into the hands of individuals, possibly now including Brooke himself, who desired the break-up of the Cdn Army, so as to use our troops as individual formations to buttress their own positions, and that I thought War Office plans might now evolve to make this result inevitable.[26]

McNaughton reiterated his position that sending a corps to Italy was not dispersion as long as there existed a 'real intent' to bring it back before the invasion. Ralston recorded that this was the first time McNaughton had attached any strings to the idea of a corps going to the Mediterranean. Here, though, he was in error, for McNaughton had explicitly done so on 29 July.[27] McNaughton went on to add: 'I definitely favoured a short time expedition for the experience it would give and the contribution it would make if additional troops were really required. I did not favour exchange with British troops, which if proposed, would indicate that no operations of importance were under contemplation.'[28]

At this point Stuart bated McNaughton, asking him what advice he would give the CWC. McNaughton replied:

> I said that as a matter of principle I was opposed to the dispersion of the

Cdn Army, but in this connection an expedition to the Middle East on the same basis as 1 Cdn Div [to return to U.K.] did not necessarily imply dispersion. If the Cdn Government decided upon dispersion, then I thought it would be wise for them to put someone in control who believed in it. I certainly did not, and did not wish the responsibility. I wanted Canada both for the war contribution it would mean, and for after the war, to end up with her Army under her own control.[29]

Clearly, this was a tense meeting, and McNaughton no doubt generated more tension when he accused Ralston of bias against the concept of an army. Ralston strongly denied that allegation, but Massey's diary for 5 August clearly supported McNaughton's contention: '[Ralston] has no patience with the army fetish. It does not now look as if the Canadian army would take the field as an army formation.'[30] The discussions of the same issues in four meetings over nine days had frayed nerves on both sides. Ralston noted that McNaughton was 'quite worked up' in general and that when the meeting was over he 'went out rather abruptly with a "good day" as we shook hands.'[31]

As frustrated as he was with Ralston and Stuart, McNaughton's faith in his own position was strengthened the next day when he met with Field Marshal Sir John Dill, Churchill's representative to the CCOS, to discuss grand strategy. McNaughton recalled: 'I gathered clearly that Field-Marshal Dill supported the American thesis, and that in his view, it was probable that the Mediterranean would soon become a minor theatre.'[32] Soon the conversation turned to McNaughton's difficulties:

> We spoke of the present situation in which pressure was being developed to put Cdns into battle and I told him personally and very frankly of my present difficulties. As regards views attributed to the War Office, he said there would be a natural tendency to agree if possible with any thesis put forward by a political representative of Canada. His own view was that the maintenance of the Cdn Army was a factor of great military importance which should not be lost unless there were most important military requirements calling for the use of our divs separately. He thought there was no case for interchange of our divs with British in the Middle East. This would waste shipping and would not add to the forces there if that were required. He thought we should await results of Washington to give general strategy and then conclusions could be taken on what was best from a strict military point of view.[33]

McNaughton left his meeting with Dill convinced of three things.

(a) The chance of a Cdn Corps and additional divs being really required in the Mediterranean is remote.

(b) Too much importance should not be placed on Mr. Ralston's statement of Brooke's views – these are possibly a mere reflection of his own ideas put forward with the semblance of a request. Naturally, Brooke would wish to agree, or at least, not to disagree, with such an important political personage.

(c) The important military advantage of maintaining the Cdn Army is an idea which is not restricted to myself, but is held also by a great military authority [Dill].[34]

On 7 August, McNaughton said goodbye to Ralston and Stuart, who were returning to Canada for the Quebec (QUADRANT) Conference. He took advantage of the opportunity to tell them of his meeting the previous day with Dill and of his view that the British were now turning to the American point of view 'rather than any large scale effort in the Mediterranean.' Ralston noted that McNaughton was 'much pleased' as a result of his talk with Dill and that his attitude 'was one of satisfaction that [the] Mediterranean show might not be gone on with.' When Ralston confirmed that he intended to push for a corps in Italy nevertheless, McNaughton stated: 'If any such proposals were put to me, I would communicate it to N.D.H.Q. I left the matter at that.'

McNaughton did, however, take Stuart aside for a one-on-one: 'I told the C.G.S. that this talk [5 August] had brought me very nearly to the breaking point. However, I had since reviewed my position, particularly in the light of my talk with Field Marshal Dill, and I felt I was absolutely right and I intended to adhere to what I had said. He [Stuart] expressed regret that the Minister should make the impression on me which he had, which he felt was not really his true intention.'[35]

Three days later Ralston and Stuart were in Quebec City. The following day Stuart and the rest of the Canadian Chiefs of Staff met with Brooke and the British chiefs at the Hotel Frontenac. Brooke suggested that once Sicily was cleared the Germans probably would not attempt to hold all of Italy, opting instead to defend the northern plain in order to prevent the Allied occupation of airfields. Brooke's deduction turned out to be wildly inaccurate. Brooke maintained that to create the conditions for a cross-Channel invasion to succeed, the German divisions in Italy – which he estimated would be thirty if Italy collapsed – would have to be held there, possibly by sixteen or seventeen Allied divisions.[36] Brooke's insistence on continuing in the Mediterranean only stiffened the resolve of Ralston and Stuart to permanently establish a corps there, just as surely as Dill's comments had buoyed up McNaughton's antithetical position. Another factor working in McNaughton's favour was the anticipated return of the 1st Canadian Division after the end of the Sicilian campaign.

Had the 1st Division returned to England as planned there would have been no Canadian corps in Italy. However, as the transition was made from operations in Sicily to invasion of the Italian mainland, Ralston, Stuart, and King all chose to conveniently forget the original role of the 1st Canadian Division. On 16 August, during the QUADRANT Conference, Churchill mentioned to King that the Canadian division in Sicily had been withdrawn for a time to prepare for the invasion of mainland Italy, code-named BAYTOWN.[37] Churchill characterized the operation as imminent and asked King if he could consider the division going into Italy as the same operation as HUSKY. King replied: 'I told him I had always understood that Sicily was simply a step to invading Italy ... I felt quite sure that the Government had always regarded the whole operation as one.' This was completely contradictory to King's record of late May, when Stuart briefed him on the fact that after the Canadian formations had gained battle experience in Sicily they would be brought back to 'impart to the other divisions in England, the experience they had gained in actual battle.'[38] Moreover, Paget had told McNaughton in June that he wanted the 1st CID to be the assault division for the cross-Channel invasion 'on account of its special training and experience.'[39] Through selective recall King had destroyed the entire logic behind sending Canadian formations to Sicily in the first place. But as Blair Neatby has noted, the prime minister had an 'infinite capacity to rationalize, to accept what he wished to believe and to reject the rest.'[40]

McNaughton gained wind of BAYTOWN simultaneously with Churchill's petition to King and cabled Stuart on 16 August for guidance. Stuart replied: 'Shortly after receipt of your message intimation was received from highest level to highest level here that exploitation may be immediate and inquired if it was in order to use Canadians in extension of operations to Italy. Reply is being made that this will be in order.'[41] The following day McNaughton met with Kennedy at the War Office to discuss the 1st Division going on into Italy. Kennedy outlined the proposed operation 'insofar as the meagre details available in London permitted.' McNaughton quickly determined that 'no effective appreciation of the practicalities of these operations could be made either by the War Office or myself, and it was agreed that, in the very brief time available, there was no other course but to leave this to the judgment of the military authorities on the spot.'[42] McNaughton had no information (as usual) because BAYTOWN had been ad hoc from the beginning and the QUADRANT Conference had not devoted time to specific planning.[43]

Montgomery sent Alexander signal after signal inquiring as to what the intent of BAYTOWN was. He was finally told that it was meant to offer a diversion to AVALANCHE, the amphibious assault on Salerno. Montgomery was never happy with the entire episode: 'It is not clear,' he noted, 'what BAYTOWN is supposed to achieve.'[44] Nevertheless, on 3 September the 1st Canadian Division along with the British 5th Division spearheaded the invasion of the toe of Italy. McNaughton had no choice but to go along with Canadian participation in BAYTOWN; nonetheless he gained the impression from Kennedy that the 1st Division would return to Britain by 1 November 1943 as part of the seven divisions that George Marshall had demanded be returned for OVERLORD. Kennedy also told McNaughton to wait upon the Quebec Conference for a decision on a Canadian corps going to Italy. Churchill was willing to explore the idea, but on 31 August he told King it would be difficult because of the hard requirement to bring back certain formations to satisfy the Americans.[45]

Things became a little clearer on 2 September, but not as McNaughton had hoped. Nye told him that the strategic plan called for the withdrawal of three British and four American divisions from the Mediterranean. In order to hold the estimated eighteen German divisions in place, the remaining forces in Italy would not be reduced. A Canadian corps 'must be considered more or less a permanent commitment, as it was highly improbable' that shipping would be available in time for it to return to Britain by the target date of 1 May 1944 and unlikely for operations by 1 September. If the Canadian corps returned it would have to be replaced by two British divisions.[46] This hardly met McNaughton's criteria for putting the Canadians in an important theatre.

Nye warned McNaughton to be very careful in his communications to NDHQ and to say that the proposal to send a Canadian corps to Italy was under discussion only. When McNaughton asked what type of division should go, Nye told him that in all likelihood it would be another infantry division since there was already too much armour in Italy. This would leave the army in England unbalanced, with two armoured divisions, one infantry division, and one independent armoured brigade, but Nye assured McNaughton that it was highly probable that British formations would be placed under its command. He was also 'very emphatic' that

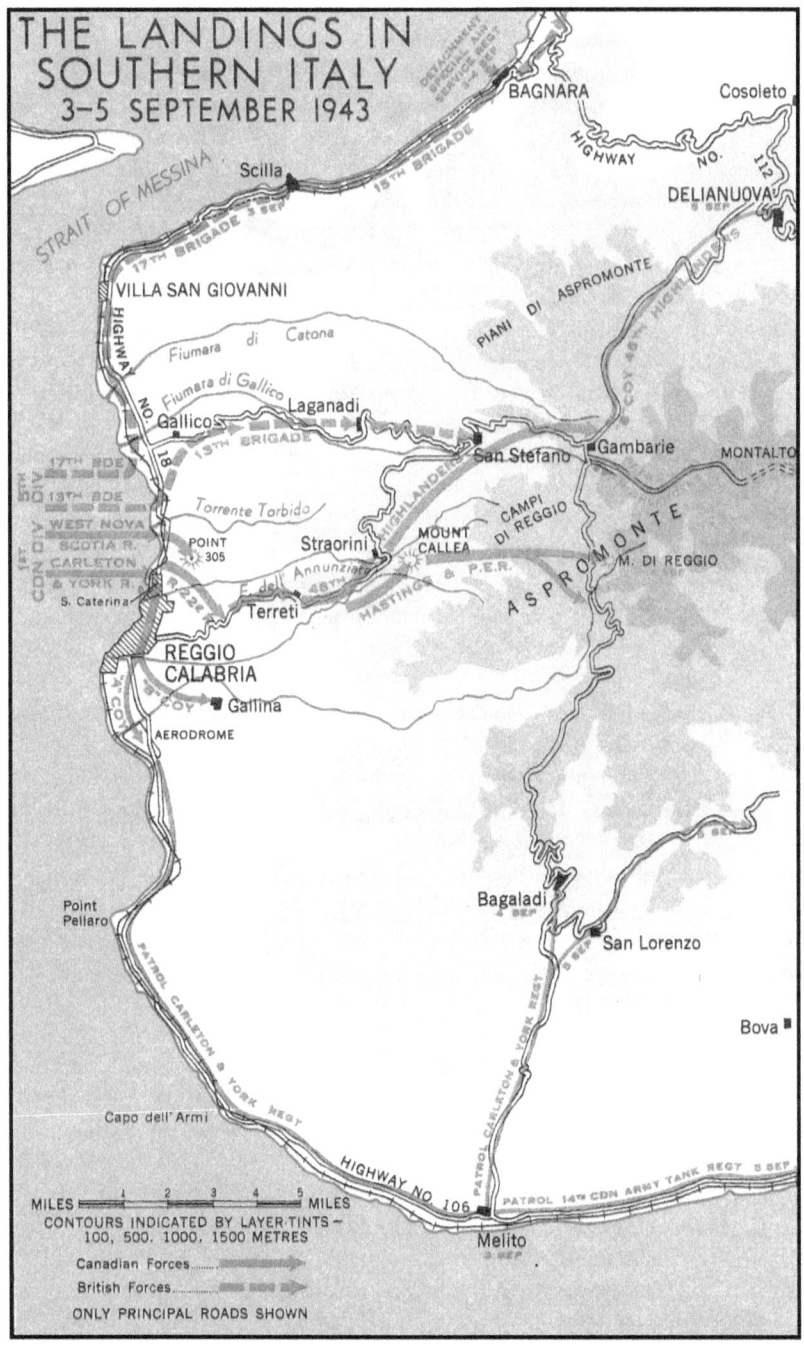

it would be most unwise to disband the Canadian Army since it was in a much more advanced stage of training and organization than the British Second Army.[47] This talk with Nye reassured McNaughton that support for an army remained fairly strong in the higher levels of the War Office. Indeed, a few weeks later Brooke told him that shipping would not be available to send a corps to Italy, which seemed to suggest that the army 'idea' was safe.[48] However, not everything Brooke had to say during their 14 September meeting was to McNaughton's liking.

By this stage of the war Brooke was certainly as fatigued as McNaughton, if not more so. He had returned from a short leave just the day before their meeting and noted that he was returning to the War Office with 'a desperate disinclination to work! I feel that nothing less than 6 months leave could really restore the necessary drive and vitality to pick up this burden again!'[49] Brooke confirmed that the 1st Canadian Division 'was now *not* included' in the three British divisions scheduled to return but 'might be considered for return as the *Eighth* Div if circumstances required the move of an additional div.'[50] The reason behind Brooke's decision is not clear, but the effect was evident: First Canadian Army remaining in Britain would have no battle-experienced division for OVERLORD. The British, in contrast, would have the benefit of an additional veteran division for the invasion, to count against the American division total. It seems beyond doubt that the British desired to ensure that the long-awaited invasion of France would be Anglo-American, not American-Canadian. This in turn strongly suggests that breaking up McNaughton's 'dagger' ultimately had as much to do with broader British geo-strategic imperatives and OVERLORD as it did with McNaughton himself. Brooke's decision to hold back the 1st CID from OVERLORD also kept the door open for the possible establishment of a Canadian corps in changed circumstances. It would not be long before that occurred.

Throughout the debate on establishing a corps in Italy, Crerar, Commander of I Canadian Corps, had been a minor player. But on 15 September he dined with Brooke and made him 'quite angry' with certain remarks regarding the War Office's failure to find employment for the Canadian army. 'I had to point out to him that night,' Brooke recorded, 'that if the Canadians had not seen more fighting up to date this was mainly due to Andy McNaughton's policy.'[51] Brooke's line was getting old, for he had told McNaughton only the day before that shipping was not available to send more Canadians to the Mediterranean. Churchill indicated the real show-stopper in a cable to King on 19 September. The

Table 12.1 Indications of the return of Canadian formations from the Mediterranean, May–August 1943

Date	Indicated by
May	Stuart (to King)
27 May	LGen Sir Archibald Nye, VCIGS (to McNaughton)
12 June	LGen Sir Archibald Nye, VCIGS (to McNaughton)
17 June	LGen Sir Bernard Paget (to McNaughton)
19 June	MGen J.N. Kennedy, DMO (to McNaughton)
28 June	Allied Force Headquarters planning memorandum
17 August	MGen J.N. Kennedy, DMO (to McNaughton)

Source: Memorandum of a Conversation, McNaughton–Nye, 27 May 1943, MP; Memorandum of Conversation, McNaughton–Nye, 12 June 1943; Memorandum of Conversation, McNaughton–Paget, 17 June 1943; Memorandum of a Discussion, McNaughton–Kennedy, 19 June 1943; Memorandum of a Discussion, McNaughton–Kennedy, 16 August 1943.

establishment of a corps in Italy was not possible, Churchill declared, because it involved 'disturbing decisions taken as recently as Quebec Conference *without any military justification* [emphasis added].'[52]

According to Brooke, Crerar refused to accept the fact that 'we should ultimately carry out operations for the liberation of Europe' and looked upon such plans as 'castles in the air of a dreamer.' It is not difficult to discern where Crerar got this attitude from. He got it from Brooke. In late May, Brooke had told Eisenhower that he was prepared to consider cancelling the cross-Channel invasion altogether because of the anticipated heavy casualties.[53] As late as November 1943 the likelihood of a cross-Channel invasion was so remote in Crerar's eyes that when Ralston asked him for his views at that time he replied: 'The possibility of a second Cdn Corps being required in the CMF [Central Mediterranean Front?] and of the setting up of [a] Canadian Army in that theatre, should not be ignored.'[54] Thus on the one hand Crerar was convinced that a cross-Channel invasion was improbable, while on the other hand McNaughton was convinced it was a certainty. The chief architect of these diametrically opposite perspectives was Brooke.

On 30 September, Massey – under instructions from Ralston – pressed Churchill to reconsider placing a Canadian corps in Italy. To add weight to the Canadian argument he passed along to Churchill a letter from King specifically stating that the reason for sending more Canadians to Italy was directly linked to the cross-Channel invasion. King indicated that Canada wanted to be 'prepared as completely as possible for the offensive [OVERLORD] when it comes. We feel that this project would be of great value for this purpose.'[55] There is no evidence that McNaughton was privy to the contents of King's letter, but he asked Massey in late October if King 'really understood the implications of the decision.' Massey replied that the matter had been considered over a long period by the CWC and King and that 'every member ... must be thoroughly familiar with the matter in all its aspects.'[56] Clearly, nothing was further from the truth. Georges Vanier, the Canadian Ambassador to France, sided with McNaughton and advised him in mid-December to lay the problem out to King in specific detail, telling him that this new policy was a complete reversal of what King had always stood for.[57] By that point, however, McNaughton's fate was sealed and he declined to do so.

By 7 October the ongoing entreaties of Ralston, Stuart, Massey, and King had finally borne fruit: the British agreed to the establishment of a Canadian corps in Italy. Brooke informed McNaughton that the entire matter had been re-evaluated on the basis of an exchange of British and Canadian personnel only. McNaughton's frustration can only be imagined, and he took pains to inform Brooke that 'I had strictly confined myself to factual accounts of the situation as he had explained it to me' and had 'neither advocated nor opposed the project.' The initiative in reopening the matter against the expressed opinions of the COS Committee, McNaughton added, 'lay there-fore [*sic*] with Canada; and I was not informed as to the policy reasons which had prompted this action.'[58]

Resigned to the inevitable, McNaughton acceded to Brooke's suggestion that an armoured division be sent. At the same time, he pressed the CIGS on the delicate matter of what was to become of First Canadian Army. Brooke indicated that a recent COSSAC plan called for one British Army of three corps in the assault in place of two armies: British Second and First Canadian. No decision had been taken, or could be taken, until the Supreme Commander had been appointed, which would be very soon. McNaughton probed for greater clarification as to the fate of the Army Headquarters. 'General Brooke made no reply,' he later noted, 'and during a long pause, he did not raise his eyes from his desk.'[59]

On 12 October, Churchill officially proposed to King that Canada establish a corps in Italy; the same day, the CWC agreed. TIMBERWOLF – the code name for the move of the 5th CAD and Headquarters I Canadian Corps – commenced immediately, to the great satisfaction of Ralston and Stuart and to the ultimate disappointment of McNaughton. With that, McNaughton's 'dagger pointed at the heart of Berlin' ceased to exist, and years of planning and organization, both in Canada and in Britain, had been undone.

Before addressing the effect that deploying a corps to Italy had on the larger Canadian war effort, it is important to understand the nature of the discussions that were going on among McNaughton, Ralston, and Stuart at the time. Their meeting of 5 August serves as perhaps the best example of unproductive interaction, for Ralston and Stuart came away from it convinced that McNaughton had unequivocally agreed with their intention of placing a corps in Italy. In his diary notes for 5 November 1943, Ralston recorded that McNaughton had answered yes, without reservations, to Stuart's direct question: 'Aren't you yourself convinced it [establishing a corps in Italy] was advisable [?]' Indeed, McNaughton did agree, but *with* reservations. In his own memorandum of the 5 November meeting he noted: 'There was some talk to the effect that I had, in our previous conversations on the Minister's last visit, indicated agreement to increasing Cdn Forces in the Mediterranean. I replied that I had, under certain conditions and with certain safeguards, but these had been swept away in the method used to obtain concurrence from the U.K. Government.'[60]

It is apparent that McNaughton, Ralston, and Stuart had tremendous difficulty recording the same version of events. McNaughton's record tends to be more consistently honest than those of Ralston or Stuart, both of whom exhibited a disturbing tendency to attribute positions to McNaughton that were not completely accurate. Stuart even skewed Paget's comments about the retention of the Army Headquarters.

As soon as TIMBERWOLF was confirmed, Paget indicated that British XII Corps would be placed under the command of First Canadian Army, which in turn would be a combined Anglo-Canadian army with a proportion of British staff officers. On 21 October, Stuart informed Ralston that he favoured Paget's proposal for a combined Anglo-Canadian army and that 'there is only one practical alternative to Paget's proposal, that is to disband H.Q. First Canadian Army and fit our 2nd Cdn Corps ... into one or more British Armies.' Stuart never even considered retaining First Canadian Army Headquarters in its present form because 'it is not a practical proposition; the British would not accept it.' Stacey would later suggest that Stuart 'read more into Paget's suggestions than McNaughton had stated' and that the ultimate retention of First Canadian Army 'proved that Stuart's judgment of the situation had been inaccurate.'[61] This was neither the first nor the last time that Stuart's interpretation of events, and his judgement, would be suspect.

The negative impact of the decision to establish a corps in Italy manifested itself immediately in the fact that none of the senior commanders in the Mediterranean wanted an untested Canadian corps. Montgomery, commanding Eighth Army in Italy, quickly cabled McNaughton that he had never heard of TIMBERWOLF; soon after, he voiced his displeasure to Alexander at the decision to establish a Canadian corps.[62] Alexander informed Brooke that the proposed move of the Canadian armoured division 'has come as a complete surprise to me [and] I do not want another Corps Headquarters at this stage.' Brooke replied that he had been subjected to 'very strong pressure' from the Canadians to have a complete Canadian corps operating in the Mediterranean. After weighing all the possibilities 'this seemed to me the best way of dealing with a very complicated problem about which it has not been possible to consult you in view of necessity for high level political consideration first.'[63]

Alexander did not confine his annoyance to Brooke. He told Major-General John F.M. 'Jock' Whiteley, Eisenhower's Deputy Chief of Staff: 'This is the first I have heard of arrival of the Canadian Armoured Div. I do not think we require more armour in Italy before we reach the valley of the Po and would have preferred an infantry div.' To Eisenhower, Alexander complained:

> Your 8448 of 14 Oct was the first news I had received from you of the proposed arrival of CANADIAN Armoured Division. It has raised all sorts of queries which I shall have to have answered. It constitutes a CANADIAN Corps and who is going to command this Corps. The Armoured Division has never been in battle and I should have preferred to have employed the battle experienced 1st Armoured Division. We are now getting too much armour out here. However since I presume they are committed to coming we must make the best use of them and I agree to their relieving 7th Armoured Division when this can be done. I request that I am consulted at the earliest possible [moment] on matters of this importance as future build-ups are concerned, and this obviously affects my order of battle and plans of campaign.[64]

Eisenhower was sympathetic to Alexander's position and apparently considered asking the CCOS to cancel TIMBERWOLF because 'our own experienced corps of which we are sufficient are better than Canadian Corps [and] we cannot rpt not guarantee to keep divisions in definite corps.'[65] In the end Montgomery, Alexander, and Eisenhower all relented in the face of the potential political difficulties; nonetheless, they made it quite clear that just because the corps showed up in Italy, that did not mean it would be used as such anytime soon.

Eisenhower informed the CCOS and the British COS that he could not say when the Canadians would be able to get into action because the equipment left behind by the returning British formations 'is already heavily depleted and almost fully mortgaged as reserves for the British forces now engaged on the mainland.' Moreover, the Canadian non-divisional troops would have to be offloaded in North Africa for administrative reasons, and the equipment of the 7th Armoured Division 'does *not* correspond' with the Canadians. He concluded: 'I would like to make it clear in advance that necessarily there will be considerable delay in equipping the Cdns and in bringing them into action.'[66] McNaughton showed this cable to Massey on 21 October; what he could *not* tell the High Commissioner (because he did not know) was that the British COS had informed the CCOS that same day: 'We are prepared to accept unavoidable delay in bringing Canadians into action.'[67]

The delay turned out to be considerable. In early November, Alexander wrote McNaughton: 'We are doing all we can to facilitate the arrival and re-equipment of the new formations and units, though the provision of vehicles is a difficulty. I am sure that soon the Canadian Corps will be adding to the successes gained by its predecessors.'[68] Yet the 7th Armoured Division's vehicles were in deplorable condition, and Montgomery's conservative estimate in late October was that neither the 5th CAD nor I Canadian Corps HQ could be employed until the new year at the very least.[69] During Ralston's visit to the Mediterranean in November, Alexander told him 'there would be no delay' in getting the Canadians into action. During this meeting Alexander had his Chief of Staff, Major-General Charles Richardson, join them. When Alexander asked Richardson how long it would be before the Canadians got into action, he replied mid-January 1944. Alexander then prompted him to speed up the schedule for the specific purpose of getting Crerar some battle experience.[70] In the interim, some units would see limited combat, but other Canadian troops would be used to run transit camps as well as for refugee control.[71]

The corps was sent to Italy in large part so that it could gain battle experience for another division and a Corps Headquarters. Yet Crerar returned to Britain on 3 March 1944 to assume command of First Canadian Army with, as Granatstein pointed out, 'little more battle experience than when he had left.'[72] As it turned out, the 3rd CID would storm the Normandy beaches as a totally raw formation. Yet its combat record in the summer of 1944 easily bears comparison with the veteran British 7th Armoured Division 'Desert Rats,' which had been brought back from the Mediterranean specifically to hedge bets against inexperienced formations. The 50th Northumbrian Division had been brought back for the same purpose, and the 3rd Canadian Division's record holds up well against this veteran infantry division as well. All of this casts a measure of doubt on the need for 'battle experience.' Brooke and Montgomery continued to insist on such experience, however, and exploited the opportunity presented by Canadian pressure for a corps to satisfy their desire for a battle-experienced armoured division and Corps Headquarters in OVERLORD.

Ultimately, it is a sad fact that I Canadian Corps, commanded by Lieutenant-General Eedson L.M. Burns, did not launch its first corps-level operation, CHESTERFIELD, against the Hitler Line until 23 May 1944. The assault on France commenced thirteen days later. In pushing for the commitment of a second division and Corps HQ to Italy, Ralston and Stuart had undermined the army program for the gain of two weeks. As far as Dan C. Spry was concerned, TIMBERWOLF was a 'waste of effort' for it created two administrative tails that exacerbated a manpower deficiency that had begun to manifest itself well before any sustained combat in Italy.[73] Stacey would agree that TIMBERWOLF was a waste of effort: 'It seems extraordinary, in retrospect, that Ralston and Stuart and others pressed the argument for dividing the Army to the extent they did, in the face of military good sense as well as Canadian tradition, which was soundly based in the close relationship between concentration of the forces and the national control of them. It had apparently become an obsession with them ... it seems evident that on this question of policy McNaughton was right and they were wrong.'[74]

A partial explanation is to be found in certain off-hand comments of Ralston and Stuart. As far as Ralston was concerned, the fact that there was a Canadian corps in Italy and a force in Europe 'identified as a Canadian Army' struck him as 'a good position.' His concern about the need for battle experience was partly valid, but there seems to have been a higher political reasoning behind sending the corps to Italy. Stuart felt the same way, telling King: 'We had done honour to our country by giving it a place in world affairs.'[75] They believed that dividing the army was a better political move than having the entire six-division (equivalent) First Canadian Army taking centre stage in Normandy.[76] That Italy would become a backwater after 6 June 1944 does not seem to have factored into their calculations.

The grand irony of their efforts to divide the army was that even as more Canadians were preparing to ship out to Italy in late 1943, Ralston was badgering Brooke about the possibility of reuniting the army. TIMBERWOLF represented the short-term political expediency of Ralston, Stuart, and King versus the long-term calculations of McNaughton. Ralston and Stuart believed that to get rid of McNaughton, they would have to destroy what he had built. However, dividing the army was not specifically necessary in order to remove McNaughton from command of First Canadian Army. This was evident by what transpired next.

13

The Final Months of McNaughton's Command

> I can no longer remain in comd of First Cdn Army responsible to any government of which he [Ralston] is a member.
>
> McNaughton to King, 10 November 1943[1]

The established view of McNaughton's ultimate relief from command of First Canadian Army holds that Ralston and Stuart acted on Brooke's advice in early 1943 to remove him. Yet the evidence suggests that Stuart voiced the opinion that McNaughton was not the right man to command the army even before Brooke did. On 20 March 1943, Stuart told King that McNaughton 'had become far too removed from the troops' and that Crerar 'was nearer to them and understood them better.'[1] Since this opinion was expressed almost two months before Brooke raised the issue of McNaughton's fitness to command in May 1943, it suggests some degree of personal motivation on Stuart's part. His open lobbying for Crerar represented a serious deterioration in the otherwise friendly relations he had enjoyed with McNaughton earlier in the war.

According to J.L. Granatstein, Stuart and McNaughton were old friends. Stuart had attended Bishop's College when McNaughton was an upperclassman and had served under him in Victoria between the wars. When Crerar took over I Canadian Corps, McNaughton supported Stuart's promotion to CGS – an act hardly conceivable had they been feuding. McNaughton told Brooke he had 'every confidence' that Stuart could be an effective CGS.[2] According to journalist Grant Dexter, however, relations between McNaughton and Stuart were not so rosy. Dexter made careful observations of many key wartime individuals and concluded that Stuart 'disliked McNaughton intensely.'[3] There do not appear to have been any open breaches between them throughout 1942, but McNaughton may have started to lose some confidence in Stuart over the simmering issue of manpower.

The animosity between McNaughton and Stuart had visibly increased by the time of the HUSKY negotiations in May 1943. The breaking point from Stuart's perspective may well have been McNaughton's personal rebuke for potentially violating security protocols before the invasion. Proving this beyond a doubt is difficult, but it is a realistic interpretation of the incident. What is not in doubt is that Stuart was especially damning of McNaughton when he wrote in his officer assessments, probably in May, that there was a 'steadily growing feeling of doubt and loss of confidence in the Army Commander both in the Army in England and with the thinking public in Canada.' While McNaughton was falling in Stuart's estimation, Crerar continued to rise. As Stacey later commented, Stuart thought that Crerar was the 'finest soldier Canada had ever produced.'[4]

According to Stuart the first hint of doubt about McNaughton's training and command abilities came from Brooke, who 'made certain remarks on his own volition about McNaughton to the Minister and myself' during the TRIDENT Conference in Washington in May 1943.[5] Dick Malone suggested that Brooke brought it up 'almost by accident,' but this is hard to accept given the long-standing doubts harboured by the CIGS.[6] Time constraints did not allow for a fuller discussion in Washington, but Brooke raised his concerns with Stuart again on 5 July in London.[7] Stuart recorded that

> the CIGS always spoke very frankly. He said that McNaughton's case had been worrying him for a considerable time ... He was more interested in the development side than in command ... He said that McNaughton did not handle his command well on exercises; he was not a good trainer and seemed to lack both the interest and ability to conduct and supervise the training of his commanders ... He was excitable and very highly strung. In other words, the CIGS stated, he lacked most of the attributes for high command ... He stated that he could not accept him as a commander in the field.[8]

As a result of this conversation with Brooke in early July, Stuart – who was just preparing to go to Sicily with McNaughton – sent one of his staff officers, Brigadier Ralph Burgess Gibson, the Assistant Chief of the General Staff (ACGS), home to report to Ralston. Gibson reported to King, Ralston, and Major-General John C. Murchie, the Vice Chief of the General Staff (VCGS), on 10 July, the very day the Canadians landed in Sicily. King was quick to ascertain that the 'real crux of the situation' was that Ralston wanted Crerar in command of the army. The minister, King noted, had 'dropped words from time to time making this clear.'[9] Ralston did suggest that 'the opinions held by the War Office ... might change, as training proceeded, and that we could not arrive at any final conclusion ... until I had had a chance to consider the matter on the ground.' Stuart echoed this: 'The Minister was reluctant to act. He felt that McNaughton's condition [performance] might improve and that Brooke and Paget might, in the course of time, change their minds.' Such comments imply a degree of objectivity on the part of Ralston and Stuart, but it is important to remember that they were written in November 1943.[10]

Now aware of the views of Brooke and Paget, Ralston began to chip away from the periphery by raising the issue of battle experience during meetings with McNaughton in London in August. Both Nye and Morgan had mentioned to McNaughton in late May that the Canadians in general needed battle experience in order to fully prepare for the invasion. On 5 August, Ralston made it clear to McNaughton that the lack of action was detrimental to his capacity to command. McNaughton noted that Ralston made specific reference to 'the lack of battle experience of our Corps and other Senior Comds, including very definite reference to myself.'[11] Nye told Ralston in mid-December that the British tended to 'depend on performance rather than taking a chance on an untried but promising appearing man'; Montgomery told him that 'nothing can substitute' for battle experience. This was precisely why Brooke would replace the inexperienced Lieutenant-General Montagu Stopford, Commander of XII Corps, with the battle-experienced Neil Ritchie in November. Stacey accepted the inherent value of such experience but argued that the British 'made a fetish of battle experience as qualification for command, exalting it even beyond its very real importance.'[12]

One historian has observed that McNaughton failed in his duty to gain battle experience during his four years in command of the army. However, this view does not acknowledge the difficulties he faced in terms of getting the army into action.[13] Circumstances undermined his efforts to command his division in action in 1940 and his corps in 1941. He had been an army commander since April 1942, but no informed historian has explained how or where he was supposed to have acquired battle experience at either of these levels of command. Even when realistic opportunities arose for Canadian commanders to gain battle experience, they could seldom take advantage because they were at Montgomery's mercy. After taking over Eighth Army in January 1944, Oliver Leese made the casual but informative comment that Montgomery 'made no effort to train Crerar during the last few months; and now refuses to have him in command of an Army, unless he has had experience and command in war.'[14] Suggesting that McNaughton did not do his duty perhaps goes too far.

There is no doubt that McNaughton recognized his limitations with respect to battle experience, for he told Ralston and Stuart on 5 August that he 'would be glad' to develop Crerar through battle experience so that he 'might take my place' in command of First Canadian Army. It was 'frequently the case that one man did the building and another got the benefit,' McNaughton observed, and it would be a case of 'to him that hath shall be given.'[15] Historians, including Stacey, have not given sufficient consideration to McNaughton's concession on this point. Indeed, it is entirely plausible that had Ralston, Stuart, and Brooke stuck to the battle experience argument, McNaughton would have accepted it as an entirely rational and legitimate position and supported Crerar's eventual succession. They did not, however, stick to this argument; instead they chose a much more personalized and negative approach to McNaughton's removal.

Nothing more came of these August meetings in London because Ralston was called back to Ottawa for the QUADRANT Conference. Then on 14 August, Brooke seized an opportunity in Quebec City to tell King that McNaughton was ill-suited to command the army. Brooke told the prime minister that McNaughton was 'more suited for planning and research than for action in the field [and that he] ought to be the head of a great research institute.' King noted: 'When I mentioned Crerar, Brooke said he had the very highest opinion of Crerar. Could not have a better man. Would trust him in command anywhere, etc. I did not in any way ask for an opinion on either of the men. All was volunteered.'[16] In this way Brooke was planting in King's mind the seed of doubt about McNaughton's abilities.

As the discussions over BAYTOWN passed into the debate over TIMBERWOLF, Stuart keyed on Paget's recommendations for a reorganized Canadian army and urged Ralston to accompany him to Britain to finalize the fate of both the army and McNaughton. They arrived in Britain on 4 November. The following day Ralston asked McNaughton whether Brooke or Paget had ever raised the issue of his lack of battle experience. McNaughton replied no and added that he was 'satisfied' that they both had confidence in him. Ralston's diary entry for 5 November gives the impression that he had actually told McNaughton as far back as August that Brooke would not accept him as an army commander; but if this was true, it certainly never registered with McNaughton, who most assuredly would have responded to the suggestion. On the contrary, McNaughton characterized his most recent meetings with Brooke over the past several months as 'very cordial.'[17]

After further meetings with Brooke, Paget, Massey, and Sir James Grigg, Ralston finally decided to inform McNaughton of the negative opinions held by the CIGS and C-in-C Home Forces. Ralston justified the decision to himself as follows: 'It is not a case where a man [McNaughton] continues at his post subject to the vicissitudes of military exigencies. In this case a conclusion has been reached by those whose opinions cannot be disregarded and it would be more fair in that case to let him know that continuance [in command] is subject to a change which those conclusions would make likely.'[18]

On 8 November, Ralston and Stuart visited McNaughton at Army Headquarters. As McNaughton recorded, after exchanging a few pleasantries Ralston 'said that he had now seen both Brooke and Paget, and after some expression of sympathy for my position, he said that both Brooke and Paget had indicated that I was no longer acceptable to them to command the Cdn Army in the Field. I made no comment.'[19] According to Ralston, a very uncomfortable silence did indeed follow: 'Nothing was said by any of us for a moment or two.' McNaughton would note: 'It seemed I was "out of harmony with the British authorities," and if so I would feel it necessary to get out of the way.' But not before he offered some defence.

> There was then some speculation on my part as to the possible cause of the trouble. I referred to my one unpleasant interview with Brooke, as CIGS, which occurred at the time 1 Cdn Div had been substituted in HUSKY for 3 British Div on an order given by the Prime Minister of the U.K. I referred also to an incident when he was C-in-C Home Forces when I had objected to his dispersion of two of 1 Cdn Div's bns in an exercise. I referred to Paget's criticism of SPARTAN and to my disagreement; also to our disagreement over Comd 2 Cdn Corps. In my view, I said that the adverse opinion on Comd 2 Cdn Corps was attributable to Wiloughby-Norrie. In any event, SPARTAN was not a fair test for Comd 2 Cdn Corps, as this Corps had been set up ad hoc purely for the exercise, and all the staff were untrained or unexperienced [sic].[20]

According to both Ralston and Stuart, McNaughton took the news 'like a soldier.' But this was superficial posturing: inside he was livid. On

9 November he went to see Paget to confront him about his long-withheld views: 'I did not offer to shake hands, and I opened the conversation by saying, "you know why I am here."' McNaughton's record of this important conversation is as follows:

> Paget replied that he wanted me to know that the conversation with Ralston and Stuart [on 8 November] was without 'premeditation' on his part; he said that he had been asked by the Minister to lunch and the question as to whether I was fit to take First Cdn Army into battle had been 'sprung on him' ... Paget then said that the Minister had taken the initiative in expressing doubt as to my fitness to command an Army. Paget said he had felt the unfairness of the suggestion, but he had in fact said in reply that there was some doubt if I would stand up to the strain. ... I had had 'much more than any other Army Comd to bear.' He said he had thought on occasion that my 'interest in the technical side might possibly be absorbing more than could properly be given without detriment to training and command' ... Stuart had asked about my handling of the Cdn Army in SPARTAN and Paget said he had replied that this had nothing to do with the matter as [the] 'Cdn Army had had less criticism than the other' British Army. He expressed great regret and chagrin at Ralston's unfair suggestions, and said something about people being ground between the stones of criticism given in conversation.[21]

McNaughton briefly changed the subject, then returned to the main issue:

> I said I had been very disturbed that there might have been something to which he ... had taken exception in our relations which was not known to me. Paget said quite the contrary ... he would welcome my continuing [in command] both while in the U.K. and also if he were to continue as Comd 21 Army Gp, in war ... He repeated with emphasis that it was Ralston and Stuart who had suggested doubts about my fitness to command, not he, and he had felt it was unfair to have pressed him for any opinion under the circumstances.

At the end of the meeting McNaughton stressed to Paget that he was still in command of the Canadian Army and that he had 'direct access to the Prime Minister of Canada' and if he pursued the matter, he 'thought some heads might fall.'[22]

After this meeting Paget telephoned Ralston to inform him that McNaughton had arrived 'very upset' and that he (Paget) had told him, to use Ralston's words, 'that there wasn't a probability that he would have that post [army commander in actual operations].'[23] McNaughton, however, had come to a radically different conclusion, and based on his interview with Paget lashed out against Ralston in a direct cable to King on 10 November:

> Today I have to report a situation which is beyond my powers to deal with. I have lost all confidence in Ralston and I must say that I can no longer remain in comd of First Cdn Army responsible to any government of which he is a member ... He ... told me that Paget had said I was unfit to comd an army in the field ... Today I called on Paget and ... Paget stated in the most emphatic terms that Ralston opened his conversation with him by expressing doubts as to my fitness to comd First Cdn Army. ... Paget then *gave me his most categorical assurance that everything had always been clear between us and he had not the least reserve that we could continue to work together both in the U.K. and in operations in the field* [emphasis added].

McNaughton added that he had 'every right' to rely on Ralston for support but that he came to England and 'by suggestions and suppositions casts doubts in the minds of senior officers of another country with whom I have to work as to my fitness. He then takes their silence or partial replies as assent to the view he has implied.' McNaughton concluded: 'I regard Ralstons [*sic*] actions as one of the meanest and most despicable of my whole experience and this is not the first time he has acted in this way to cause me very great anxiety for the welfare of the Cdn Army.'[24]

When Ralston saw McNaughton's message he cabled King the following day that the allegations were 'utterly groundless.' Ralston attempted to take the moral high ground by pointing out that 'since these conclusions [as to fitness] had been reached I could not in fairness to him allow him to go on with the implication that he would command in the field.' This, naturally enough, was defensive writing on Ralston's part, in that he had allowed McNaughton to 'go on with the implication' for six months. On top of this, Ralston equated McNaughton's justifiable anger over the whole situation with 'temperamental instability not in keeping with the judgment required of an army commander.'[25] Equating anger or frustration with some sort of mental instability was a tactic that Crerar would later employ against Simonds in Italy.[26]

On 11 November, Paget wrote to McNaughton that

> some of your statements in this telegram do not tally with my own notes of

the conversation you and I had ... It was later in our talk that the Minister and Stuart raised the question of your fitness to command the Army during active operations. I gathered they were both doubtful about this, and I told them frankly that I was too ... I was concerned that you should know that I had not pre-meditated expressing to them an opinion on your fitness to Command in the field, but ... when asked to do so, I gave my opinion frankly that you were not ... I was not referring to Command during active operations.[27]

Paget ended his letter: 'I deeply regret that there should have been these misunderstandings between us.' In his reply on 12 November (also sent to King in a telegram), McNaughton insisted that their accounts differed 'only in some omissions and in some details which having regard to the conditions of stress under which we met is perhaps not unexpected.' Indeed, Paget had reiterated that Ralston and Stuart had raised the issue of his fitness first. The sequence was tangential. Moreover, on the specific issue of remaining in command during actual operations, McNaughton told Paget: 'On this I am absolutely positive that my record is correct for it was indeed the sole purpose of my going to see you to ascertain this, to me, vital information on which my whole future course of action depended.'[28] McNaughton's logic here seems difficult to challenge.

King's first instinct on receiving McNaughton's 10 November cable and Ralston's subsequent replies was to suggest that both men return to Canada to settle the matter. It is conceivable that had that happened, McNaughton would have been in a much stronger position to defend himself. Ralston, however, told King that it was more important that he visit Canadian troops in Italy rather than return to Canada. In this way he maintained his strong hand. Faced with Ralston's refusal to return, King appealed to both men's sense of loyalty and duty to ensure that some compromise or solution was found to mitigate the 'appalling series of consequences which otherwise would be inevitable' if the dispute made it into the public domain.[29] It does seem that Ralston attempted to carry out King's request in good faith, for on 12 November he informed Brooke of King's concerns, adding: 'We [Canadian government] would be glad if he could accept him [McNaughton].' Brooke, however, replied that he 'wouldn't want to.' When Ralston asked what the reaction would be 'if we insisted,' Brooke replied that he 'would make stronger representations' against McNaughton. When Ralston asked once again why Brooke did not consider McNaughton fit for command, the CIGS replied:

> One thing – not a trainer. Should have [had] schools for senior officers.

Didn't stick to it. Would have taken in hand when C in C Home Forces if [he] had same control as of U.K. forces.

Another – not making familiar with personnel. Like mechanics better than men. Always on equipment rather than knowing and sizing up his officers.

Another – High strung. Not capable of sound judgment in tight places. Army Cmdr must take losses. He couldn't be collected under those circumstances and make plans. If found going to France 'jolly well take steps to stop it.'[30]

The following day, Paget, Stuart, and McNaughton met at Paget's headquarters at St Paul's School. Paget began the meeting by saying that they 'had all known one another for a long time' and that there was 'no need for finesse.'[31] Paget then gave McNaughton a note that stated: 'I do not consider you [McNaughton] fitted to command the 1st Canadian Army in active operations.' What was still not clear was the role of Ralston and Stuart. Paget insisted: 'It was not General Stuart nor Mr. Ralston who inspired this opinion; it is one to which the CIGS and I have been coming [to] for some time.' This was reasonable, but Paget managed to muddle the entire issue once more when he added: 'I did not know when I met Mr. Ralston and General Stuart that the CIGS had already given his opinion to them, nor did I know that I was going to be asked for mine; but that is why I was asked for my opinion.'[32] McNaughton remained unconvinced, telling Guy Turner after the meeting: 'It was quite evident Paget had not been truthful ... My account of our first meeting was absolutely correct ... Paget was probably acting under instructions of higher authority.' McNaughton remained in a combative mood but decided to stand down because 'my future action had to derive from the Prime Minister's appeal and Paget's two statements which, while I do not believe them, I must accept.'[33]

McNaughton might well have stayed in command of the army into early 1944 until such time as Crerar gained sufficient battle experience in Italy, but by 7 December he had informed Ralston that based on recent medical check-ups he needed an extended rest. A week later Nye informed Brooke that McNaughton 'is genuinely ill and has said he must give up command almost at once.'[34] It is quite apparent, however, that McNaughton had simply had enough of the intrigue. He gave up command of First Canadian Army effective 26 December and began his convalescence.

Epilogue

The problem of finding a replacement for McNaughton reared its head only two days after he began his medical leave. Ralston appointed Stuart Acting Army Commander, but the decision did not sit well with Crerar. According to Ralston, Crerar advocated giving the army to a British general rather than to Stuart.[1] In his last hours as Army Commander, McNaughton sided with Crerar, arguing that appointing Stuart was 'most unsound.'[2] Ralston appointed Stuart anyway as an interim measure until Crerar returned from Italy. Crerar came back to England in mid-March 1944 without having gained any battle experience. McNaughton's relief never solved the problem of battle experience for the Commander of First Canadian Army.

When McNaughton returned to Canada at the beginning of 1944 he told reporters at the Seigniory Club in Quebec City in February that illness had nothing to do with his recall. 'There is nothing wrong with me,' he insisted. 'It will be up to those who made statements about my health to explain them.'[3] On 30 September he retired from the army at the age of fifty-seven as a full general after thirty-four years of military service. Only Currie and Sir William Otter before him had attained that rank. Crerar would be promoted to full general in mid-November.

Though officially retired, McNaughton could not resist the natural compulsion to defend his decisions while Army Commander. 'I still think I was right,' he told a reporter in early October. 'It was a terrible mistake to break up the army. That was a political decision. They thought they couldn't hold the patience of the country. I know Canadians and I know we could have held their patience and the Army's as long as necessary.' His pent-up anger still palpable, he added: 'In two minutes I could start a controversy the like of which you have never seen, but I'm not going to do it ... I have given every report, every memorandum and every scrap of paper to the historians. Let them write about it 25 years from now.'[4]

Soon thereafter, in early November, at the height of the manpower crisis, Mackenzie King accepted Ralston's resignation as Minister of National Defence. At King's request, McNaughton replaced Ralston and once more entered the fray. One of his first executive decisions as minister was to exorcise the personal enemy immediately within his reach: he fired Stuart, who by this time was serving as Chief of Staff at CMHQ.[5] As personally satisfying as it must have been for McNaughton to replace Ralston and fire Stuart, nothing could have alleviated the disappointment he felt over the fate of First Canadian Army.

Time magazine announced in early January 1944 that because of the move of a corps to Italy the Canadian forces in Northwest Europe 'will be something less than a full, tactical, all-Canadian Army.'[6] This quasi-Canadian army performed decently in Northwest Europe. It played a major role in the Normandy campaign, fought a gruelling series of battles up the coast, capturing numerous coastal fortresses, successfully cleared the Scheldt Estuary to make the critical port of Antwerp fully operational, liberated Holland, and helped clear the Rhineland for the final push into Germany. The army made a significant contribution to the war in Northwest Europe, but McNaughton had envisioned even more for Canada's first ever field army.

As minister, McNaughton oversaw the return of I Canadian Corps from Italy and was shocked when he read Ralston's files, which revealed that he had pushed for the idea in 1943 and again at the Second Quebec Conference in September 1944. The British were certainly not prepared to weaken Commonwealth forces in Italy until it suited them, but by early February 1945, Operation GOLDFLAKE had commenced, and by 15 March, Lieutenant-General Charles Foulkes's I Canadian Corps was operational under First Canadian Army in the Netherlands. The Canadian Army was at long last complete, but it was a fleeting accomplishment. It served as a unified force for a period of fifty-one days at the very end of the war, when the defeat of Germany was assured. This window dressing was a sad end to McNaughton's big idea of a powerful, self-contained Canadian shock army maximizing the effects of an 'all-Canadian' effort against the enemy.

Conclusion

If ever a man was unfit to command an army in the field it was Andy, and I said so to Alanbrooke in no uncertain voice.

Montgomery to Trumbell Warren, 1969

Looking back over these painful discussions, it is clear now that a crisis in connection with General McNaughton's command could hardly have been avoided.

Massey commenting on diary entry for 21 October 1943

This study has attempted to trace the reasons why Andrew McNaughton, a brilliant scientist and outstanding soldier from the First World War, failed as commander of First Canadian Army. Three lines of investigation have been followed: McNaughton's apparent refusal to sanction the division of the army; his training and command deficiencies; and the effects of personality. The purpose of the multiple lines of investigation has been to compare the current interpretation of McNaughton's relief against the historical record.[1] This fresh analysis has revealed that the historical record has much more to say about McNaughton's wartime difficulties and ultimate failure than has generally been appreciated or acknowledged by historians.

McNaughton was certain that he was finally removed as a result of his clash with Ralston and Stuart over the deployment of the Canadian Army. He explained his thinking to Frederick Morgan on 12 December 1943 in the following terms: 'I had stood for the maintenance of the Cdn Army as a whole for employment in the decisive theatre of N.W. Europe, at the decisive time. I was prepared to exercise patience until that time arrived, and I was confident I could hold the morale of the troops until it did. Meanwhile, as a short term arrangement, I was prepared to use units and formations in acceptable operations in other theatres, partly for the training and experience they would gain, and partly for the contribution to operations they would make.'[2]

McNaughton's position has not had much currency with historians. Instead, the dominant interpretation is best summed up by J.L. Granatstein's statement that 'the British raised various ideas for employment (North Africa, Norway?), and the Canadian commander usually rejected them.'[3] However, context is important.

The British did not want or need Canadian formations in North Africa, and McNaughton was never invited to study possible Canadian participation in TORCH. Fortunately, he rejected JUPITER, Churchill's high-risk Norway adventure. But he did agree to ten other operations, including HAMMER, GAUNTLET, ABERCROMBIE, RUTTER, JUBILEE, SLEDGEHAMMER, TONIC, BRIMSTONE, HUSKY, and BAYTOWN, between 1940 and 1943. He would have been much more supportive of TIMBERWOLF had Ralston and Stuart agreed to support the return of I Canadian Corps in time for OVERLORD.

McNaughton has been blamed for intentionally holding back the army from deployment outside Britain, but except for North Africa, there was no place to fight a full-scale campaign between mid-1940 and the summer of 1943. The fact that the Allies did not begin the invasion of Fortress Europa until July 1943 spoke to the dearth of British strategic options after the loss of France, Norway, and Greece. Except for HUSKY and BAYTOWN, the other operations to which McNaughton agreed reflected that bankruptcy. HAMMER, RUTTER, JUBILEE, and SLEDGEHAMMER were fraught with considerable risk, and the last one would certainly have led to the complete destruction of the Canadian formations involved had it actually been executed.

Since there was literally no place to fight, creating opportunities for Canadian formations to gain battle experience was beyond McNaughton's control. Consistent exposure to combat realities came only with HUSKY – an operation that he supported. Though the War Office talked about the need for Canadian commanders to gain battle experience, Brooke resisted Canadian participation in HUSKY. This fact points to a measure of duplicity on the part of the British. Brooke's excuse for resisting – that McNaughton was against it – does not hold up under close examination. It had far more to do with the War Office's desire to keep Canadian formations in Britain so that British formations could deploy to acquire their own battle experience. The Canadian Army experienced considerable difficulty getting into action because it was the victim of unique circumstances. Events as they unfolded did not mirror those of the First World War. Deprived of a large, stable theatre of operations, the Canadian Army had to wait until Allied strategy created the conditions for its full deployment. Having no say in that strategy, McNaughton could only react and consider derivatives of the main effort, a cross-Channel invasion.

Another line of investigation has been McNaughton's professional military competence as a trainer and as a commander. The current interpretation justifiably highlights his professional weaknesses but minimizes the enormous challenges he faced in training the army in Britain. Widespread equipment shortages, a lack of skilled commanders and staff officers, the spatial limitations of training areas in Britain, the lack of opportunity for large-scale exercises, commitment to various operations, a defensive mindset until 1942, an unrefined offensive doctrine thereafter, and a rapidly expanding Canadian army resulting in the consistent dilution of staff and command talent were not minor impediments to training.

Perhaps the most important limitation was McNaughton's own rate of promotion relative to his expanding training responsibilities. That training improved under Crerar was due less to his skill as a trainer than to the gradual replacement of many commanders by younger talent who simply had not been ready to step up between 1940 and 1942. McNaughton always understood that the original cadre of senior commanders were a 'cover crop' and that weeding out would be both inevitable and necessary.[4] What he did not understand was that the time involved in this process would count against him with Brooke.

As far as McNaughton's abilities as a commander were concerned, the breaking point for Brooke was SPARTAN. Yet McNaughton's performance was not the debacle portrayed by historians, especially considering that it was his first attempt at commanding an army in the field with a host of new faces at all levels of command and staff. D.C. Spry has argued that SPARTAN was 'an unfair test, but I think that the British were absolutely delighted to find an excuse like that to put pressure on the Canadian government and say that we want a new commander – hoping that it would be a British commander put in to command the colonials.'[5] Spry was correct in that Paget actually suggested to Ralston and Stuart on 11 December 1943 that Leese be appointed to command the Canadian Army if Crerar was not available.[6] Montgomery even suggested that Dempsey be given the army.

It is clear that Brooke wanted Crerar to take over First Canadian Army. Yet it is by no means apparent that Crerar was an improvement over McNaughton as a commander. On 23 December 1943, prior to actual command in combat, Montgomery made this cumulative assessment of Crerar: 'The more I think of Harry Crerar, the more I am convinced that he is quite unfit to command an army in the field at present. He has much to learn.'[7] An element of personal friction certainly coloured Montgomery's view, but the core of his observation was that while Crerar generally looked good on exercise as a corps commander, he had considerable doubts about his ability to transition to commanding an army.

After observing Crerar in action in Normandy, Montgomery did not revise his impression. Indeed, after the war he told Major Lionel F. Ellis, the British official historian, that Crerar was a 'staff officer with no command experience [and] completely at sea when it came to exercising the functions of command in battle. He improved slightly as the Campaign in North-West Europe progressed, but he was never really any good. But, of course, I had to "carry" him.' On another occasion after the war Montgomery declared emphatically that Crerar was 'quite unfit to command troops.'[8]

Given the opportunity, McNaughton might have been a diamond in the rough. Had he commanded the army in the static, attrition-based fighting in Normandy or the Rhineland he might have been highly effective, especially in the use of artillery to serve Montgomery's large set-piece battle operational technique. After the war McNaughton stated that he had wanted to use artillery 'in a systematic way so that, when we wanted to support the infantry, in place of using the artillery as a lot of scatter guns shooting at hell's half acre and not hitting anything, if we had our intelligence right, we could bring it down and absolutely obliterate the enemy's resistance without the troops having to do much in the way of an attack on the main centres.'[9]

Indeed, McNaughton and Simonds were both imaginative (McNaughton more so) and searched for solutions to practical problems on the battlefield. McNaughton also understood tactical airpower. As the Canadian official historians of the RCAF in the Second World War noted, his ideas about army–air cooperation 'went deeper than those of any British general except, perhaps, Montgomery.'[10] McNaughton and Simonds might have made a good command pairing. Simonds certainly liked him, far more than he did Crerar: 'I have the most tremendous admiration for him, for his drive and energy, his imagination, and for his very quick mind.'[11]

Another often minimized aspect of the difference between McNaughton and Crerar relates to leadership. McNaughton was somewhat weak on command technique but strong on the intangibles, whereas Crerar was better on technique but very weak on the leadership intangibles. McNaughton possessed a magnetic personality – an attribute Crerar certainly lacked. As Douglas Fisher once observed: 'Crerar? What a cold, dull fellow!'[12] When asked, most First Canadian Army veterans said that they fought under Montgomery, not Crerar – a profound commentary on the latter's inability to connect with the men. Farley Mowat, a veteran of the Royal Canadian Regiment, would recall that during his time as Army Commander, McNaughton was 'looked upon by every soldier with the affection that belongs to a good father.'[13] Elliot Rodger declared that 'whether he would have been the right man to take the forces into battle I'm not competent to say. He had many real attributes of a Field Commander. He was colourful, he was decisive, and made himself loved by the troops.'[14]

Brooke repeatedly expressed concern about placing British troops under McNaughton's command, but a lack of battle experience was clearly not the show-stopper, for he allowed Crerar to take the army into action, and Crerar had no battle experience either. The stakes were high enough in OVERLORD with an untried commander like Crerar, and Brooke felt that the probability of failure was even higher with McNaughton, whose command technique was questionable.

The most compelling aspect of this entire story is the effect of personality. Here the fundamental question is why Brooke, Paget, Ralston, or Stuart did not go to McNaughton immediately on having doubts about his abilities. Brooke's excuse was that 'it wasn't the same as if he were a British General.' According to Ralston, Brooke 'didn't feel like working so directly ... The task of informing McNaughton ... was, in the circumstances, not his responsibility but was that of the Canadian Government.'[15] Ottawa needed input from Brooke to make any decision. He could have provided that input much earlier but felt constrained by the imperfect chain of command as well as by the circumstances of Canadian sovereignty.

When McNaughton was finally told he was out, Brooke was actually surprised that he 'seemed to find it quite impossible to realize that he was quite unsuitable for Command of an Army.'[16] Up to that point, Brooke could not – or perhaps would not – muster the moral courage to tell McNaughton that he was not measuring up in command technique. It is impossible to know how a simple 'Andy, we need to talk about your handling of SPARTAN' might have played out, but it is a good bet that had it been presented in an appropriate manner, he would have responded favourably and tried to address Brooke's concerns. Yet Forrest C. Pogue was right when he described Brooke as 'highly voluble' and impatient in discussion, with the effect that he offended colleagues and Allies 'by his brusqueness and short temper.' Unfortunately, it is highly unlikely that he could have approached McNaughton effectively with constructive criticism.[17] The total absence of any effort on Brooke's part to address the problems up front leads one to suspect that he wanted McNaughton to fail. In addition to this, there simply was no one to replace him between 1940 and 1943.

Paget, McNaughton's immediate superior after the end of 1941, also refrained from voicing his concerns. As he told Ralston, he, like Brooke, 'wanted to keep strictly regular in not interfering with McNaughton's relations with his Government.' Stuart was largely justified in telling Paget in early November 1943 that 'you have given us a dirty deal in going on as if McNaughton would command the show.'[18] Yet Stuart never told McNaughton either. Instead he chose the indirect route, informing Foulkes, McNaughton's BGS, of Brooke's reservations during the visit to Sicily in July. Foulkes quite properly refused to do Stuart's dirty work.[19] It is not clear what Stuart hoped to achieve by such a questionable approach. One can only imagine McNaughton's reaction had he been made aware of Stuart's concerns by his own BGS. On 29 November, McNaughton told Stuart that 'I thought it most unfortunate that he had not been frank with me ... I had never had any idea that this interchange of adverse views on myself was taking place behind my back. I think that Stuart then, perhaps for the first time, realized the wrongness and unethical character of his action, for he again said very defensively, that having these views, he had felt it was his duty as CGS to press them on the Canadian Government.'[20]

The very fact that Stuart chose to wait so long, and to recruit numerous unnecessary individuals into the conspiracy of silence, negates Granatstein's belief that Stuart 'unquestionably acted in the best interests of the army throughout [and] behaved in exemplary fashion.' Stacey was correct in concluding that keeping McNaughton in the dark for six months 'is surely the most doubtful aspect of their proceedings in the whole affair.'[21]

Several authors, including Stacey, suggest that after their clash in Sicily, McNaughton could not have served under Montgomery during OVERLORD.[22] McNaughton would have clashed with Montgomery.

Table C.1 Time line regarding McNaughton's fitness to command

Date	Raised by	Addressed to	Location
March 1943†	Stuart/Brooke	King	Ottawa
May 1943	Brooke	Pope	Washington
May 1943†	Brooke	Stuart	Washington
18 June 1943?	Brooke	Crerar	London
5 July 1943	Brooke	Stuart	London
10 July 1943	Brigadier Gibson	King	Ottawa
10 July 1943	Stuart	Ralston/King	Cable
10 July 1943	Ralston	King	Ottawa
14 July 1943	Brooke	Massey	London
30 July 1943	Crerar	Ralston	London
3 August 1943	Ralston	Brooke	London
5 August 1943*	Ralston	McNaughton	London
14 August 1943†	Brooke	King	Ottawa
14 August 1943†	Ralston/Stuart	King	Ottawa
2 November 1943	Brooke	Churchill	London
10 November 1943†	Paget	McNaughton	London

*Specific mention of battle experience.
†Fitness to command

That seems a near certainty, for as he told Ralston in November 1943: 'We do not like one another.'[23] Yet it is not apparent that McNaughton would have had any more difficult a time with Montgomery than Crerar eventually did. Crerar's persistent difficulty with Montgomery strongly suggests that McNaughton was not possessed of a unique ability to engender discord.

Crerar was incensed by Montgomery's harsh public criticism of Basil Price during a post-exercise conference. Instead of discussing the matter with Montgomery he went directly to his old friend Brooke. Bypassing the chain of command in this fashion certainly did not raise Montgomery's estimation of Crerar. There was a reason why Crerar referred to the first half of 1942 as 'the most trying and difficult' of his professional life. Montgomery felt much the same way: 'What I suffered from that man!'[24]

Crerar was as strong a Canadian nationalist as McNaughton – indeed, one could debate who was more passionate about it. In the final stages of preparation for RUTTER, the raid on Dieppe, Montgomery attempted to deny Crerar access to the operation's command and control facilities at 11 Group Headquarters of the RAF. Crerar felt he had no choice but to raise the 'constitutional' card as McNaughton had done with Brooke after the VICTOR episode in 1941. Crerar bluntly told Montgomery he was 'quite wrong' in his interpretation. Canadian autonomy, he added, was a 'complicated problem ... involving national policies and Imperial Constitutional relations.' He even told Montgomery that 'if this attitude was maintained I was quite certain that the issue would be raised to the highest political levels and that the decision would go against ... himself.'[25]

Crerar took Canadian nationalism to such an extent that he displayed annoyance when Eisenhower, the Supreme Allied Commander, visited the 3rd CID prior to D-Day without apparently consulting the Commander of First Canadian Army. Crerar thought that the act 'completely ignores the political situation and the fact that I happen to be the "National Representative" in the field.' Montgomery was greatly annoyed, telling Brooke that 'the idea that commanders cannot pay friendly visits anywhere they like without advising the whole place is dreadful' and pointing out that 'Harry took Mackenzie King to visit 3 Cdn Div on 18 May, and never told Dempsey!!'[26] That Crerar could generate such annoyance over what was essentially a very minor issue suggests a constitutional sensitivity at least the equal of McNaughton's. Indeed, it seems far more petty.

Crerar would remember that throughout the rest of the campaign in Northwest Europe he had many other difficulties with Montgomery, several of which were 'of considerable importance from a Canadian point of view.'[27] They clashed heatedly in early September 1944 when Montgomery instructed Crerar to meet with him the following day to discuss future operations. Crerar chose instead to visit the 2nd Division at Dieppe to take part in a ceremony marking the ill-fated raid. When Crerar finally arrived, Montgomery severely scolded him. Crerar noted: 'I received the impression ... that the C-in-C was out to eliminate forcefully, from my mind that I had any other responsibilities than to him.'[28] It has been suggested that Crerar possessed the necessary tact, character, and personality to be an army commander. His ability to survive clashes with Montgomery and numerous errors of judgement can only be attributed to the fact that he enjoyed the confidence and friendship of Brooke.[29]

During the war McNaughton enjoyed very good personal relations with Ironside, Dill, Nye, Kennedy, Morgan, Paget, and King. Unfortunately, except for Paget the rest were peripheral players (or did not stay around long) whose influence was not decisive. It was difficult for McNaughton to overcome the handicap of his poor personal relations with Brooke, the most powerful individual in the British Army after December 1941.[30] The depth of feeling – at least from McNaughton's perspective – was captured by Brooke after the war when he stated: 'I regret that he has ever since this time borne me some form of grudge. He has refused ever to have any communication with me.'[31]

The tension with Brooke was compounded by personal clashes with Ralston and Stuart. On 14 November 1943, Ralston and McNaughton talked at length about their relationship: 'I had known him on and off for more than 25 years and ... had always credited him with the best and proper motives, but his methods were not mine, nor his conclusions. We were both very strong willed and perhaps naturally clashed. I would do my best to get on with him in the interim, but then we would probably clash again.'[32] As a result of all this McNaughton had no one senior enough either in Canada or in Britain to run interference for him or support him. King performed this function to a limited extent behind the scenes, but ultimately pulled the rug from under him when the political winds shifted.

The one person whose support could have been decisive in securing McNaughton's position was Crerar, but he actively worked with Brooke behind McNaughton's back. Crerar coveted the position of Army Commander and ultimately succeeded in getting it because he enjoyed a strong personal relationship with Brooke. McNaughton intentionally shielded Crerar from political controversy, in part because Crerar was supposed to be focusing on training the formations. The result, however, was that Crerar felt let down by McNaughton, who had not taken him into his confidence. McNaughton's decision to shield Crerar in this way meant that an opportunity was gradually lost for shoring up a close ally as a counterpoint to his friction with Brooke.

McNaughton was right on many important issues. He was justified in confronting Brooke over the VICTOR episode and over the need to build the army slowly to match anticipated missions with what Canada could actually sustain. Deferring to the British COS on matters of strategy was the only sensible course to follow. He was also right on numerous points of organization, carefully weighing manpower considerations, all the while working from Crerar's, then Stuart's, suspect numbers. McNaughton was most certainly correct in trying to keep the army intact for the cross-Channel invasion, and he understood better than anyone else that the big show would be Northwest Europe, not Italy. In this sense, he was more attuned to the American approach to strategy than to the British one. Crerar's idea of a big army was fine, and it served the national interest, but McNaughton understood the problems of fitting it into British operations prior to the beginning of the invasion of Europe in mid-1943.

But in the end, McNaughton failed to build a consensus for his positions. Without that consensus he could not see his fundamentally sound policies carried to fruition. The evidence suggests that he never felt he had to build up personal alliances. He assumed that everyone was behind him (at least in Canada) and in general agreement on the value of perpetuating the 'Currie Doctrine.' McNaughton had left Canada in 1939 with the firm conviction that he was doing the government's will – that is, by making the Canadians as autonomous as possible – but he did not fully understand that things were different in the current war. Fundamental divisions in the Canadian Cabinet had created an inconsistent application of the 'Currie Doctrine.'[33] Clearly, the decisions taken against McNaughton's wishes signified that Canada had taken a step back from what Currie and the Canadian Corps had achieved a generation earlier. Because McNaughton consistently misread the political currents, he never attempted to change the beliefs of those who openly opposed him. He never convinced Crerar, Ralston, or Stuart of the importance of keeping the army intact. Their ideas developed along antithetical lines. In hindsight, the critical factor was the death of Norman Rogers in June 1940, which brought Ralston once more into close working proximity to McNaughton. Perhaps Rogers would have gone to McNaughton with Brooke's concerns much earlier.

McNaughton's efforts to keep the army together for OVERLORD and his training and command deficiencies were real issues to Ralston, Stuart, and Brooke. Yet the evidence presented here strongly suggests that personality was the most important factor in McNaughton's ultimate failure as Commander of First Canadian Army. McNaughton's personality had a definite edge to it, and he had to work with Brooke, Ralston, and Stuart, who were also difficult personalities. Eisenhower and Montgomery clashed on different occasions during the campaign in Northwest Europe, but Eisenhower remained committed to a conciliatory, workable approach even while Montgomery remained committed to confrontation and was capable of the most appalling errors in judgement, as evidenced by his disastrous press conference during the Battle of the Bulge.[34] As Burns declared in his comments on the official history dealing with his own troubles as commander of I Canadian Corps in Italy: 'There is a two-way responsibility for friendly relations.'[35]

No one represented the conciliatory side in the relations among McNaughton, Brooke, Ralston, and Stuart. They talked past one another throughout the war, thereby aggravating problems where a friendlier discourse would have yielded positive results. It was a recipe for intellectual, professional, and personal gridlock. No discussion of McNaughton's ultimate failure as Commander of First Canadian Army can begin to approach the truth without first considering the effect of personality. In the end, it doomed McNaughton's command of the army and his 'dagger pointed at the heart of Berlin.'

Appendices

Appendix A: Principal Canadian Wartime Positions, 1939–43

Minister of National Defence
Norman McL. Rogers   19 September 1939–10 June 1940
J.L. Ralston   5 July 1940–2 November 1944

Chief of the General Staff
Maj.-Gen. T.V. Anderson   21 November 1938–21 July 1940
Lt.-Gen. H.D.G. Crerar   22 July 1940–23 December 1941
Lt.-Gen. K. Stuart   24 December 1941–26 December 1943

Vice Chief of the General Staff
Maj.-Gen. H.D.G. Crerar   6 July 1940–21 July 1940
Maj.-Gen. K. Stuart   13 March 1941–23 December 1941
Maj.-Gen. M.A. Pope   24 December 1941–14 February 1942
Maj.-Gen. J.C. Murchie   15 February 1942–2 May 1944

Senior Combatant Officer (Overseas)
Maj.-Gen. H.D.G. Crerar   17 October 1939–5 July 1940.
Maj.-Gen. P.J. Montague   6 July 1940–25 December 1943

GOC-in-C First Canadian Army
Lt.-Gen. A.G.L. McNaughton 6 April 1942–26 December 1943

GOC-in-C Canadian Corps
Lt.-Gen. A.G.L. McNaughton 25 December 1940–14 November 1941
Maj.-Gen. G.R. Pearkes   15 November 1941–23 December 1941

GOC-in-C I Canadian Corps
Lt.-Gen. H.D.G. Crerar   23 December 1941–19 March 1944

GOC-in-C II Canadian Corps
Lt.-Gen. E.W. Sansom   15 January 1943–29 January 1944

GOC-in-C 1st Canadian Infantry Division
Maj.-Gen. A.G.L. McNaughton 17 October 1939–19 July 1940
Maj.-Gen. G.R. Pearkes   20 July 1940–1 September 1942
Maj.-Gen. H.L.N. Salmon   8 September 1942–29 April 1943

Maj.-Gen. G.G. Simonds   29 April 1943–31 October 1943
Maj.-Gen. C. Vokes   1 November 1943–30 November 1944

GOC-in-C 2nd Canadian Infantry Division
Maj.-Gen. V.W. Odlum   20 May 1940–6 November 1941
Maj.-Gen. J.H. Roberts   7 November 1941–12 April 1943
Maj.-Gen. G.G. Simonds   13 April 1943–28 April 1943
Maj.-Gen. E.L.M. Burns   6 May 1943–10 January 1944

GOC-in-C 3rd Canadian Infantry Division
Maj.-Gen. E.W. Sansom   26 October 1940–13 March 1941
Maj.-Gen. C.B. Price   14 March 1941–7 September 1942
Maj.-Gen. R.F.L. Keller   8 September 1942–8 August 1944

GOC-in-C 4th Canadian Armoured Division
Maj.-Gen. L.F. Page   10 June 1941–24 December 1941
Maj.-Gen. F.F. Worthington   2 February 1942–29 February 1944

GOC-in-C 5th Canadian Armoured Division
Maj.-Gen. E.W. Sansom   14 March 1941–14 January 1943
Maj-Gen. C.R.S. Stein   15 January 1943–18 October 1943
Maj-Gen. G.G. Simonds   1 November 1943–29 January 1944

Appendix B: Exercises in Britain, 1940–41

1941

VICTOR (January)	1st Canadian Infantry Division.
FOX (February)	1st Canadian Infantry Division road move to concentration area to practise its anti-invasion role.
DOG (26–28 February)	2nd Canadian Infantry Division practice of its anti-invasion role.
HARE (9–11 April)	1st Canadian Infantry Division and Corps Troops practise their counter-attack role.
BENITO (16–19 April)	2nd Canadian Infantry Division road move.
WATERLOO (14–16 June)	First major exercise where McNaughton commands the entire Canadian Corps. The scenario is a counter-attack against invasion, with 8th Armoured Division participating.
ALBERT (31 July–2 August)	1st Canadian Infantry Brigade acts as enemy force for 55th Division.
BUMPER (29 September– 3 October)	Largest exercise ever held in Britain. Two armies, four corps, and twelve divisions (250,000 men) practise the anti-invasion role.

1942

MUSKRAT (28–29 January)	1st Canadian Infantry Division signals exercise to practise the passing of information.
BEAVER II	Skeleton force exercise to practise commanders and staff (3–5 February) of Canadian divisions in the anti-invasion role. BEAVER I was also a skeleton HQ exercise.
LANCELOT (14 March)	South-Eastern Command skeleton exercise to study the handling of a corps in the attack and breakthrough battle.
RAM (9 April)	5th Canadian Armoured Division skeleton exercise to practise commanders and staff in the exercise of command, practise communications with 414th Sqn RAF, and familiarize staffs in the use of maps.
FLIP, FLOP, FLAP (April)	2nd Canadian Infantry Division.

BEAVER III (22–24 April)	1st Canadian Infantry Division acts as invading force and attacks 2nd Canadian Infantry Division.
BEAVER IV (10–13 May)	2nd Canadian Infantry Division in offensive role versus 3rd Canadian Infantry Division.
TIGER (19–30 May)	South-Eastern Command exercise involving I Canadian Corps with a heavy emphasis on physical conditioning.
RAM III (8–11 June)	5th Canadian Armoured Division skeleton exercise.
YUKON (11 June)	2nd Division rehearsal for Dieppe Raid.*
YUKON II (23 June)	2nd Division rehearsal for Dieppe Raid.*
CHEDDAR (July)	First full-scale divisional exercise for Guards Armoured Division.
FREEHOLD (11 July)	Exercise conducted at Camberley for division commanders.
KITTEN II (15–16 July)	1st Canadian Infantry Division HQ exercise to practise the control of brigade HQs by wireless with air and tank support.
HAROLD (25–31 July)	3rd Canadian Infantry Division acts as an invasion force to practise the 46th Division in the defence.
CRACKER (31 August)	2nd Canadian Infantry Brigade exercise designed to determine the best method of moving heavy equipment across numerous water obstacles after a landing.
MARCONI (16 September)	5th Canadian Armoured Division skeleton exercise.
MAPLE 9 (9 October)	5th Canadian Armoured Division skeleton exercise.
CAVENDISH (October)	Designed to rehearse troops for a seven-division invasion of the continent. 1st Canadian Infantry Division takes part.
BLACKBOY (November)	Anti-invasion refresher training conducted by 1st Canadian Infantry Division.

1943

HAMMER I (January)	Interdivisional exercise between the 3rd and 53rd Divisions.

Appendices

BLACKMORE (February)	Guards Armoured Division exercise.
SPARTAN (4–12 March)	Largest exercise since BUMPER. Designed to test First Canadian Army in its ability to break out of an established bridgehead. First time that 5th Canadian Armoured Division practises as a complete formation.
HOTSPUR II (6 April)	4th Canadian Armoured Division exercise to study a squadron seizing a pivot of manoeuvre.
BRICKBAT (April)	4th Canadian Armoured Division exercise to practise R.T. procedure, use of codes and C3.
GUNBUSTER I (28 April–5 May)	Tactical handling of 5th Canadian Armoured Division's artillery.
HOTSPUR III (4 May)	4th Canadian Armoured Division exercise to practise a squadron in the advance to a pivot of manoeuvre and dispositions on the objective.
FLANKEM (11–12 May)	5th Canadian Armoured Division armoured regiments practise tactical movement.
QUATTUOR (15 May)	First exercise where large portions of 4th Canadian Armoured Division work together. Consists of an all-arms break-in of a defended position.
TROJAN (31 May–1 June)	I Canadian Corps signals exercise to train replacement officers.
COUGAR III (24–25 June)	4th Canadian Armoured Division's first practice of its recce regiment as an advance/rear guard.
HARDTACK (26–28 July)	5th Canadian Armoured Division practises anti-invasion role and harbouring at night.
GRIZZLY I (1–4 August)	Full-scale 5th Canadian Armoured Division test of its recce regiment in the advance to contact and the proper use of armour and infantry in the attack.
SNAFFLE (7–11 August)	5th Canadian Armoured Division and its recce regiment practise the delay and the approach march by armoured and infantry brigades.
ATTACK (13–20 August)	5th Canadian Armoured Division.
GRIZZLY II (22–24 Aug)	First time the 4th Canadian Armoured Division exercises as a complete formation. Full-scale attack on a semi-prepared position.

HARLEQUIN (August–September)	5th Canadian Armoured Division road movement to embarkation points.
BRIDLINGTON (August)	11th Armoured Division exercise.
LINK (13 September)	II Canadian Corps HQ controls 1st Polish Armoured 61st Division.
TAKEX (25–28 September)	4th Canadian Armoured Division (– 4th Armoured Brigade) exercises as a covering force.
DITTO (9–12 September)	Last tactical exercise by 5th Canadian Armoured Division in Britain.
VICTOR (27 September)	5th Canadian Armoured Division artillery exercise.
BRIDOON (2–5 October)	4th Canadian Armoured Division versus 9th Armoured Division.
BLAST I (3–4 October)	Canadian artillery exercise at Sennybridge.
PLODDER (23 October)	2nd Canadian Infantry Division.
LIMBER (28–31 October)	HQ First Canadian Army movement and operations in the field with HQ 83 Group RAF.

Appendix C: Allied Operations, Proposed and Executed, 1940–44

1940
ANGEL MOVE
Move of 1st Canadian Infantry Brigade to Dunkirk.^
ANKLET
Against the island of Vaagso in Norway in December.*
BRITTANY REDOUBT
A second BEF with the 1st Canadian Infantry Division to establish a defensive line in Brittany from St Nazaire through Rennes to Pontorson. (Started but soon cancelled.)^
COMPASS
O'Connor's offensive against the Italian Tenth Army (9 December 1940–7 February 1941). First British offensive in the Western Desert and a complete success.*
DYNAMO
Evacuation of BEF from France via Dunkirk.*
HAMMER
British assault on Trondheim in April 1940. Canadian participation consisted of the 2nd Infantry Brigade but the operation was cancelled.^
HURRY
Reinforcement of Malta in July.^
MAURICE
Original operation against Trondheim.*
RUPERT
Operation to expel Germans from Narvik in mid-April.*

1941
AJAX
Churchill-backed operation to capture Trondheim as a means of relieving pressure on the Russians. Officially proposed in late September.^
ARCHERY
Successful raid on Vaagso in December.
BARITONE
Raid. Planning taken over by I Canadian Corps.^

---
* Denotes operations actually carried out.
^ Denotes operations cancelled.
Bold represents operations involving Canadian forces.

BATTLEAXE
Wavell's offensive against Rommel's newly arrived panzer force, 15 June.*
BREVITY
British counter-attack against Rommel in May.*
CRUPPER
Raid. Planning taken over by I Canadian Corps.^
CRUSADER
Brief victory over Rommel on Libyan coast in November.*
EXPORTER
British occupation of Syria.*
GAUNTLET
Occupation of Spitzbergen Island by elements of 2nd Canadian Infantry Brigade in August.*
INFLUX
Invasion of Sicily in late 1941.^
PILGRIM
Proposed occupation of the Canary Islands in anticipation of German invasion of Spain.^
PUMA
Proposed occupation of the Canary Islands.^
TIGER
Tank reinforcement sent to Egypt in April.*
TIGER No. 2
Second tank reinforcement sent to Egypt in July.*
TRUNCHEON
Raid on Italian coast in December.*
VELVET
Air and military aid to southern Russia. Authorized to commence in September.^
WHIPCORD
Plan for the invasion of Sicily. Abandoned in November.^
WORKSHOP
Attack on Pantellaria in January.*

1942
ABERCROMBIE
Raid on Hardelot involving a small number of Canadian troops to test coastal defenses south of Boulogne in April. The raid was attempted but failed in execution.*

Appendices

AFLAME
Continuing operation designed to draw the *Luftwaffe* into attritional battles in the West.*
BLAZING
Raid planned for August but cancelled in May.^
BOLERO
Buildup of U.S. forces and supplies in Britain.*
CHARIOT:
Raid on St Nazaire in January.*
FRANKTHON
Small raid against the docks at Bordeaux in October.*
GUNNER
Operation to establish a bridgehead on the Brest peninsula in November.^
GYMNAST
Brooke's North African project.^
IMPERATOR
Super-raid of 4 divisions to stay in France up to a week during the fall.^
IRONCLAD
Capture of Diego Suarez, near Madagascar, in March.*
JUBILEE
Remounted raid on Dieppe on 19 August.*
JUPITER:
Operation to be carried out principally by Canadian forces to seize aerodromes in Norway to support Russian convoys.^
MAGNET
Buildup of U.S. forces in Northern Ireland and Iceland.^
MYRMIDON
Raid on Bayonne.^
OVERTHROW
Deception scheme to hold German forces in France during TORCH.*
RUTTER
Raid on Dieppe in July.^
SLEDGEHAMMER
Proposed invasion of France with 2 American and 10 British divisions if the Russians appeared to be on the verge of collapse.^
SOLO I
Deception operation designed to fix German attention on Norway during TORCH.

SUPER-GYMNAST
Proposed invasion of French North Africa with American and British forces.^
TONIC
Proposed invasion of the Canary Islands by I Canadian Corps.^
TORCH
Invasion of French North Africa by U.S. and British forces in early November.*
WETBOB
Operation to gain a permanent foothold on the continent in the Cherbourg area in the autumn.^

1943
AVALANCHE
Combined Anglo-American amphibious assault on Salerno in September.*
BAYTOWN
Invasion of mainland Italy on 3 September with the 1st Canadian Infantry Division participating.*
BRIMSTONE
Proposed invasion of Sardinia with a Canadian division to participate.^
BUTTRESS
Original British plan for the invasion of mainland Italy after HUSKY.^
COCKADE
Elaborate camouflage and deception scheme devised by COSSAC to pin the Germans in the West consisting of three sub-operations.
HADRIAN
Proposed cross-Channel invasion plan calling for an assault on the Cherbourg peninsula.^
HERCULES
Proposed plan to capture the Isle of Rhodes.^
HIRES
Proposed Anglo-American invasion of Sicily in March.^
HUSKY
Invasion of Sicily on 10 July.*
POINTBLANK
Combined U.S.–British strategic bombing offensive on German industrial facilities commencing in June.*

RANKIN C
Operation to invade if the Germans withdrew from France or became seriously weakened.^
ROUNDUP
Cross-Channel invasion originally scheduled to take place in the spring with 30 American and 18 British divisions. It represented a sacrificial attempt to support the Russians if they were considered on the verge of collapse.^
ROUNDHAMMER
Modified ROUNDUP invasion plan.^
SKYSCRAPER
Cross-Channel assault proposed in 1943 but never officially presented to the British COS.^
STARKEY
Pre-invasion deception rehearsal conducted in September consisting of an amphibious feint from southern England against the Pas de Calais. Part of COCKADE.*
TIMBERWOLF
Move of I Canadian Corps HQ and 5th Canadian Armoured Division to Italy.*
TINDAL
Threat generated from Scotland against Norway as part of COCKADE.*
WADHAM
Threat generated by U.S. forces in southwest England against French ports as part of COCKADE.*

1944
CHESTERFIELD
I Canadian Corps' first corps-level operation, conducted against the HITLER LINE in Italy in May.*

1945
GOLDFLAKE
Move of I Canadian Corps to Northwest Europe to rejoin First Canadian Army in February.*

Appendix D: Montgomery's Assessment of Canadian COs, 1942

Commander	Unit	Comments
LCol Tweedie	Carleton & York	Inexperienced but will do well.
LCol Bernatchez	Royal 22nd	First class.
LCol Ernst	W. Nova Scotia	Very poor, not a good trainer.
LCol Tait	Seaforths	Runs a good show.
LCol Lindsay	PPCLI	Very good, very keen.
LCol Wilson	Edmonton	Away sick for two months.
LCol Cantlie	Black Watch	Quite unfit for command.
LCol MacLaughlan	Calgary Highlanders	Totally unfit.
LCol Roche	Reg de Maisoneuve	Knows little of training a battalion.
LCol Basher	Royal Reg of Canada	Very good despite age.
LCol Jasperson	Essex Scottish	Will do very well.
LCol Labatt	R. Hamiton L.I.	Weakest CO in the 4th Inf Bde.
LCol Macpherson	H.L.I.	Good solid stuff.
LCol Murdock	N. Nova Scotia H.	Cautious, runs a good show.
LCol Rutherford	S.D. & Glengarry H.	First class, age 34.
LCol Sharpe	Regina Rifles	Should do well with considerable guidance.
LCol Gibson	Winnipeg Rifles	Should make a first class CO.
LCol Kingham	Canadian Scottish	Should be replaced.
LCol Grenier	Fus. Mount Royal	A very poor CO.
LCol Gostling	Camerons	Very keen with sound ideas.
LCol Lett	S. Sask R.	An excellent CO. Doing very well.
LCol MacKenrick	Q.O.R. of Canada	Will do well with firm guidance.
LCol Calkin	N. Shore Reg.	Will do well with firm guidance.
LCol Chouinard	R. de Chaudiere	Teachable, but requires firm guidance.

LCol Snow	R.C.R.	The best CO I have met and first class.
LCol Graham	Hastings & P.E.R.	Very good.
LCol Hendrie	48 Highlanders	Quite unfit to command a battalion.

Source: Montgomery's Notes on Inf. Bdes of Canadian Corps, CP, Vol. 2, File 958C.009(D182).

Notes

Introduction

1 Charles P. Stacey, 'Canadian Leaders of the Second World War,' *Canadian Historical Review* (hereafter *CHR*) 66, no. 1 (1985): 67. Stacey remained devoted to McNaughton, remembering him past his death in July 1966 with 'deep affection and admiration.' Even with his faults of judgement, Stacey considered him a 'great man and the most compelling personality it has ever been my fortune to encounter.' See *A Date with History: Memoirs of a Canadian Historian* (Ottawa: Deneau, 1983), 79.
2 J.L. Granatstein, *The Generals: The Canadian Army's Senior Commanders in the Second World War* (Toronto: Stoddart, 1993), 82.
3 It is interesting that the whole thrust of Granatstein's study of Canadian wartime commanders was not to inquire as to why so many failed, but rather 'why so many succeeded.' Ibid., 261.
4 General Sir Archibald Wavell, *Allenby: A Study in Greatness* (London: George G. Harrap, 1940), 16.
5 Dominick Graham, 'Stress Lines and Grey Areas: The Utility of the Historical Method to the Military Profession,' in *Military History and the Military Profession*, ed. David A. Charters, Marc J. Milner, and J. Brent Wilson (Westport: Praeger, 1991), 148–9.
6 Sir Ian Jacob, 'Statesmen and Soldiers in War,' *Foreign Affairs* 38, no. 4 (1960): 657.
7 John Buchan, *Augustus* (London: Hodder and Stoughton, 1937), 9; David Curtis Skaggs, 'Michael Howard and the Dimensions of Military History,' *Military Affairs* 49, no. 4 (1985): 180.

1. Early Life and the Crucible of the First World War

1 John Swettenham, *McNaughton*, 3 vols., *1887–1939* (Toronto: Ryerson 1968), I:7–8.
2 Sir Brian Horrocks, *A Full Life* (London: Collins, 1960), 13.
3 Sir Bernard Law Montgomery, *The Path to Leadership* (New York: Putnam, 1961), 14–15.
4 J.L. Granatstein, *The Generals: The Canadian Army's Senior Commanders in the Second World War* (Toronto: Stoddart, 1993), 6. Over 90 per cent of American army generals from 1898 to 1940 were Protestant. Morris Janowitz, *The Professional Soldier: A Social and Political Portrait* (New York: Free Press, 1960), 78.
5 Donald G. Creighton, *Canada's First Century, 1867–1967* (Toronto: Macmillan, 1970), 78. For an excellent discussion of English Canadians as keen imperial federalists, see Ronald G. Haycock, *Sam Hughes: The Public Career of a Controversial Canadian, 1885–1916* (Ottawa: Wilfrid Laurier University Press, 1986).
6 Carl Berger, *The Sense of Power: Studies in the Ideas of Canadian Imperialism, 1867–1914* (Toronto: University of Toronto Press, 1970), 9, 258.
7 Creighton, *Canada's First Century*, 92.
8 William L. Morton, ed., *The Shield of Achilles: Aspects of Canada in the Victorian Age* (Toronto: McClelland and Stewart, 1968), 317.
9 Richard A. Preston, Sydney F. Wise, and Herman O. Werner, *Men in Arms: A History of Warfare and Its Interrelationships with Western Society* (New York: Praeger, 1968), 219.
10 A healthy literature aimed specifically at the youth of the empire existed at the time. Lawrence James, *The Rise and Fall of the British Empire* (New York: St Martin's, 1994), 207.
11 Swettenham, *McNaughton*, I:11–12.
12 McNaughton was particularly taken with Matthew Flinders's *Voyages to Australia* as a child and Swettenham concluded that his world view was built almost exclusively on 'the deeds of Britain and the great figures of the past.' Swettenham, *McNaughton*, I:12.
13 William G.F. Jackson, *Alexander of Tunis as Military Commander* (London: Batsford, 1971), 9; Brian Holden Reid, 'Alexander,' in *Churchill's Generals*, ed. John Keegan (New York: Grove Weidenfeld, 1991), 105. Nigel Nicolson, however, has argued that Alexander 'did not intend to be a soldier, except temporarily. He wanted to paint, professionally. This was a serious ambition, not a passing caprice.' *Alex: The Life of Field Marshal Earl Alexander of Tunis* (London: Weidenfeld and Nicolson, 1973), 20–1. Horrocks stated: 'There was never any question of my entering a profession other than the

army.' *A Full Life*, 13. One of the few British army commanders of the Second World War who did not attend Sandhurst or Woolich was Oliver Leese. He spent six-and-a-half years at Eton prior to the First World War. He was commissioned into the Coldstream Guards directly from the Eton Officer Training Course. Roland Ryder, *Oliver Leese* (London: Hamish Hamilton, 1987), 14.

14 There was no guarantee that he would have chosen to remain in the Permanent Force upon graduation because the trend was that most graduates took their education into civilian life. Richard A. Preston, *Canada's RMC: A History of the Royal Military College* (Toronto: University of Toronto Press, 1969), 185, 192.

15 Swettenham, *McNaughton*, I:17.

16 Montgomery was shot in the chest at Meteren and was awarded the DSO. Patton was shot through the thigh while leading a dismounted advance. Alexander was wounded first at Ypres in 1915 and again at Passchendaele in October 1917. MacArthur was wounded while conducting reconnaissance near the Cote-de-Chataillon, and Leese was hit by shrapnel.

17 Swettenham, *McNaughton*, I:44–5.

18 Nicolson, *Alex*, 33. Valentine Williams served with Alexander and observed that he was 'the only man I knew who appeared to enjoy the Battle of the Somme.' Norman Hilsman, *Alexander of Tunis: A Biographical Portrait* (London: W.H. Allen, 1952), 23. Horrocks described his first experience of being shelled as 'a thrill' and saw war in a romantic light. When he saw the horse artillery 'galloping into action' he declared that it was 'just what I had always imagined war would be like.' *A Full Life*, 16.

19 So too did Alexander. His first wound was serious enough that he was invalided back to Britain. He spent two months in hospital and, like McNaughton, had to prove to the army doctors that he was fit for service. To do so, he walked sixty-four miles and reported his feat to the War Office. Nicolson, *Alex*, 31.

20 McNaughton had the 41st, 44th, and 46th Batteries (18 pounders) and the 29th (Howitzer) Battery.

21 Swettenham, *McNaughton*, I:62.

22 Swettenham never suggested that this letter, date unknown, was a sign of temporary mental degradation, but he did call it 'strange.' Indeed, it had no introduction and almost reads like poetry. Yet even poetry has recognizable separations. The letter was essentially a run-on sentence, something one would clearly not expect from someone as highly educated as McNaughton. *McNaughton*, I:63.

23 James S. Finan and W.J. Hurley, 'McNaughton and Canadian Operational

Research at Vimy,' *Journal of the Operational Research Society* 48, no. 1 (1997): 14.
24 Gerald W.L. Nicholson, *The Gunners of Canada: The History of the Royal Regiment of Canadian Artillery*, 2 vols, *I: 1534–1919* (Toronto: McClelland and Stewart, 1967), 282; Pierre Berton, *Vimy* (Toronto: McClelland and Stewart, 1986), 163–5. McNaughton did not invent the counter-battery technique. Lieutenant-Colonel A.G. Haig, Douglas Haig's cousin, was so successful in developing the technique in V Corps that it quickly spread throughout the BEF. McNaughton was receptive to Haig's ideas, especially the technique of sound ranging. Though he quickly adopted and refined Haig's ideas, McNaughton always acknowledged his 'very deep gratitude' to him. Swettenham, *McNaughton*, I:70.
25 Ibid., I:117. McNaughton had indicated to Currie before the battle that morale was suffering and that the 'effectiveness of the artillery was played out.' Quoted in Daniel G. Dancocks, *Legacy of Valour: The Canadians at Passchendaele* (Edmonton: Hurtig, 1986), 100, 106.
26 Swettenham, *McNaughton*, I:149.
27 Ibid., I:136.
28 McNaughton had direct control of three Counter Battery Groups totalling 111 guns ranging from 60 pounders to 9.2-inch howitzers. He could also call on the super-heavies of the 26th Heavy Artillery Group, which consisted of one 15-inch and four 12-inch howitzers. Nicholson, *The Gunners of Canada*, I:295.
29 Montgomery was promoted to captain after suffering his wound but did not return to France until 1916 as the Brigade-Major of the 104th Infantry Brigade, 35th Infantry Division. In 1917 he was GSO 2 of the 33rd Infantry Division and GSO 2 of IX Corps. As for Brooke, it should not be supposed that he never oversaw artillery operations. He arrived late in France with his troop and was promoted to Staff-Captain, Royal Artillery, in the 2nd (Indian) Cavalry Division in January 1915. When his brigade was combined with two others for the assault on Neuve Chapelle in early March 1915, Brooke supervised fifty-four guns. David Fraser, *Alanbrooke* (London: Collins, 1982), 63.
30 Archibald Wavell, *Allenby: A Study in Greatness* (London: Harrap, 1940), 294. This idea was firmly embedded in German training manuals before the Second World War. See Bruce Condell and David T. Zabecki, eds., *On the German Art of War: Truppenführung* (Boulder: Lynne Rienner, 2001), 17.
31 Ernest G. Black, *I Want One Volunteer* (Toronto: Ryerson, 1965), 136–9.
32 Paul D. Dickson, *A Thoroughly Canadian General: A Biography of General H.D.G. Crerar* (Toronto: University of Toronto Press, 2007), 60. One of the

few criticisms of McNaughton's skill as a gunner comes, oddly enough, from Crerar, who questioned the former's organizational and tactical abilities. 'Questions for a Programme on Artillery Tactics,' Crerar Papers (hereafter CP), MG30 E157, vol. 19, file 958C.009(D304), LAC, Ottawa. See also Crerar to GOC, CCHA, 9 December 1919, Records of the Department of Militia, RG9 III, vol. 3922, LAC.

33 Bill Rawling, *Surviving Trench Warfare: Technology and the Canadian Corps, 1914–1918* (Toronto: University of Toronto Press, 1992), 111–12, 190; Daniel G. Dancocks, *Spearhead to Victory: Canada and the Great War* (Edmonton: Hurtig, 1987), 33; John A. English, *The Canadian Army and the Normandy Campaign: A Study of Failure in High Command* (New York: Praeger, 1991), 154n6; Swettenham, *McNaughton*, I:74; Sir Frederick Pile, *Ack-Ack* (London: Harrap, 1949), 183.

34 Samuel P. Huntington, *The Soldier and the State: The Theory and Practice of Civil–Military Relations* (Cambridge, MA: Harvard University Press, 1957), 13–14.

35 Nicholson, *The Gunners of Canada*, I:317.

36 Robert H. Berlin, 'United States Army World War II Corps Commanders: A Composite Biography,' *Journal of Military History* (herafter *JMH*) 53, no. 2 (1989): 155.

37 John Connell, *Wavell: Scholar and Soldier* (London: Collins, 1964), 145.

38 Swettenham, *McNaughton*, I:95.

39 Despite being a newly appointed corps commander, Currie immediately went to Horne and said, 'If we were to fight at all, let us fight for something worth having.' Horne passed the objections on to Haig, who was subsequently convinced of the correctness of Currie's arguments. Incredibly, when Haig ordered the operation to start around 4 August, Currie again resisted, requesting more time to prepare properly. John Swettenham, *To Seize the Victory: The Canadian Corps in World War I* (Toronto: Ryerson, 1965), 174–5.

40 Quoted in A.M.J. Hyatt, *Sir Arthur Currie: A Military Biography* (Toronto: University of Toronto Press, 1987), 66. The first occasion on which Currie balked at the prospect of unnecessary casualties was at Festubert in May 1915. He was ordered to go forward and reconnoitre the ground. He had been given virtually no time to prepare. He could not even locate his objective, designated 'K.5.,' on a totally inadequate map. So he pressed Lieutenant-General E.A.H. Alderson for a postponement. A. Fortescue Duguid, *Official History of the Canadian Forces in the Great War, 1914–1919*, 2 vols. (Ottawa: Patenaude, 1938), I:463.

41 McNaughton later echoed Currie's hesitation. 'It isn't that we didn't want

to fight,' he said, 'but when you fight you like to fight under reasonable conditions, particularly when you've got a good mechanism in which you have confidence.' Hyatt, *Sir Arthur Currie*, 79, 81. Currie was not the only commander to balk at fighting at Passchendaele. Both General Sir Herbert Gough and Lieutenant-General Sir H.C.O. Plummer, commanders of the Fifth and Second Armies respectively, made their feelings known to Haig. During the fighting at the Somme in 1916, General Sir Henry Rawlinson, commander of the Fourth Army, challenged Haig's offensive strategy. Dominick Graham and Shelford Bidwell, *Coalitions, Politicians, and Generals: Some Aspects of Command in Two World Wars* (London: Brassey's, 1993), 94–5.

42 Currie's engineering preparations took longer than expected. He relented when Haig pressed for an attack date of 26 October. The Canadians took Passchendaele at the cost of 2,238 casualties.

43 Robert Blake, ed., *The Private Papers of Douglas Haig, 1914–1919* (London: Eyre and Spottiswoode, 1952), 303–4. Lieutenant-General Sir John Monash, commander of the Australians, felt that the policy of keeping the corps together produced inestimable advantages of mutual knowledge and confidence among all commanders and arms. Monash, *The Australian Victories in France in 1918* (London: Hutchinson, 1920), 5.

44 R. Craig Brown, 'Sir Robert Borden, the Great War, and Anglo-Canadian Relations,' in *Character and Circumstance: Essays in Honour of Donald Grant Creighton*, ed. John S. Moir (Toronto: Macmillan, 1970), 211–12; James Eayrs, *The Art of the Possible: Government and Foreign Policy in Canada* (Toronto: University of Toronto Press, 1961), 6. Borden never actually sent this letter but it clearly demonstrates his state of mind. With Borden searching for ways to increase Canadian influence on the strategic level, and Currie fighting to keep his divisions, Ottawa gradually but effectively achieved autonomy over the administration of the CEF. See Desmond Morton, *A Peculiar Kind of Politics: Canada's Overseas Ministry in the First World War* (Toronto: University of Toronto Press, 1982).

45 Charles P. Stacey, *Canada and the Age of Conflict: A History of Canadian External Policies, I: 1867–1921* (Toronto: Macmillan, 1977), 238. The Australians fought to keep their units together from the beginning of the war. See Charles E.W. Bean, *Official History of Australia in the War of 1914–1918, The Story of ANZAC, I: From the Outbreak of the War to the End of the First Phase of the Gallipoli Campaign, May 4, 1915* (Sydney: Angus Robertson, 1921), 30–1.

46 Hyatt, *Sir Arthur Currie*, 13–14. Currie joined the militia in 1897 and was commissioned from the ranks in 1901. By 1906 he was a major and second-in-command of the 5th Regiment of Artillery. Three years later he assumed command of the regiment but retired after five years at the age of thirty-

eight. Shortly thereafter he took on the role of Commanding Officer of the newly established 50th Regiment. Upon the outbreak of war the Minister of Militia, Sir Sam Hughes, promised him command of one of the provisional brigades in the contingent then forming for overseas duty. Currie was plagued by personal financial troubles throughout the war. See R. Craig Brown and Desmond Morton, 'The Embarrassing Apotheosis of a 'Great Canadian': Sir Arthur Currie's Personal Crisis in 1917,' *Canadian Historical Review* (hereafter *CHR*) 60, no. 1 (1979): 41–63.

47 Winston S. Churchill, *Great Contemporaries* (London: Thornton Butterworth, 1937), 197; David Lloyd George, *War Memoirs of David Lloyd George* (London: Ivor, Nicholson, and Watson, 1936), VI:3424. Haig served with Lord Kitchener in the 1898 Sudan campaign as a brevet major and commanded the 17th Lancers in the final stages of the Boer War. He served as Director of Military Training and Chief of the General Staff in India before returning home to command at Aldershot. He commanded I Corps at Mons, the Marne, Aisne, and Ypres and took command of the First Army in 1915 when the BEF expanded to an army group. Denis Winter, however, has presented significant evidence that Haig advanced not through any true professional skill but through good fortune, powerful patrons, and skilled diplomacy. *Haig's Command: A Reassessment* (New York: Viking, 1991), 33–4.

48 Sir Bernard Law Montgomery, *The Memoirs of Field Marshal Montgomery of Alamein* (London: Collins, 1958), 80. The most questionable of Currie's actions was ordering two of his battalions to retire during the German gas attack at Ypres. See Tim Travers, 'Currie and 1st Canadian Division at Second Ypres, April 1915: Controversy, Criticism, and Official History,' *CMH* 5, no. 2 (1996): 7–15.

49 Major Shane B. Schreiber, *Shock Army of the British Empire: The Canadian Corps in the Last 100 Days of the Great War* (Westport: Praeger, 1997), 140–1; John A. English, 'Lessons from the Great War,' *CMJ* 4, no. 2 (2003): 56. Currie worked hard to understand what was possible. The argument has been made that he was a blank slate and that this allowed him to avoid the pitfalls of the 'professionals,' who could not see 'outside the box.' Currie's classroom was the battlefield. Captain Roger A. Barrett, 'General Sir Arthur William Currie: A Common Genius for War,' *ADTB* 2, no. 3 (1999): 53.

2. The Road to High Command

1 The Otter Committee, under the chairmanship of Major-General Sir William Otter of Boer War fame, was constituted in the spring of 1919. In mid-

April of that year Patton reported to Washington and began to formulate tank regulations. In order to do so he went to the Rock Island Arsenal in Illinois and the Springfield Armory in Massachusetts to inspect tank production. At the time the Mark VIII tank was in production. Martin Blumenson, ed., *The Patton Papers, Vol. I: 1885–1940* (Boston: Houghton Mifflin, 1972), 705–6.

2 David Fraser, *Alanbrooke* (London: Collins, 1982), 110.

3 Brian Bond, *The Victorian Army and the Staff College, 1854–1914* (London: Eyre Methuen, 1972), 12. See also Lieutenant-Colonel F.W. Young, ed., *The Story of the Staff College* (Camberley: Staff College, 1958). The year before, the *Journal of the Royal United Services Institute* was established to further the study of war. Brian Holden Reid, *The American Civil War* (London: Cassell, 1999), chapter 1 on the Crimean War. The Staff Colleges (including Quetta in India, Greenwich for naval officers, and Andover for air force officers) trained officers in their respective services but also provided the means for 'building up a liaison between the services.' Major D.H. Cole, *Imperial Military Geography: General Characteristics of the Empire in Relation to Defence* (London: Sifton Praed, 1935), 38.

4 Sir Brian Horrocks, *A Full Life* (London: Collins, 1960), 69; Roland Ryder, *Oliver Leese* (London: Hamish Hamilton, 1987), 31. On the American side, there is a limited debate concerning the utility of the Command and General Staff School at Fort Leavenworth and other American service schools in preparing commanders for the Second World War. Omar Bradley, who finished first out of eighty students in the Leavenworth class of 1915, considered the instruction methods to be less than stellar but thought the courses were intellectually stimulating. Joseph L. Collins considered instruction methods to be first-rate. Timothy K. Nenninger, 'Leavenworth and Its Critics: The U.S. Army Command and General Staff School, 1920–1940,' *Journal of Military History* (hereafter *JMH*) 58, no. 2 (1994): 204–5.

5 Those officers in the Senior Division underwent a one-year course while those in the Junior Division stayed for two years. Oddly, Montgomery in his memoirs offers little direct evidence of his performance at Camberley. He noted: 'I believe I got a good report, but do not know as nobody ever told me if I had done well or badly: which seems curious.' Sir Bernard Law Montgomery, *The Memoirs of Field Marshal Montgomery of Alamein* (London: Collins, 1958), 39. Nigel Hamilton has little to add, other than that Montgomery was disappointed by the quality of instruction. Hamilton, *Monty: The Making of a General, 1887–1842* (London: Hamish Hamilton, 1981), 151–3. Montgomery's brother, Brian, recalled that the Directing Staff considered him a 'bloody menace.' Brian Montgomery, *A Field Marshal in the Family* (London: Constable, 1973), 179.

6 John Swettenham, *McNaughton*, 3 vols., *I: 1887–1939* (Toronto: Ryerson, 1968), 191; National Personnel Records Centre (NPRC), Ottawa, A.G.L. McNaughton, Confidential Report, 21 December 1921. Anderson had been a major-general on Sir Henry Horne's First Army staff during the war.
7 The brigade, with nine battalions, actually had the strength of a division and a large area of responsibility. Brian Montgomery, *A Field Marshal in the Family*, 181.
8 MacBrien commanded the 12th Infantry Brigade during the war.
9 Cole, *Imperial Military Geography*, 37.
10 Quoted in Swettenham, *McNaughton*, I:236.
11 J.L. Granatstein, *The Generals: The Canadian Army's Senior Commanders in the Second World War* (Toronto: Stoddart, 1993), 57.
12 Norman Hillmer and William J. McAndew, 'The Cunning of Restraint: General J.H. MacBrien and the Problems of Peacetime Soldiering,' *Canadian Defence Quarterly* (hereafter *CDQ*) 8, no. 4 (1979): 43; Stephen J. Harris, 'The Canadian General Staff and the Higher Organization of Defence, 1919–1939: A Problem of Civil–Military Relations,' *War and Society* 3, no. 1 (1985): 89. McNaughton acted much like General Sir Pierre van Ryneveld, South Africa's senior military officer, who treated the service commanders like staff officers. Lieutenant-General H.J. Martin and Colonel Neil D. Opren, *South African Forces World War II, Vol. VII: South Africa at War: Military and Industrial Organization and Operations in Connection with the Conduct of the War, 1939–1945* (Cape Town: Purnell, 1979), v.
13 John A. English, *The Canadian Army and the Normandy Campaign: A Study of Failure in High Command* (New York: Praeger, 1991), 42. For a good analysis of McNaughton's role in the camps, see Roy R. Maddocks, 'A.G.L. McNaughton, R.B. Bennett, and the Unemployment Relief Camps, 1932–35,' unpublished paper, 1973, file 74/795, Directorate of History and Heritage (hereafter DHH), Department of National Defence, Ottawa.
14 Stephen J. Harris, 'Or There Would Be Chaos: The Legacy of Sam Hughes and Military Planning in Canada, 1919–1939,' *Military Affairs* 46, no. 3 (1982): 122. McNaughton's decision to scrap Defence Scheme No. 1 and his personal and professional conflict with Brown are covered in Richard H. Gimblett, 'Buster Brown: The Man and His Clash with "Andy" McNaughton,' BA thesis, Royal Military College, 1979, 40–4.
15 Quoted in Stephen J. Harris, *Canadian Brass: The Making of a Professional Army, 1860–1939* (Toronto: University of Toronto Press, 1988), 179.
16 Major-General Andrew G.L. McNaughton, 'The Defence of Canada: A Review of the Present Situation, 28 May 1935,' file 112.3M2009(D7), DHH; Charles P. Stacey, *Official History of the Canadian Army in the Second World War, I: Six Years of War: The Army in Canada, Britain, and the Pacific* (Ottawa:

Queen's Printer, 1957), 7. In late October, after Bennett's Tory government was replaced by King's Liberals, McNaughton, no longer in a position of authority, nevertheless pressed O.D. Skelton to make sure the new prime minister saw his memo 'so that he might be under no misapprehension as to the situation.' It seems that Skelton, Major-General E.C. Ashton, the new CGS, and Loring Christie read McNaughton's memo towards the end of May 1935. However, King apparently never found the time to read it until August 1936. Swettenham, *McNaughton*, I:315; James Eayrs, *In Defence of Canada: Appeasement and Rearmament* (Toronto: University of Toronto Press, 1965), 135. Lieutenant-Colonel Eedson L.M. Burns summed up the deplorable situation in 1936: 'Our forces are poorly trained and equipped.' Burns, 'The Defence of Canada,' *CDQ* 13, no. 4 (1936): 379.

17 Address to the United Services Institute, 10 December 1921, quoted in Swettenham, *McNaughton*, I:228–9. He repeated this belief years later when he became CGS; see 'Trend of Army Development,' Address to Canadian Military Institute, Toronto, 2 May 1929, McNaughton Papers (hereafter MP), MG30 E133, vol. 347, LAC. See also 'Canada's Land and Air Forces,' An Address by Maj.-Gen. A.G.L. McNaughton, C.M.G., D.S.O., Ottawa, 2 May 1929, *The Empire Club of Canada Speeches 1929* (Toronto: Empire Club of Canada, 1930), 189–99.

18 McNaughton, 'The Defence of Canada'; Correlli Barnett, *Britain and Her Army 1509–1970: A Military, Political, and Social Survey* (New York: Allen Lane, 1970), 411.

19 Douglas MacArthur, *Reminiscences* (New York: McGraw-Hill, 1964), 99. The financial limitations of British experimentation in the interwar period are analysed in 'Tanks, Votes, and Budgets: The Politics of Mechanization and Armored Warfare in Britain, 1919–1939,' in *The Challenge of Change: Military Institutions and New Realities, 1918–1941*, ed. Harold R. Winton and David R. Mets, (Lincoln: University of Nebraska Press, 2000), 74–107.

20 Desmond Morton, 'Changing Operational Doctrine in the Canadian Corps, 1916–1917,' *ADTB* 2, no. 4 (1999): 39. In 1936 the Militia underwent a significant reorganization. Four cavalry regiments and six infantry regiments were converted to armoured car and tank units respectively. As for the Permanent Force, a tank school was not established until 1938. For a good survey of the changes, see Major W. Alexander Morrison, 'Major-General A.G.L. McNaughton, The Conference of Defence Associations, and the 1936 N.P.A.M. Re-organization: A Master Military Bureaucratic Politician at Work,' file 82/470, DHH, Department of National Defence, Ottawa.

21 English, *The Canadian Army and the Normandy Campaign*, 29; Special Correspondent, 'The Army Manoeuvres,' *The Fighting Forces* 12 (October 1935): 355. For an overview of what the British had achieved in the mid-1920s, see Major-General James H. MacBrien, 'The British Army Manoeuvres September 1925,' *CDQ* 2 (January 1926): 132–50.
22 Harris, *Canadian Brass*, 193.
23 Heinz Guderian, *Panzer Leader* (New York: Dutton, 1952), 22–3; Major Eedson L.M. Burns, 'A Step Towards Modernization,' *CDQ* 12, no. 3 (1935): 305.
24 James Eayrs, *In Defence of Canada: From the Great War to the Great Depression* (Toronto: University of Toronto Press, 1964,) 258, 260.
25 Major-General A.G.L. McNaughton, 'The Military Engineer and Canadian Defence,' *CDQ* 7, no. 2 (1930): 151, 154. Harris has made the case that the mainstay of interwar education at RMC was engineering, and most Canadian officers probably thought as McNaughton did. Harris, *Canadian Brass*, 192–209.
26 Swettenham, *McNaughton*, I:229; Major-General Andrew G.L. McNaughton, 'The Development of Artillery in the Great War,' *CDQ* 6, no. 2 (1929): 171.
27 J.F.C. Fuller, 'The Application of Recent Developments in Mechanics and Other Scientific Knowledge to Preparation and Training for Future War on Land,' *Royal United Services Institute Journal* 65 (May 1920): 240–1, 252–6.
28 McNaughton, 'The Development of Artillery in the Great War,' 170; idem, 'The Capture of Valenciennes: A Study in Co-ordination,' *CDQ* 10, no. 3 (1933): 279. English has suggested that he perhaps 'fell into the trap of believing that the only experience that counts is one's own.' English, *The Canadian Army and the Normandy Campaign*, 50. For a good discussion of Guy Simonds's faith in, and ultimate reliance on, the timed use of artillery, see William J. McAndrew, 'Fire or Movement? Canadian Tactical Doctrine, Sicily – 1943,' *Military Affairs* 51, no. 3 (1987): 144–5.
29 Shelford Bidwell and Dominick Graham, *Firepower: British Army Weapons and Theories of War, 1904–1945* (London: Allen Unwin, 1982), 138–9.
30 Shane B. Schreiber, *Shock Army of the British Empire: The Canadian Corps and the Last 100 Days of the Great War* (Westport: Praeger, 1997), 142. Harris correctly observed that as a 'scientist-engineer,' McNaughton could not help but be 'acutely aware of the implications of modern weapons on the battlefield.' Harris, *Canadian Brass*, 201. However, knowing that mechanization would change warfare is a long way from understanding precisely how it would be changed.
31 William A. Stewart, 'Attack Doctrine in the Canadian Corps, 1916–1918,' MA thesis, University of New Brunswick, 1984, 213. At Amiens, tank–infantry cooperation was 'practically impossible' due to the poor visibility.

32 Fraser, *Alanbrooke*, 119. Fraser added that 'at no point' had he 'yet given evidence that he had a real insight into the strategic and operational opportunities, as opposed to the tactical advantages.' In 1944 a small pamphlet titled 'British Commanders' was in circulation that gave brief biographical sketches of the principal commanders. The section on Brooke claimed that he was 'regarded as Britain's greatest expert on mechanization.' This was no more true in 1944 than it had been in 1939. File 000.2(D29), DHH. Basil Liddell Hart recalled that while Brooke was teaching at Camberley in 1925 he had 'set forth the attack method he was teaching, which made the action of tanks subordinate and subservient to the requirements of a well-organized barrage and timetable, while distributing the tanks in a string of small packets to aid the barrage-following infantry.' Liddell Hart, *The Memoirs of Captain Liddell Hart*, 2 vols. (London: Cassell, 1965), I:101.

33 James H. Lutz, 'Canadian Military Thought, 1923–1939: A Profile Drawn from the Pages of the old *Canadian Defence Quarterly*,' *CDQ* 9, no. 2 (1979): 48; Serge M. Durflinger, 'The *Canadian Defence Quarterly* 1933–1935: Canadian Military Writing of a Bygone Era,' *CDQ* 20, no. 6 (1991): 46.

34 Bernard Law Montgomery, 'The Major Tactics of the Encounter Battle,' *Army Quarterly* 36, no. 2 (1938): 272. Montgomery perfectly anticipated various dilemmas that would affect both the Allies and the Germans in Normandy. He recognized that in the defence, armoured divisions 'must be replaced by normal divisions' to regain mobility. However, an actual relief would be hard because armoured divisions 'should not allow themselves to become static.' As for air power, he argued for its employment against enemy concentrations rather than rearward objectives. In effect, he favoured the interdiction of units on the move to the main battle area: 'The slowing up and the disorganization of the enemy's main armies might have far-reaching results for the side which could achieve this object.' As for offensive tactics, he suggested that divisions should 'push on without any idea of keeping alignment' so as to 'drive in the advance elements without delay' in order to confirm the contours of the enemy's main position. This is what the 3rd Canadian Infantry Division did on D-Day, only to have its lead elements isolated and decimated by flanking attacks of the 12th SS Panzer Division.

35 Bernard Law Montgomery, 'The Problem of the Encounter Battle as Affected by Modern British War Establishment,' *CDQ* 15 (October 1937): 13–25; see also Stuart's editorial in same issue, 1. Montgomery was by no means a prolific article writer, but he did publish a five-part series titled 'The Growth of Modern Infantry Tactics' in *The Antelope* during 1924–5 and 'Some Prob-

lems of Mechanization' in the same journal in 1926. See also 'Letter of Advice to a Newly Appointed Adjutant in the TA' (the editors changed his original title, 'Training in the Territorial Army'), *Army Quarterly* (Autumn 1924).

36 Burns was certainly studying. He wrote sophisticated reviews of works by Liddell Hart and Fuller. See Captain Eedson L.M. Burns, 'The Principles of War: A Criticism of Colonel J.F.C. Fuller's Book "The Foundation of the Science of War,"' *CDQ* 4, no. 2 (1927): 168–75; idem, 'The Remaking of Modern Armies: A Review,' *CDQ* 5, no. 1 (1927): 115–17.

37 Steve E. Dietrich, 'The Professional Reading of General George S. Patton, Jr.,' *JMH* 53, no. 4 (1989): 390. In 1919 Patton gave a presentation to the officers of his tank brigade titled 'The Obligation of Being an Officer' in which he stated: 'Few are born Napoleons, but any of us can be good company commanders if we study ... I earnestly advise you all to read military subjects 3½ hours a week.' Blumenson, *The Patton Papers*, I:724. Omar Bradley reflected that while teaching at West Point 'I began to seriously read – and study – military history ... learning a great deal from the mistakes of my predecessors.' In particular he studied Sherman, 'a master of the war of movement.' Bradley, *A General's Life* (New York: Simon and Schuster, 1983), 53-54.

38 Brigadier-General R.J. Orde, quoted in Interview Transcript, 'Andrew McNaughton – Canadian,' as broadcast on 11 September 1966, MG 31, vol. 3, box E42, LAC.

39 Address to the United Services Institute, 10 December 1921, quoted in Swettenham, *McNaughton*, I:228–9.

40 Stewart, 'Attack Doctrine in the Canadian Corps,' 250.

41 John J. Pershing, *My Experiences in the World War*, 2 vols. (New York: Stokes, 1931), II:2.

42 Stewart, 'Attack Doctrine in the Canadian Corps,' 256.

43 Swettenham, *McNaughton*, I:315–16; John Bryden, *Best-Kept Secret: Canadian Secret Intelligence in the Second World War* (Toronto: Lester, 1993), 44; Wilfrid Eggleston, *Scientists at War* (London: Oxford University Press, 1950), 13.

44 Mel Thistle, ed., *The Mackenzie–McNaughton Wartime Letters* (Toronto: University of Toronto Press, 1975), xiii; Lieutenant-General Elliot Rodger, quoted in Interview Transcript, 'Andrew McNaughton – Canadian,' 14; Mackenzie quoted in Bernd Horn and Stephen J. Harris, eds., *Warrior Chiefs: Perspectives on Senior Canadian Military Leaders* (Toronto: Dundurn, 2001), 74.

45 MacArthur, too, was called out of retirement to resume command. He had resigned from the active list on 31 December 1937, but President Roosevelt wanted him as Senior Commander in the Orient. On 26 July 1941 he was

reappointed major-general in the army; the following day he was promoted to lieutenant-general. William Manchester, *American Caesar: Douglas MacArthur 1880–1964* (New York: Little, Brown, 1978), 189.
46 John W. Pickersgill, ed., *The Mackenzie King Record, I: 1939–1944* (Toronto: University of Toronto Press, 1960), 38; King Diary, 6 October 1939, quoted in Granatstein, *The Generals*, 61; Granatstein, *Canada's War: The Politics of the Mackenzie King Government, 1939–1945* (Toronto: Oxford University Press, 1975), 25.
47 Patton was known to break down at the thought of his soldiers dying. Montgomery related a similar incident about Brooke at Dunkirk when the latter had been ordered to return to Britain: 'I saw at once that he was struggling to hold himself in check; so I took him a little way into the sand hills and then he broke down and wept – not because of the situation of the B.E.F. ... but because he had to leave us all to a fate which looked pretty bad. He, a soldier, had been ordered to abandon his men at a critical moment – that is what disturbed him.' Montgomery, *The Path to Leadership* (New York: Putnam, 1961), 127–8.
48 Pickersgill, *The Mackenzie King Record*, I:39.
49 Victor Odlum, a militia officer, was fifty-nine and had not commanded anything since the 23rd Infantry Brigade in 1925. In the intervening years he had pursued business and politics. Halfdan F.H. Hertzberg, a Permanent Force officer, was fifty-five in 1939 and was serving as the Quartermaster General in Ottawa. Forty-nine-year-old E.W. Sansom, a future corps commander in England, was only a colonel.
50 During the 'Canada and the Second World War' conference held in London, England, in June 2004, Desmond Morton scoffed at the idea that Pearkes could have commanded the 1st Division. Pearkes had an interest in military history and was always keen to extract lessons. See George R. Pearkes, 'The 1914 Campaign in East Prussia,' *CDQ* 8, no. 2 (1932): 248-254; and idem, 'The Winter March of a Brigade of Guards through New Brunswick, 1862,' *CDQ* 11, no. 1 (1934): 100–10.
51 Reginald Roy, *For Most Conspicuous Bravery: A Biography of Major-General George R. Pearkes, V.C. through Two World Wars* (Vancouver: UBC Press, 1977), 129. Pearkes managed to get himself appointed as Assistant Chief Umpire for the three-week exercise. The 1st Division was commanded by Major-General C. Armitage; the 2nd Division was commanded by Major-General Henry 'Jumbo' Wilson. Pearkes recalled: 'There was very little "air" recce and the battle like so many manoeuvres in peace time became deadlocked. I was very impressed by the night move of H.Q.s and could quite understand the thrill or romance of a mechanized force. The speed and distance

of the move were remarkable at that time (of course we did the same sort of thing many times a few years later but I had never seen anything like it in Canada).' Pearkes Papers (hereafter PP), University of Victoria Archives, MacPherson Library, Special Collections, box 7.

52 Reginald Roy, *1944: The Canadians in Normandy*, Canadian War Museum, Publication no. 19 (Toronto: Macmillan, 1984), 2.

53 Brooke's brigade consisted of four battalions, and Fraser has suggested that handling them on exercise 'presented him with few difficulties,' specifically because he had studied 'voraciously' for the task. Brooke commanded the Mobile Division from the end of 1937 to July 1938. Fraser, *Alanbrooke*, 108, 117, 121. Even Paget had interwar command experience. He had commanded the Quetta Brigade in 1936–7.

54 Donald E. Houston, *Hell on Wheels: The 2nd Armoured Division* (San Rafael: Presidio, 1977), 69. The umpires called it the most magnificent road move ever made by tanks. Patton specifically focused on repeatedly attacking the enemy's rear area. In Louisiana he pushed his division two hundred miles in twenty-four hours. Christopher R. Gabel, *The US Army GHQ Maneuvers of 1941* (Washington: Office of the Chief of Military History, 1991), 52, 108, 110. Through study and actual field manoeuvres, Patton evolved a belief that tanks should not be used en masse. This differed fundamentally from Guderian's concept of 'Nicht klechern, sondern klotzen' (Not a drizzle, but a downpour). Major-General Frederick W. von Mellenthin, *German Generals of World War II: As I Saw Them* (Norman: University of Oklahoma Press, 1977), 89. The Tennessee Manoeuvres exercised corps-level formations only. In Louisiana, Lieutenant-General Ben Lear's Second Army (160,000 men) attacked Lieutenant-General Walter Krueger's Third Army (240,000 men). Twenty-seven divisions took part in the Louisiana Manoeuvres. The Carolina Manoeuvres followed in mid-November.

55 Burns, 'A Step Towards Modernization,' 301.

3. A Willingness to Fight, 1940–1941

1 J.L. Granatstein, *Canada's Army: Waging War and Keeping the Peace* (Toronto: University of Toronto Press, 2002), 176, 213–14; Desmond Morton, *A Military History of Canada*, rev. ed. (Edmonton: Hurtig, 1990), 203, 210; George F.G. Stanley, *Canada's Soldiers: A Military History of an Unmilitary People*, 3rd ed. (Toronto: Macmillan, 1974), 366; Daniel J. Dancocks, *The D-Day Dodgers: The Canadians in Italy, 1943–1945* (Toronto: McClelland and Stewart, 1991), 17; David J. Bercuson, *Maple Leaf against the Axis: Canada's Second World War* (Toronto: Stoddart, 1995), 65; Dominick Graham, *The Price of Command:*

A Biography of General Guy Simonds (Toronto: Stoddart, 1993), 105; Denis Whitaker, *Dieppe: Tragedy to Triumph* (Toronto: McGraw-Hill Ryerson, 1993), 60; Paul D. Dickson, 'The Hand That Wields the Dagger: Harry Crerar, First Canadian Army Command, and National Autonomy,' *War and Society* 13, no. 2 (1995): 114.

2 Quoted in Charles P. Stacey, *Arms, Men, and Governments: The War Policies of Canada, 1939–1945* (Ottawa: Queen's Printer, 1970), 20. King had achieved an almost unanimous declaration of war in the House of Commons on 10 September, but mainly because of his stand against conscription. The very idea of war troubled him. He made it quite clear in a pre-war article that any sort of overt alliance, even with Britain, that held out the prospect of war, was anathema to his world view. King, 'Canada's Foreign Policy,' *CDQ* 15, no. 4 (1938): 380–401.

3 John W. Pickersgill, ed., *The Mackenzie King Record, I: 1939–1944* (Toronto: University of Toronto Press, 1960), 40–1. By mid-December, King had approved the ambitious British Commonwealth Air Training Plan (BCATP) and had even extracted a guarantee from Chamberlain that a statement would be included in the final draft declaring the program 'the most essential and effective form of military co-operation open to Canada.' Spencer Dunmore, *Wings for Victory: The Remarkable Story of the British Commonwealth Air Training Plan in Canada* (Toronto: McClelland and Stewart, 1994), 36. Military correspondent Hanson Baldwin would declare a year later that the BCATP 'is probably the only Canadian effort that may have a major influence upon the outcome of the war.' Baldwin, *New York Times*, 10 October 1940. King eventually found himself in a difficult position after the fall of France. 'Ironically,' stated Desmond Morton, 'the total war concept movement was fed by King's own propaganda appeals for greater sacrifices and by his own decisions.' Morton, *Canada and War: A Military and Political History* (Toronto: Butterworths, 1981), 117.

4 Quoted in John A. Munro, ed., *Documents on Canadian External Relations, VI: 1936–1939* (Ottawa: Department of Foreign Affairs, 1972), 1269.

5 As far as Rogers was concerned, McNaughton offered no such warning, but it definitely sounded like something McNaughton would say. Norman Rogers, 'Conversations with Major-General McNaughton,' 4 October 1939, Norman Rogers Papers (hereafter NRP), Queen's University Archives, Kingston. During the BCATP discussions in late September 1939, the Chief of the General Staff (hereafter CGS), Major-General T.V. Anderson, also expressed the view that the Canadian public would not be satisfied with an exclusive air effort, no matter how significant in scale. Minutes of the Cabinet War Committee (hereafter CWC) (Emergency Council), 28 September 1939, RG 2, vol. I, C4653A, LAC.

6 Winston S. Churchill, *Their Finest Hour* (London: Cassell, 1949), 460.
7 Memorandum, 17 January 1940, McNaughton Papers (hereafter MP), MG30 E133, vol. 251, LAC. Ottawa also acted quickly to prevent the kind of difficulties that had arisen during the First World War from an administrative standpoint and established Canadian Military Headquarters (hereafter CMHQ) in London in late September 1939.
8 Pickersgill, *The Mackenzie King Record*, I:76.
9 War Cabinet, Confidential Annex, 2 January 1940, in *The Churchill War Papers, I: At the Admiralty, September 1939–May 1940*, ed. Martin Gilbert (London: Heinemann, 1993), 598. The Admiralty had identified the value of Canadian participation in any Norwegian operations three months earlier. The Chiefs of Staff had generated a report which indicated that a force of 5,000 to 7,000 Canadians 'could be organized for operations in Northern Scandinavia in March' and that the Secretary of State for Dominion Affairs 'understood from General McNaughton ... that the Canadians had no troops at present trained to work on skis.'
10 The Visiting Forces (British Commonwealth) Acts of 1933 stipulated that Canadians were either 'serving together' or 'in combination' with British forces. Under the first stipulation the forces were independent of each other; but under the second, Canadian forces were under the operational authority of a British commander.
11 Winston S. Churchill, *The Gathering Storm* (London: Cassell, 1948), 608; Sir Adrian Carton de Wiart, *Happy Odyssey: The Memoirs of Lieutenant-General Sir Adrian Carton de Wiart* (London: Jonathan Cape, 1950), 167. General Hastings Ismay, soon to be Churchill's representative to the Chiefs of Staff Committee, concluded that the resources available were 'inadequate and unsuitable.' Lord Ismay, *The Memoirs of General the Lord Ismay* (London: Heinemann, 1960), 120–1.
12 Charles P. Stacey, *Official History of the Canadian Army in the Second World War, I: Six Years of War: The Army in Canada, Britain, and the Pacific* (Ottawa: Queen's Printer, 1957), 261.
13 J.L. Granatstein, *The Generals: The Canadian Army's Senior Commanders in the Second World War* (Toronto: Stoddart, 1993), 63.
14 Charles P. Stacey, *A Date With History: Memoirs of a Canadian Historian* (Ottawa: Deneau, 1983), 76; Lester B. Pearson, *Mike: The Memoirs of the Right Honourable Lester B. Pearson, I: 1897–1948* (Toronto: University of Toronto Press, 1972), 164–5. McNaughton knew the Statute of Westminster intimately, having taken part in the deliberations at the Imperial Conference held in London in October 1930. He had been sent by Prime Minister R.B. Bennett as the government's sole military adviser and principal representative to the committee on constitutional relations.

15 Pearson, *Mike*, I:164.
16 McNaughton responded that it was the assistance of Stanley's officers in the Aldershot Command who made the move so quick. Oliver Stanley to McNaughton, 23 April 1940, and McNaughton to Stanley, 26 April 1940, E.W. Sansom Papers (hereafter SaP), MG30 E537, vol. 1, file 10, LAC; Dewing Diary, 16 April 1940, General Richard Dewing Papers (hereafter DP), Liddell Hart Centre for Military Archives (hereafter LHCMA).
17 Ironside to McNaughton, 25 April 1940, and McNaughton to Ironside, 27 April 1940, SaP, vol. 1, file 10.
18 Eric Hutton, 'A Scientist Commands Canada's First Division,' *Star Weekly*, 9 December 1939.
19 John Swettenham, *McNaughton, II: 1939–1943* (Toronto: Ryerson, 1969), 30–2; Interview Transcript, 'First and Second World War: McNaughton 1965,' file 1, session 3, tape 2, 9, MG31, vol. 3, box E42, LAC. Crerar, however, did warn McNaughton to tone down comments about Ironside's ignorance of constitutional issues. H.D.G. Crerar Papers (hereafter CP), MG30 E157, War Diary (hereafter WD), 31 March 1940, vol. 15, file 958C.009(D271), LAC.
20 On 17 April, Vincent Massey had sent an explanatory cable about the impending operation to Ottawa. Since King was in the United States and Rogers was at that time still en route to Britain, the CWC never met to discuss the Norwegian crisis. Stacey, *Arms, Men, and Governments*, 32.
21 McNaughton acted under the authority granted him in PC 3391, promulgated on 2 November 1939. It stated that he could place Canadian units 'in combination' with the British 'when such action is necessitated by military exigencies.' 26 February 1940, MP, vol. 252, file 'War Orders.'
22 Pearson asked Crerar prior to the request how he thought McNaughton would respond. Crerar 'felt sure' that he would act on his own. Pearson recorded: 'If the whole thing turns out badly and they cannot drive the Germans out of Norway, then I would not wish to be in McNaughton's shoes if he agrees to the use of Canadians in this way without prior authorization from Ottawa.' Pearson, *Mike*, I:168–9; H.A. Dyde to Richard Malone, 7 January 1969, Malone Papers (hereafter MaP), Queen's University Archives, Kingston.
23 Record of visit to United Kingdom of Canadian Minister of National Defence (Mr Norman McL. Rogers) April 18th to May 9th 1940, NRP, box 8, Queen's University Archives; Vincent Massey, *What's Past Is Prologue: The Memoirs of the Right Honourable Vincent Massey* (Toronto: Macmillan, 1963), 321–2; Stacey, *Six Years of War*, 262.
24 John Robinson Campbell, 'James Layton Ralston and Manpower for the

Canadian Army,' MA thesis, Wilfrid Laurier University, Waterloo, 1984, 13–14.
25 In a speech to the Canadian Club of Vancouver on 12 December 1928, McNaughton, then DCGS, advocated a 'large force in embryo.' When Ralston heard of it he came down hard on McNaughton, declaring in no uncertain terms that he should take the 'utmost care' that 'nothing approaching policy is discussed publicly without consultation and approval.' Quoted in James Eayrs, *In Defence of Canada: From the Great War to the Great Depression* (Toronto: University of Toronto Press, 1964), 258. McNaughton was certainly not innocent in such things; indeed, he was constantly demonstrating a willingness to exceed his authority. For example, when Ian Mackenzie, Minister of National Defence, read McNaughton's correspondence with the War Office concerning the dispatch of a Canadian expeditionary force in time of war, he concluded that such speculative intrusion into policy making was 'fraught with danger.' Quoted in James Eayrs, *In Defence of Canada: Appeasement and Rearmament* (Toronto: University of Toronto Press, 1965), 83.
26 Claude Bissell took an entirely different view, stating that Ralston's military experience 'gave him confidence in dealing with professional soldiers.' Bissell, *The Imperial Canadian: Vincent Massey in Office* (Toronto: University of Toronto Press, 1986), 141.
27 Quoted in R. MacGregor Dawson, *The Conscription Crisis of 1944* (Toronto: University of Toronto Press, 1961), 29.
28 Interview Transcript, 'First and Second World War,' 7; Swettenham, *McNaughton*, II:116. Rogers never wanted to be Minister of National Defence because he neither liked military things nor had any aptitude for them. McNaughton no doubt found it easier to work with him because – as Chubby Power indicated – Rogers, a Rhodes Scholar, was a 'quiet, unassuming man of gentle disposition [and] far and away the most cultured member of the cabinet.' Norman Ward, ed., *A Party Politician: The Memoirs of Chubby Power* (Toronto: Macmillan, 1966), 184, 186.
29 Martha Ann Hooker interview with D.C. Spry, 14 March 1985, in 'In Defence of Unity: Canada's Military Policies, 1935–1944,' MA thesis, Carleton University, Ottawa, 1985, 152. According to Chubby Power, Ralston 'manifested a vast reluctance to take over National Defence' because he felt that 'his health was none too good, and that the strain of carrying on the routine of the department in wartime would be very heavy.' Ward, *A Party Politician*, 187.
30 Pearson Diary, 24 December 1939, Lester Pearson Papers (hereafter LPP), MG26 N1, LAC.
31 The 1st Canadian Division originally served under Major-General Claude

Auchinleck's IV Corps. Eventually, however, it became a self-contained entity known as 'Canada Force.'
32 Stacey, *Six Years of War*, 268–9.
33 Dewing to McNaughton, 27 May 1940, quoted in Stacey, *Six Years of War*, 271. McNaughton recorded that any attempt to reinforce Dunkirk 'did not appear to be [a] practical operation of war.' MP, vol. 238.
34 Dill claimed that he was not even aware of the plan to hold out in Brittany; Brooke, however, later commented that it was 'one of my blackest days' when Dill informed him he was to take a second BEF back to France. Brooke also bluntly informed Sir Anthony Eden, Secretary of State for War, that the mission had 'no possibility of accomplishing anything.' Alex Danchev and Daniel Todman, eds., *War Diaries, 1939–1945: Field Marshal Lord Alanbrooke* (London: Weidenfeld and Nicolson, 2001), 74; Swettenham, *McNaughton*, II:111.
35 For Freyberg's authority, see *Documents Relating to New Zealand's Participation in the Second World War, 1939–45*, vol. I (Wellington: War History Branch, Department of Internal Affairs, 1949), 32.
36 Dill to McNaughton, 21 June 1940, and McNaughton to Dill, 29 June 1940, quoted in Stacey, *Six Years of War*, 284.
37 Dewing to McNaughton, 27 May 1940.
38 David Fraser, *And We Shall Shock Them: The British Army in the Second World War* (London: Sceptre, 1983), 113. The divisions were the 1st, 2nd, and 7th Armoured Divisions and the 70th Infantry Division.
39 Lieutenant-General Maurice A. Pope, *Soldiers and Politicians: The Memoirs of Lt.-Gen. Maurice A. Pope* (Toronto: University of Toronto Press, 1962), 153.
40 Bercuson, *Maple Leaf against the Axis*, 65; Jacques Mordal, *Dieppe: The Dawn of Decision* (Toronto: Stoddart, 1962), 99; Ronald Atkin, *Dieppe 1942: The Jubilee Disaster* (London: Macmillan, 1980), 22; Brian Loring Villa, *Unauthorized Action: Mountbatten and the Dieppe Raid* (Oxford: Oxford University Press, 1989), 218.
41 He made no mention of sending Canadians but specifically cited the need for Australians and Indians. Churchill stated that only the air force could win the war. 'In no other way at present visible can we hope to overcome the immense military power of Germany ... The Air Force and its action on the largest scale must, therefore, subject to what is said later, claim the first place over the Navy or the Army.' Churchill, *Their Finest Hour*, 458–9.
42 Study Group of the Royal Institute of International Affairs, *Political and Strategic Interests of the United Kingdom: An Outline* (London: Oxford University Press, 1939), 100.
43 Ibid., 5.

44 Interview Transcript, 'First and Second World Wars,' session 3, tape 2, p. 6. Crerar's personal motives are explained in Paul D. Dickson, *A Thoroughly Canadian General: A Biography of General H.D.G. Crerar* (Toronto: University of Toronto Press, 2007), 136.
45 Francis H. Hinsley, *British Intelligence in the Second World War, I: Its Influence on Strategy and Operations* (London: HMSO, 1979), 482. It was estimated that the Germans would reach Moscow in three to four weeks and with perhaps six weeks of preparation after that could attempt invasion. A case can be made, however, that Churchill always knew there was a low probability of German invasion during 1940–1. See Anthony Cave Brown, *'C': The Secret Life of Sir Stewart Graham Menzies, Spymaster to Winston Churchill* (New York: Macmillan, 1987), 300, 769.
46 See Dickson, *A Thoroughly Canadian General*, 154.
47 Eedson L.M. Burns, *General Mud: Memoirs of Two World Wars* (Toronto: Clarke, Irwin, 1970), 107–8; Swettenham, *McNaughton*, II:179–80.
48 Pickersgill, *The Mackenzie King Record*, I:239, 244, 261. One historian has recently implied that Churchill intended to send the Canadians to North Africa once the Luftwaffe was defeated in the Battle of Britain. This, however, is an inaccurate claim. Gordon Corrigan, *Blood, Sweat, and Arrogance and the Myths of Churchill's War* (London: Weidenfeld and Nicolson, 2006), 394.
49 According to Burns, Nye suggested in early 1942 that 'many more troops, in all probability, would have been dispatched to different theatres had resources in shipping permitted.' Burns Diary, 25 February 1942, Burns Papers (hereafter BP), MG31 G6, LAC.
50 Stacey, *Arms, Men, and Governments*, 41. The argument in favour of having the Canadians stay in Britain was only reinforced when Ralston returned to Canada in January 1941 and told the CWC that he had not realized until then 'how great was the need for men' there. Minutes of the CWC, 24 January 1941, RG 7C, vol. 4, C4653A, LAC. King was heavily influenced by Ralston's statement and recorded in his own diary on the same day that the Canadians were needed to defend Britain. King Diaries, 24 January 1941, King Papers (herafter MKP), MG26 J13, Diaries, 1938–1944, LAC.
51 During a 21 October 1941 meeting with Crerar, Ralston, Massey, and Macready, Sir Archibald Sinclair, Secretary of State for Air, emphasized the need to keep strong forces in Britain. CP, vol. 15, WD.
52 For a detailed order of battle, see Barrie Pitt, *Crucible of War: Western Desert 1941* (London: Jonathan Cape, 1980), 488–9.
53 Martin Gilbert, *Finest Hour: Winston S. Churchill 1939–1941* (London: Heinemann, 1983), 1191–2, 1196. Churchill was particularly concerned with Auchinleck's decision to send the British 50th Division to Cyprus 'in what

looks like a safe defensive role.' It actually made strategic sense to send Australians and New Zealanders, not Canadians, to the Middle East. Williamson Murray, 'British Military Effectiveness in the Second World War,' in *Military Effectiveness, III: The Second World War*, ed. Alan R. Millett and Williamson Murray, (Boston: Allen Unwin 1988), 105.

54 At this time Churchill was thinking of Canada more as a strategic producer of supply and training. Churchill to King, 8 December 1940, in Gilbert, ed., *The Churchill War Papers*, I:1189.

55 Ralph Bennett, *Ultra and Mediterranean Strategy* (New York: William Morrow, 1989), 78. In November 1941 the Afrika Korps consisted of the 15th and 21st Panzer Divisions and the 90th Light Division.

56 Roosevelt to Churchill, 3 May 1941, in Francis L. Loewenheim, Harold D. Langley, and Manfred Jonas, eds., *Roosevelt and Churchill: Their Secret Wartime Correspondence* (New York: Saturday Review Press/Dutton, 1975), 141. Roosevelt was at that moment sending thirty ships loaded with supplies, and more would follow 'until there is a final decision in the Mediterranean.' On 8 June 1941, Australian, Indian, and Free French forces invaded Syria and quickly secured the country. Churchill stated that this success 'greatly improved our strategical position in the Middle East.' Winston S. Churchill, *The Grand Alliance* (London: Cassell, 1950), 331.

57 King Diaries, 1 October 1940.

58 Stacey, *Arms, Men, and Governments*, 40. Gardiner had reported to the CWC on his conversation with Eden on 4 December.

59 Stacey, *Six Years of War*, 323n. Stacey has suggested that Crerar was 'doubtless' influenced to ask Dill such a question because of a Canadian report in the *Ottawa Evening Journal* that Canadian troops 'may be thrown into Britain's increasingly important campaign in the Near East.' Stacey, *Arms, Men, and Governments*, 39.

60 Pickersgill, *The Mackenzie King Record*, I:156. The telegram was drafted by King's anglophobe minister, O.D. Skelton.

61 Ralston met with Churchill on 17 and 22 December. Regarding the later meeting, Ralston wrote: 'I repeated our position regarding troops in the Middle East and that we assumed that it would be regarded as a matter for Canada and would be left to our suggestion. He said that was so.' Ralston Diary, 17, 22 December 1940, Ralston Papers (hereafter RP), MG27 III BII, vol. 41, LAC; Stacey, *Arms, Men, and Governments*, 41.

62 Pickersgill, *The Mackenzie King Record*, I:156.

63 When Churchill was told on 7 July 1940 that Canadian troops were serving in Iceland, he was aghast that such 'fine troops should be employed in so distant a theatre,' and he stated: 'We require two Canadian Divisions

to work as a Corps as soon as possible.' Churchill to Eden, 7 July 1940, in *The Churchill War Papers, Volume II: Never Surrender, May 1940–December 1940*, ed. Martin Gilbert (London: Heinemann, 1994), 488. According to King, Norman Rogers had 'fought violently' for a corps at the 2 April 1940 CWC meeting. Rogers's reasoning, according to King, was that 'he could maintain the stand of two divisions, whereas if this were not done we might have a third division on our hands.' Pickersgill, *The Mackenzie King Record*, I:76; J.L. Granatstein, *Canada's War: The Politics of the Mackenzie King Government, 1939–1945* (Toronto: Oxford University Press, 1975), 93. In October 1939, McNaughton only raised the possibility of a corps. In February 1940, he again mentioned it but did not press it with any vigour. The government declined to address the issue until after the federal election. Granatstein has suggested that McNaughton viewed this as a rebuff and that he took it 'badly,' but there is little evidence to support this assertion. King Diaries, 6 October 1939; Granatstein, *The Generals*, 63.

64 Letters of 2nd Division, 20 April 1940, Victor Odlum Papers (hereafter OP), MG30 E300, vol. 26, LAC. The consensus was that 'a balanced Canadian corps of two divisions and ancillary troops might prove to be the maximum which Canada could maintain by voluntary recruiting in a war of long duration.' In accordance with the Militia Service Plan, Anderson had indicated in his 29 August 1939 memorandum that the army's contribution would be a two-division corps.

65 McNaughton pointed out that there was a considerable deficiency of qualified staff officers on the 'G' and 'A' side. Forming the corps would mean the retention of 25 to 30 per cent British staff officers. Memorandum of a Meeting with the Minister of National Defence, 5 December 1940, MP, vol. 251. McNaughton and Ralston also clashed over the formula for paying for British non-divisional units that would still need to serve with the corps. See Pope, *Soldiers and Politicians*, 155–6. The CWC had approved the establishment of a corps on 17 May.

66 Paul D. Dickson, 'The Politics of Army Expansion: General H.D.G. Crerar and the Creation of First Canadian Army, 1940–1941,' *Journal of Military History* (hereafter *JMH*) 60, no. 2 (1996): 280. On 8 September 1940 the Land Forces Committee recommended that the aim be to equip fifty-five divisions by the end of September 1941. This would break down to thirty-two British divisions, fourteen from the dominions, four from India, and five reserve divisions. Where the reserve divisions were to come from is not clear. The principal obstacle to achieving this goal was equipment. Even if the scales were cut drastically, noted James R.M. Butler, there simply was not enough equipment 'for anything like fifty-five divisions.' Butler, *History of the Second*

World War, Grand Strategy, II: September 1939–June 1941 (London: HMSO, 1957), 31–3.
67 Crerar to McNaughton, 9 September 1940, CP, vol. 1, file 958C.009(D12). In late September, Crerar submitted his Army Programme for 1941 based on a corps of three divisions and an armoured brigade group, with a fourth division to follow in 1941. Crerar also suggested that the public was getting anxious, and he concluded: 'I am wondering whether the next few months will not find you out of the U.K. and in some other area of operations where the Canadian Corps [it was not even formed at this time] can better demonstrate its fighting power. The next few weeks should give the answer.' What Crerar was implying is unclear. Dickson has indicated that Crerar sent this letter on 8 August, but Crerar makes it clear that it was not sent until 9 September.
68 Crerar to McNaughton, 11 August 1941, CP, vol. 1, file 958C.009, 'Pers Corr Lgen McNaughton.'
69 Stacey, *Six Years of War*, 94; *Globe and Mail*, 29 September 1941.
70 Power cabled Ralston in July: 'McNaughton specifically stresses that before embarking on new adventures ... we should be certain that we will be in a position to carry out to the full, even in a long war, the military commitments already undertaken and supply reinforcements already pledged.' Quoted in Swettenham, *McNaughton*, II:160.
71 In late October 1941, Ralston, Crerar, and McNaughton discussed future operations. McNaughton told Ralston that other than raids, no other operations involving Canadians were envisioned. Crerar recorded: 'It was agreed that the detachment of a Canadian formation ... would inevitably result in a lowering of the morale of the troops comprising the balance of the Corps.' CP, vol. 15, WD, 25 October 1941.
72 Dickson, 'The Politics of Army Expansion,' 281.

4. From ROUNDUP to TORCH

1 Charles P. Stacey, *Arms, Men, and Governments: The War Policies of Canada, 1939–1945* (Ottawa: Queen's Printer, 1970), 190. The Combined Chiefs of Staff (hereafter CCOS) was formed exclusively by the senior service chiefs of the United States and Britain. It consisted of General George C. Marshall, Admiral Ernest J. King, Admiral William D. Leahy, and General Henry H. Arnold on the American side; and Brooke, Air Chief Marshal Sir Charles Portal, Admiral of the Fleet Sir Dudley Pound, and Lieutenant-General Hastings Ismay on the British side.

2 On the Canadian inability to penetrate the CCOS, see Stanley W. Dziuban, *U.S. Army in World War II: Special Studies: Military Relations between the United States and Canada, 1939–1945* (Washington: Center of Military History, 1990), 85. According to Charles P. Stacey, King was unwilling to press for greater influence because he 'did not care to risk disturbing the relationship [between himself and Roosevelt and Churchill] ... which he considered so valuable to him politically.' *Canada and the Age of Conflict: A History of Canadian External Policies, II: 1921–1948* (Toronto: Macmillan, 1981), 275, 334. King and Churchill also had completely different backgrounds. In December 1941, Lord Moran, Churchill's private physician, suggested that he 'is not really interested in Mackenzie King. He takes him for granted. I cannot help noticing Winston's indifference to him after the wooing of the President at the White House.' Lord Moran, *Churchill Taken from the Diaries of Lord Moran: The Struggle for Survival 1940–1965* (London: Constable, 1966), 21.
3 McNaughton was content to allow the COS to decide how best to use the Canadians. By contrast, Freyberg had suggested to his prime minister in early January 1940 that 'the committing of the New Zealand Expeditionary Force to a theatre of war should still be decided by the Government of New Zealand, with my emergency powers fully retained.' *Documents Relating to New Zealand's Participation in the Second World War, 1939–45*, vol. II (Wellington: War History Branch, Department of Internal Affairs, 1951), 4.
4 John Swettenham, *McNaughton, II: 1939–1943* (Toronto: Ryerson, 1969), 112, 185. McNaughton stated: 'The relations became very difficult and I thought that I had to be exceedingly careful.' He did not want Canadian units broken up by 'some commander who thought he could run around as he might have done in the days of the Zulu War.' Brooke had not been Churchill's first choice for CIGS. That distinction belonged to Nye.
5 Brooke's biographer, David Fraser, concluded that there was no evidence that he bore any prejudice against McNaughton from the First World War. *Alanbrooke* (London: Collins, 1982), 188n. Brooke made no mention of any such friction in his original diary or in his later notes on his diary. This does not prove, however, that prejudice did not exist.
6 Alanbrooke, 'Notes for My Memoirs,' vol. 1, file 5/2/1, Alanbrooke Papers, Liddell Hart Centre for Military Archives (hereafter LHCMA), King's College, London.
7 John Swettenham, *McNaughton, I: 1887–1939* (Toronto: Ryerson, 1968), 89.
8 File # 1, Interview Transcript, 'First and Second World War: McNaughton 1965,' session 2, tape 7, p. 3, MG31, vol. 3, box E42, LAC. McNaughton stated that 'it's clear that Brook[e] was told that I felt that he'd let me down

completely.' McNaughton investigated, and in his letter to Dill of 21 June he placed the blame on Brooke and continued to inquire as to his culpability. On 6 August, McNaughton spoke with Major-General T.R. Eastwood, Brooke's COS, and asked him who had interfered with the 1st Brigade's movement. Eastwood said that Brooke knew nothing of the order until after the event and that it probably came from Movement Control at the War Office. McNaughton, memorandum of conversation with T.R. Eastwood, 6 August 1940, McNaughton Papers (hereafter MP), vol. 227.

9 Interview Transcript, 'First and Second World War,' session 2, tape 7, p. 3; Swettenham interview with McNaughton, 17 June 1965, in *McNaughton*, II:109–11. The arrangement McNaughton referred to seemed to be that the Canadians would be allowed to concentrate northeast of the port of Brest upon landing. Charles P. Stacey, *Official History of the Canadian Army in the Second World War, I: Six Years of War: The Army in Canada, Britain, and the Pacific* (Ottawa: Queen's Printer, 1957), 279.

10 Interview Transcript, 'First and Second World War,' session 2, tape 7, p. 4. The result was that McNaughton made a special provision in all subsequent orders detailing Canadian forces 'in combination with' the British, making it effective only for as long as he directed. 'This was done,' he stated, 'because of Brooke's attitude.' McNaughton reflected: 'Actually, we would have carried [out] every lawful and proper order without any question but we were not going to have the repetition of the kind of experience particularly through 1917 and 1918.' Brooke made no mention of this incident in his diaries.

11 J.L. Granatstein interview with General W.A.B. Anderson, 21 May 1991, Granatstein interviews, Directorate of History and Heritage (herafter DHH), Department of National Defence; Interview Transcript, 'Andrew McNaughton: Canadian,' as broadcast on 11 September 1966, p. 16, MG31, vol. 3, box E42, LAC.

12 Interview Transcript, 'First and Second World War,' session 3, tape 2, p. 8. Lieutenant-Colonel A.S. Price stated that McNaughton 'couldn't compromise … Black was black and white was always white with Andy.' Transcript, 'Andrew McNaughton: Canadian,' p. 19.

13 Sir Bernard Law Montgomery, *The Memoirs of Field Marshal Montgomery of Alamein* (London: Collins, 1958), 535; Alex Danchev and Daniel Todman, eds., *War Diaries, 1939–1945: Field Marshal Lord Alanbrooke* (London: Weidenfeld and Nicolson, 2001), 248. The loss of his brother, Victor, in 1915 in France was followed ten years later by the death of his first wife, Janey, in an automobile accident. The mental torture of his wife's death was compounded by the fact that he had been driving at the time. When his second wife's health faltered in mid-April 1942 he penned in his diary that it was a 'deep source of anxiety.' Fraser, *Alanbrooke*, 60, 92.

14 Denis Whitaker, *Dieppe: Tragedy to Triumph* (Toronto: McGraw-Hill Ryerson, 1992), 40; James Leasor, *War at the Top: The Experiences of General Sir Leslie Hollis* (London: Michael Joseph, 1959), 11; Fraser, *Alanbrooke*, 202.
15 Forrest C. Pogue interview with Lieutenant-General Walter Bedell Smith, 8 May 1947, 'Pogue Interviews,' United States Army Military History Institute (hereafter USAMHI), Carlisle Barracks, Pennsylvania. Bedell Smith was particularly aroused against Brooke because he complained to Eisenhower about his attitude.
16 Charles P. Stacey, Memorandum on VICTOR Exercise, January 1941, RG 24, vol. 10431, file 210.051, LAC. After the war he wrote in his memoirs that McNaughton had been right about VICTOR. *A Date with History: Memoirs of a Canadian Historian* (Ottawa: Deneau, 1983), 75. However, the episode does reinforce the idea that McNaughton had a tendency to exaggerate such moments. During the crisis over the unemployment relief camps in July 1934, McNaughton had suggested to Bennett that closing the camps could plunge the country into chaos and possibly revolution. James Eayrs, *In Defence of Canada: From the Great War to the Great Depression* (Toronto: University of Toronto Press, 1965), 131.
17 Danchev and Todman, *War Diaries*, 137. Brooke's postwar comments were more cutting. The VICTOR episode represented McNaughton's 'warped outlook concerning the sanctity of this Charter. He had not sufficient strategic vision to realize that under conditions of extreme emergency no Charter could be allowed to impede the employment of troops from a purely strategic role.' 'Notes for My Memoirs,' Alanbrooke Papers, vol. 1, 5/2/1, p. 252, LHCMA.
18 McNaughton to Brooke, 1 February 1941, MP, vol. 252, CC7/3-2, War Orders. Ironically, Brooke would vent his frustration against the War Office in late October 1941 for dispatching the 6th Armoured Division to the Middle East. 'When will the WO learn,' he ranted, 'not to break up formations which it has taken months to build up!' Danchev and Todman, *War Diaries*, 194–5. After the war, Burns specifically challenged the assertion that separated Canadian formations fought less well than they might have had the army remained united. Major-General Eedson L.M. Burns, *Manpower in the Canadian Army, 1939–1945* (Toronto: Clarke, Irwin, 1956), 32.
19 McNaughton to Brooke, 1 February 1941, and Brooke to McNaughton, 5 February 1941, MP, vol. 252, War Orders; Stacey, *A Date with History*, 74–5.
20 Danchev and Todman, *War Diaries*, 214.
21 Ibid., 213.
22 Ibid.
23 Quoted in Lord Ismay, *The Memoirs of General the Lord Ismay* (London: Heinemann, 1960), 247.
24 Memorandum [meeting between McNaughton, Paget, and Crerar] 25 De-

cember [27 December] 1941, Crerar Papers (hereafter CP), MG30 E157, vol. 1, PA 3-6 (C), file 958C.009(D18). When Ralston heard of this he cabled McNaughton: 'This involves a somewhat imposing expansion in overhead and did not understand that it had been advocated by you.' Quoted in Stacey, *Six Years of War*, 96. David A. Wilson has suggested that Crerar intentionally minimized McNaughton's involvement in formulating the 1942-3 Army Programme. 'Close and Continuous Attention: Human Resources Management in Canada During the Second World War', PhD diss., University of New Brunswick, 1997, 148. Indeed, Crerar carried on numerous talks with the British on the issue. On 10 October 1941 he met with the ACIGS, Major-General Macready, who indicated that he favoured establishing a Corps HQ. CP, vol. 15, War Diary (hereafter WD).

25 Crerar, Department of National Defence, Army, The Minister, Army Programme 1942-3, 18 November 1941, CP, vol. 1, file 958C.009(D1).
26 Memorandum, 14 January [16 January] 1942, CP, vol. 1, file 958C.009(D18); Danchev and Todman, *War Diaries*, 220.
27 Visit of Lieutenant-General A.G.L. McNaughton to Washington, [11] March 1942, MP, vol. 248, PA 3-6, WD.
28 For the spring 1943 cross-Channel invasion, the American planners estimated that they would have 30 divisions and 3,250 combat aircraft in Britain. Larry I. Bland, ed., *The Papers of George Catlett Marshall, III: 'The Right Man for the Job', December 7, 1941–May 31, 1943* (Baltimore: Johns Hopkins University Press, 1991), 158; Charles P. Stacey, *Official History of the Canadian Army in the Second World War, III: The Victory Campaign: The Operations in Northwest Europe, 1944–1945* (Ottawa: Queen's Printer, 1960), 4.
29 Danchev and Todman, *War Diaries*, 248.
30 Dominick Graham and Shelford Bidwell, *Coalitions, Politicians, and Generals: Some Aspects of Command in Two World Wars* (London: Brassey's, 1993), 158-9. Brooke indicated that seven infantry divisions and two armoured divisions would take part. Gordon A. Harrison, *Cross-Channel Attack* (Washington: Office of the Chief of Military History, 1989), 17. King had voiced his concerns about Canadian participation in SLEDGEHAMMER to Roosevelt as early as April. See Stacey, *Six Years of War*, 320-1. American strategic thinking regarding SLEDGEHAMMER was questionable. See Leasor, *War at the Top*, 184-7.
31 Winston S. Churchill, *The Grand Alliance* (London: Cassell, 1950), 660; Dwight D. Eisenhower, *Crusade in Europe* (New York: Doubleday, 1948), 70-1. Arguments in favour of a cross-Channel invasion in 1943 are few. See John Grigg, *1943: The Victory That Never Was* (London: Eyre Methuen, 1980), 3, 73.

32 Winston S. Churchill, *The Hinge of Fate* (London: Cassell, 1951), 434. Ismay felt that the future misunderstanding would have been alleviated had the British 'expressed their views more frankly' in the beginning. *The Memoirs of General the Lord Ismay*, 249–50.
33 Alan F. Wilt, *War From the Top: German and British Military Decision Making during World War II* (Bloomington: Indiana University Press, 1990), 256.
34 Joseph L. Strange, 'The British Rejection of Operation SLEDGEHAMMER, an Alternative Motive,' *Military Affairs* 46, no. 1 (1982): 6–14.
35 Churchill told Dill that if Roosevelt decided against GYMNAST both countries would 'remain motionless in 1942 and all will be concentrated on Roundup in 1943.' Quoted in Trumball Higgins, *Soft Underbelly: The Anglo-American Controversy over the Italian Campaign, 1939–1944* (New York: Collier-Macmillan, 1968), 32.
36 Bland, *The Papers of George Catlett Marshall*, III:279. For an analysis of the encircling, defensive strategy devised at ARCADIA, see Richard M. Leighton, 'OVERLORD Revisited: An Interpretation of American Strategy in the European War, 1942–1944,' *American Historical Review* 68, no. 4 (1963): 926–7.
37 On 20 February 1942 the Army Council issued the directive for the establishment of the Expeditionary Force Planning Staff. The Combined Commanders consisted of the C-in-C Dover (Navy), the C-in-C Fighter Command, a representative from the United States, and apparently McNaughton representing the Canadian Army. The committee's first meeting had already taken place on 15 May. On 22 June McNaughton even cabled Stuart that for the first time he would be able to stay ahead of operational planning. Yet McNaughton was never invited to play an official role in the Combined Commanders. Stacey, *Six Years of War*, 321–2; McNaughton to Ralston, 23 September 1944, quoted in Swettenham, *McNaughton*, II:246.
38 Gerald W.L. Nicholson, *Official History of the Canadian Army in the Second World War, II: The Canadians in Italy, 1943–1945* (Ottawa: Queen's Printer, 1957), 24. When Brooke asked him what he knew of TORCH, McNaughton replied: 'I had never been told anything about it formally.' Memorandum [meeting between McNaughton, Paget, and Brooke] 17 October [19 October] 1942, MP, vol. 135, WD, PA 1-11-1.
39 Stacey, *Six Years of War*, 323. On 5 October 1942, while in Britain, Churchill had apparently told him that he understood that McNaughton 'did not fancy taking troops out for Torch' – a statement that Brooke confirmed to Ralston the next day. James L. Ralston, 'English Trip 1942,' Lunch with Churchill, 5 October 1942, Ralston Papers (hereafter RP), MG27 III BII, vol. 41, LAC.
40 Ralston, 'Overseas Trips in 1943,' RP, vol. 59, entry for 30 November 1943.

41 Churchill declared that 'every American division which crossed the Atlantic gave us freedom to send one of our matured British divisions out of the country to the Middle East.' *The Grand Alliance*, 684.
42 Brian Horrocks, *A Full Life* (London: Collins, 1960), 103. Lieutenant-General Kenneth Anderson's First Army expanded from Force 110, originally trained for amphibious operations. Gregory Blaxland, *The Plain Cook and the Great Showman: The First and Eighth Armies in North Africa* (London: Purnell Book Services, 1977), 39; David Fraser, *And We Shall Shock Them: The British Army in the Second World War* (London: Sceptre, 1988), 213.
43 Harrison, *Cross-Channel Attack*, 32.
44 Ibid., 27.
45 Memorandum of a Discussion by Lt.-Gen. McNaughton with C.-in-C., Home Forces, on 8 July [11 July] 1942, MP, vol. 248, WD, PA 3-6, Appendix 'H'; Memorandum, Discussion with VCIGS, War Office, on 3 August [4 August] 1942, MP, vol. 249, WD, PA 5-3-8; Historical Report no. 182, Historical Section CMHQ, 'The Strategic Role of First Canadian Army, 1942–1944,' DHH. ROUNDUP was not taken very seriously in its initial stages. Major-General J.A. Sinclair, who served originally with the Combined Commanders and then with Home Forces, stated: 'Clearly the commanders who put up ROUNDUP weren't very keen on it. Thought it a poor show.' Hughes-Hallett called it 'idiotic ... Just played around with it.' Forrest C. Pogue interview with Major-General J.A. Sinclair, 21 February 1947, and Hughes-Hallett, 12 February 1947, 'Pogue Interviews.' ROUNDUP generated all kinds of offshoots, including HADRIAN, a plan to take and hold the Cherbourg Peninsula.
46 Burns Diary, 25 February 1942, Burns Papers (hereafter BP), MG31 G6, vol. 5, LAC; Danchev and Todman, *War Diaries*, 227–8.
47 For the expansion of McNaughton's raiding authority, see Stacey, *Arms, Men, and Governments*, 209. Dill told McNaughton that the Canadian performance in GAUNTLET 'calls for nothing but praise.' McNaughton also authorized I Canadian Corps to take over the planning for two additional raids, BARITONE and CRUMPER. Neither plan, however, was executed. In April 1942, McNaughton authorized Crerar to undertake ABERCROMBIE, a raid on Hardelot. See Stacey, *Six Years of War*, 306–7.
48 On Crerar's machinations, see Paul D. Dickson, 'The Limits of Professionalism: General H.D.G. Crerar and the Canadian Army, 1914–1944,' PhD diss., University of Guelph, 1993, 406–07. McNaughton was not informed by Montgomery until 30 April, well after Crerar had solicited participation behind McNaughton's back. As commander of South East Command, Montgomery was the responsible army officer for RUTTER. According

to Mountbatten, Brooke 'spoke to me personally about the desire of the Canadians to be brought into a raiding operation as soon as possible.' Lieutenant-General G.G. Simonds to Lord Louis Mountbatten, 22 January 1969, and Mountbatten to Simonds, 4 February 1969, Admiral John Hughes-Hallett Papers, MG30 E463, LAC.

49 Tim Cook, *Clio's Warriors: Canadian Historians and the Writing of the World Wars* (Vancouver: UBC Press, 2006), 100; Quentin Reynolds, *Dress Rehearsal: The Story of Dieppe* (New York: Random House, 1943).

50 McNaughton War Diary, 14 July 1942, MP, vol. 248; Stacey, *Six Years of War*, 341.

51 Memorandum, Operation JUBILEE, 16 July [25 July] 1942, MP, vol. 248, WD, file 8-3-5/Ops, Appendix 'O.'

52 Swettenham, *McNaughton*, II:202–3; Stacey, *Six Years of War*, 317. According to General Albert C. Wedemeyer, Marshall 'continued to advance the idea of SLEDGEHAMMER simply to restrain wild and diversionary efforts' proposed by Churchill. *Wedemeyer Reports!* (New York: Henry Holt, 1958), 135.

53 Harrison, *Cross-Channel Attack*, 15–16.

54 McNaughton, Memorandum 'Operation JUBILEE,' 25 July 1942, MP, vol. 248, WD, file 8-3-5/Ops, Appendix 'K.'

55 Marshall had told the British COS on 24 July that the United States intended to withdraw fifteen groups of various types of aircraft from BOLERO for use in the Pacific specifically because of the adoption of a defensive posture in Europe. Bland, *The Papers of George Catlett Marshall*, III:279.

56 Harrison, *Cross-Channel Attack*, 28–9.

57 Sources supporting this position include Fraser, *Alanbrooke*, 294; Lieutenant-General Sir James Wilson, 'Dieppe: Vindication,' *AQDJ* 124, no. 1 (1994): 69; Whitaker, *Dieppe*, 57, 284–5; Nigel Hamilton, *Monty: The Making of a General, 1887–1942* (London: Hamish Hamilton, 1981), 556; Philip Ziegler, *Mountbatten: A Biography* (New York: Alfred A. Knopf, 1985), 183; Richard Lamb, *Churchill as War Leader* (New York: Carroll and Graf, 1991), 170–1; and Brian Loring Villa, *Unauthorized Action: Mountbatten and the Dieppe Raid* (Oxford: Oxford University Press, 1989), 74–94. The idea of JUBILEE as a psychological substitute can be found in Helmut Heiber and David M. Glantz, eds., *Hitler and His Generals* (New York: Enigma, 2002), 925n866.

58 Sholto Douglas, *Years of Command* (London: Collins, 1966), 173.

59 Brooke diary entries for 10, 17, and 28 March 1942; Danchev and Todman, *War Diaries*, 238; Sir John N. Kennedy, *The Business of War: The War Narrative of Major-General Sir John Noble Kennedy* (London: Hutchinson, 1957), 223; Alfred D. Chandler, Jr, *The Papers of Dwight D. Eisenhower: The War Years* (Baltimore: Johns Hopkins University Press, 1970), I:370; Stacey, *Six Years of*

War, 388. When Churchill later expressed some anxiety a month after the raid that some thought it had been launched for this purpose, McNaughton said, 'This was in fact not the case.' Memorandum of Conversation between McNaughton and Churchill, 19 September 1942 [22 September], MP, vol. 248.

60 William G.F. Jackson, *Overlord: Normandy 1944* (London: Davis-Poynter, 1978), 60.
61 Kennedy, *The Business of War*, 264.
62 See Richard W. Steele, *The First Offensive 1942: Roosevelt, Marshall, and the Making of American Strategy* (Bloomington: Indiana University Press, 1973), 146. The idea that the raid was remounted to gain experience is clearly expressed in Bernard Fergusson, *The Watery Maze: The Story of Combined Operations* (London: Collins, 1961), 168.
63 Anthony Cave Brown, *Bodyguard of Lies* (New York: Harper and Row, 1975), 81.
64 John W. Pickersgill, ed., *The Mackenzie King Record, I: 1939–1944* (Toronto: University of Toronto Press, 1960), 417. There has been a long-standing debate about whether British deception operations such as OVERTHROW intersected with the Dieppe Raid. For a good discussion, see Nigel West, *Unreliable Witness: Espionage Myths of the Second World War* (London: Weidenfeld and Nicolson, 1984), 85–98.
65 For the background to the convoy problems, see Captain Stephen W. Roskill, *Official History of the Second World War, The War at Sea, Volume II: The Period of Balance* (London: HMSO, 1956), 136.
66 Churchill, *The Hinge of Fate*, 436. Churchill presented his initial memorandum on the operation to the COS on 2 June 1942.
67 McNaughton later told Churchill that his part in JUPITER 'had caused some acidity in my relations with the Chiefs of Staff.' Memorandum [meeting between McNaughton and Churchill], 19 September [22 September] 1942, MP, vol. 248, WD, PA 1-7-1. However, McNaughton's War Diary entry for 9 July stated: 'It was clearly evident that no bitterness whatever existed in the minds of the Chiefs of Staff on account of an outsider being called into consultation on this subject.'
68 Danchev and Todman, *War Diaries*, 278. Memorandum of Lt.-Gen. McNaughton's Visit with the Prime Minister on 12 July [15 July] 1942, MP, vol. 248, WD, Appendix 'I.' On 12 July, Churchill presented the situation to McNaughton at Chequers, describing the predicament of the Russian convoys, and said that studying JUPITER in no way committed the Canadians to attacking Norway. Churchill had been agitating for a Norway offensive for some time and refused to give up JUPITER easily. See Gerald Pawle, *The War and Colonel Warden: Based on the Recollections of Commander C.R. Thompson,*

Personal Assistant to the Prime Minister, 1940–45 (London: Harrap, 1963), 167, 187, 291.

69 On 19 September Churchill confronted McNaughton about his appreciation. On 22 September, Churchill asked King for permission to send McNaughton to Moscow. King declined two days later, adding: 'I need not assure you that the views expressed ... should not be construed as in any way modifying our fixed policy that Canadian forces are to be available to be used wherever they can best serve the common cause.' Historical Report no. 182, 68.

70 Alanbrooke, 'Notes for My Memoirs,' 5/2/18, 536. He criticized McNaughton 'very forcibly' for not having been more emphatic with Churchill regarding JUPITER. Brooke suggested that JUPITER was so nonsensical that if it had been carried out he would have resigned. See Memorandum [conversation between McNaughton, Paget, and Brooke], 17 October 1942, MP, vol. 249, WD, PA 1-11-1. There is an entry for 21 September 1942 in Brooke's diary indicating that McNaughton visited him that day. Arthur Bryant, ed., *Turn of the Tide, 1939–1943* (London: Collins, 1957), 501–2. Swettenham argued that Brooke fabricated it, for McNaughton's War Diary clearly shows he was not even in London on 21 September. *McNaughton*, II:259. Danchev and Todman cited the same quote but incorrectly placed it after an entry for 16 October 1941. The text is italicized, indicating it was Brooke's post-war comments on his original diary entries. *War Diaries*, 191.

71 Churchill to Roosevelt, 27 July 1942, in Francis L. Loewenheim, Harold D. Langley, and Manfred Jonas, eds., *Roosevelt and Churchill: Their Secret Wartime Correspondence* (New York: Saturday Review Press/Dutton, 1975), 226. On British deception efforts at this time, see Donal J. Sexton, 'Phantoms of the North: British Deceptions in Scandinavia, 1941–1944,' *Military Affairs* 47, no. 3 (1983): 109–14.

72 Memorandum, 2 September [3 September] 1942, MP, vol. 248, WD, PA 3-6-1, Appendix 'B.' McNaughton believed that OVERTHROW included 'planning for the invasion of the Continent with a force of about 7 Divs. This is the part of the larger project "ROUNDUP" on which studies are to be directed for the time being.' CAVENDISH was an exercise with troops rehearsing the part of OVERTHROW from D-10 to D-1 with D-10 being 5 October 42. For a description of OVERTHROW and its role in the larger TORCH operation, see Michael Howard, *British Intelligence in the Second World War, V: Strategic Deception* (London: HMSO, 1990), 56–8.

73 Memorandum [meeting between McNaughton and Brooke], 17 September [20 September] 1942, MP, vol. 135, WD, PA 1-7-1. McNaughton had only discovered the change through informal conversation with Paget.

McNaughton's meeting with Churchill at Chequers on 19 September had also started off in confusion, for when Churchill suggested that the army was busy 'preparing for Arctic War,' McNaughton replied that 'this was *not* so' and that the army was still focused on ROUNDUP. Churchill 'evidenced some surprise but the matter was not pursued.'
74 Memorandum, 17 September 1942.
75 Memorandum, Minutes of a Meeting held in General McNaughton's office at 1720 hrs, 3 October [4 October] 1942, MP, vol. 249, WD, PA 5-3-1, Appendix 'B1.' McNaughton indicated that SLEDGEHAMMER envisioned employing six divisions, including two Canadian, against the Pas de Calais, but the plans were now 'dormant.' McNaughton told Ralston that the time had now come to request greater say in the future planning of the war; however, he urged a sympathetic approach because of the 'many difficulties' under which the War Office was labouring. Days later, Ralston and Stuart visited Sir James Grigg, Secretary of State for War. Stuart noted that he was 'very appreciative of the way in which the Canadian Army had accepted its allotted role in the defence of the U.K., and it appeared to the C.G.S. that they were well aware of the many difficulties which had been faced and were yet to be faced by the Canadian Army during this period of watching and waiting.' Memorandum, Discussion of Army Comd with Lt.-Gen. Stuart (CGS) at CMHQ, 1600 hrs, 15 October [16 October] 1942, MP, vol. 249, PA 5-3-1, Appendix 'E.'

5. Practical Operations of War

1 Alex Danchev and Daniel Todman, eds., *War Diaries, 1939–1945: Field Marshal Lord Alanbrooke* (London: Weidenfeld and Nicolson 2001), 327, 329.
2 Memorandum, [meeting between McNaughton and Stuart], 17 October [18 October] 1942, McNaughton Papers (hereafter MP), MG30 E133, vol. 249, War Diary (hereafter WD), PA 1-11-1, LAC.
3 Occupying the Canaries was not a new idea. Throughout 1940–1 the War Office had reserved an expeditionary element to execute PILGRIM, as it was then called.
4 Memorandum [conversation between McNaughton, Paget, and Brooke], 17 October [19 October] 1942, MP, vol. 249, WD, PA 1-11-1. During the initial portion of the meeting Paget went to great lengths to point out that they were taking 'very serious risks' regarding the defence of Britain. After TORCH there would only be twenty-five British and Canadian divisions whereas thirty were required. McNaughton noted: 'I agreed with all he had said and reiterated my opinion that it was necessary to maintain a strong

force in the British Isles.' Brooke, however, thought that the invasion threat was 'not serious at present' and that they would have at least three months forewarning.

5 Memorandum [meeting between McNaughton and Major-General Watson et al.], 8 January [16 January] 1942, Crerar Papers (hereafter CP), MG30 E157, vol. 1, file 958C.009(D18), PA 5-3-2, LAC.
6 Memorandum of conversation, McNaughton – Paget – Brooke, 17 October 1942; Brian P. Farrell, 'Yes, Prime Minister: Barbarossa, Whipcord, and the Basis of British Grand Strategy, Autumn 1941,' *Journal of Military History* (hereafter *JMH*) 57, no. 4 (1993): 614.
7 McNaughton to Brooke, 23 October 1942, MP, vol. 249, WD, PA 1-11-1. McNaughton had issued Crerar the planning directive for the operation on 4 November. McNaughton to Crerar, 'TONIC' Directive, 4 November 1942, MP, vol. 249, WD, PA 1-11-1. Brooke replied: 'I note with pleasure that you have nominated' Crerar as commander. Brooke to McNaughton, 29 October 1942, MP, vol. 136.
8 McNaughton to Ralston, 23 September 1944, quoted in John Swettenham, *McNaughton, II: 1939–1943* (Toronto: Ryerson, 1969): 246.
9 Memorandum of Conversation, McNaughton – Paget – Brooke, 17 October 1942.
10 Handwritten draft of letter, Crerar to McNaughton, 19 October 1942, CP, vol. 2, file 958.009(D21); MP, vol. 240. McNaughton's anger at Crerar was generated when he saw a transcript of a conversation between Churchill and Ralston in which Ralston advocated the employment of Canadians at the earliest opportunity. Two days later McNaughton recanted somewhat. McNaughton to Crerar, 21 October 1942, MP, vol. 240; Dominick Graham, *The Price of Command: A Biography of General Guy Simonds* (Toronto: Stoddart, 1993), 60.
11 J.L. Granatstein, *The Generals: The Canadian Army's Senior Commanders in the Second World War* (Toronto: Stoddart, 1993), 75; Paul D. Dickson, 'The Limits of Professionalism: General H.D.G. Crerar and the Canadian Army, 1914–1944,' PhD diss., University of Guelph, 1993, 357, 433.
12 Paul D. Dickson, 'The Hand That Wields the Dagger: Harry Crerar, First Canadian Army, and National Autonomy,' *War and Society* 13, no. 2 (1995): 124.
13 Memorandum [meeting between McNaughton and Brooke], 19 November [20 November] 1942, MP, vol. 249, PA 5-0-3, Appendix 'P.'
14 Ibid. On 30 October, Brooke noted in his diary: 'We finally finished off our policy for the conduct of the war in 1943 at this morning's COS.' Danchev and Todman, *War Diaries*, 337.

15 McNaughton to Stuart [To Defensor from Canmilitary], 21 November 1942, MP, vol. 249, WD. McNaughton added that the War Office 'have no present thought' of employing the army or formations elsewhere.
16 Paget to McNaughton, 11 December 1942, MP, vol. 249.
17 Danchev and Todman, *War Diaries*, 346, 350. The operation was to be commanded by Lieutenant-General Frederick E. Morgan. A few days later Brooke let McNaughton know that he would update the proposal in about two weeks. Memorandum of conversation, McNaughton and Brooke, 6 January 1943, MP, vol. 249.
18 Memorandum of conversation, General McNaughton – General Brooke, CIGS, War Office, held at the War Office at 1700 hrs, 31 December 1942 [4 January 1943], MP, vol. 249, WD, Appendix 'O.' Bedell Smith had met with Brooke to discuss Sardinia on 13 November. Danchev and Todman, *War Diaries*, 341. Brooke also promised to ensure that McNaughton was better informed about operations. McNaughton let Hollis know that he needed copies of all relevant papers for any proposed operation 'as expeditiously as possible as otherwise I find myself at a grave disadvantage.' McNaughton to Hollis, 2 January 1943, MP, vol. 136.
19 McNaughton to Stuart, 2 January 1943, MP, vol. 249.
20 Minutes of a Conference held at H.Q. First Cdn Army at 1130 hrs, 10 January 1943, MP, vol. 249, PA 5-0-3, Appendix 'M.' McNaughton indicated to the press that morale was good because the Canadians realized they were 'serving a cause greater than themselves.' McNaughton press conference, 23 January 1943, MP, vol. 249.
21 John W. Pickersgill, ed., *The Mackenzie King Record, I: 1939–1944* (Toronto: University of Toronto Press, 1960), 502.
22 Memorandum of conversation by Lt.-Gen. A.G.L. McNaughton with CIGS and Lt.-Gen. K. Stuart CGS NDHQ, 9 February [10 February] 1943, MP, vol. 131, WD, PA 1-0-4. Brooke mentioned Sicily as part of the 'Southern Task Force.'
23 Danchev and Todman, *War Diaries*, 381.
24 Alanbrooke, 'Notes for My Memoirs,' 5/2/20, Alanbrooke Papers (hereafter AlP), Liddell Hart Centre for Military Archives (hereafter LHCMA), King's College, London.
25 Memorandum of conversation Lt-Gen McNaughton – Gen Sir Bernard Paget, C-in-C Home Forces, 10 February 1943, MP, vol. 249, WD, PA 1-0-2.
26 Charles P. Stacey, *Official History the Canadian Army in the Second World War, III: The Victory Campaign: The Operations in Northwest Europe, 1944–1945* (Ottawa: Queen's Printer, 1960), 29. Reinforcements had already become an issue, and McNaughton told Stuart that he would 'keep out of the order of

battle sufficient formations to ensure that the reinforcements then available were adequate for ... the forces actually committed to battle.' Memorandum of conversation, McNaughton and Stuart, 12 February [15 February] 1943, MP, vol. 249.

27 Pickersgill, *The Mackenzie King Record*, I:496. It was proposed to reduce Canada's shipping allotment by some 37,000 soldiers to expedite the transport of American air crews for POINTBLANK, the bombing of Germany. On 6 April, Ismay expressed the chiefs' 'profound apologies' for the 'awkward and unsatisfactory position' as a result of the decision to prioritize American air crews. Ismay to McNaughton, 6 April 1943, MP, vol. 132.

28 Desmond Morton, *A Military History of Canada* (Edmonton: Hurtig, 1990), 188–90. J.L. Granatstein pointed out that the anti-King sentiment reached its highest point between November 1941 and February 1942 in the wake of the devastating series of Allied reverses. *The Politics of Survival: The Conservative Party of Canada, 1939–1945* (Toronto: University of Toronto Press, 1967), 82.

29 Pickersgill, *The Mackenzie King Record*, I:440. Indeed, King was not happy with the existing Bureau of Public Information in the Department of National War Services. For an excellent description of what Canada had achieved, see *Debates of the Senate of the Dominion of Canada, 1943–1944, Official Report* (Ottawa: Edmund Cloutier, 1944), 1213.

30 Quoted in Marc Milner, 'The Royal Canadian Navy and 1943: A Year Best Forgotten?' in *1943: The Beginning of the End*, ed. Paul D. Dickson (Waterloo: Wilfrid Laurier University, 1995), 126.

31 W.A.B. Douglas and Brereton Greenhous, *Out of the Shadows: Canada and the Second World War*, rev. ed. (Toronto: Dundurn, 1995), 131.

32 Quoted in Gerald W.L. Nicholson, *Official History of the Canadian Army in the Second World War, II: The Canadians in Italy, 1943–1945* (Ottawa: Queen's Printer, 1957), 24; Stuart to McNaughton, 18 March 1943, MP, vol. 249.

33 Pickersgill, *The Mackenzie King Record*, I:500. Based on what Ralston told him after the war, Richard Malone concluded that the Defence Minister was the driving force behind the deployment. He was 'haunted by the risk of having the entire Canadian Army launched across the beaches of Normandy without any real battle experience.' *A Portrait of War 1939–1945* (Toronto: Collins, 1983), 144.

34 McNaughton to Stuart, 20 March 1943, MP, vol. 132. Stuart did not agree entirely. 'Clause 3 if it were laid down as an exclusive rule of procedure,' he stated, 'might be construed too rigidly. I think it ought to be clear that it would not exclude information being given in advance to the CIGS regarding any factors which would help him in deciding as to how Canadians

might best be employed.' On 16 March, McNaughton told Nye that he needed 'timely and adequate information.' Stuart was also frustrated with the lack of information, declaring that 'the British staff in Washington appear to know more about projected plans than I do.' Stuart to McNaughton, 31 March 1943, MP, vol. 132. McNaughton subsequently explained that he never intended to abdicate the responsibility of making suggestions. McNaughton to Stuart, 5 April 1943, MP, vol. 132, WD, PA 1-3-2. The problem, as McNaughton saw it, was that the system 'under which I have relied on C.I.G.S. personally to keep me sufficiently informed' had broken down. He also said that neither he nor the army 'relish prolonged waiting for action.'

35 Pickersgill, *The Mackenzie King Record*, I:498. The friction between Ralston and McNaughton had only increased. After the Hong Kong disaster McNaughton told reporters off the record in Canada that Ralston was 'completely unfitted for his job' and was simply 'unable to distinguish between policy and detail.' Quoted in Frederick W. Gibson and Barbara Robertson, eds., *Ottawa at War: The Grant Dexter Memorandum* (Winnipeg: Manitoba Record Society, 1994), 284ff.

36 Pickersgill, *The Mackenzie King Record*, I:498, 500. King added: 'My mind keeps going back to McNaughton's words to me that he did not intend to allow life to be unnecessarily sacrificed.'

37 Memorandum, Conversation General McNaughton with General Sir Alan Brooke, C.I.G.S., War Office, at 1530 hrs, 5 April [8 April] 1943, MP, vol. 132, WD, PA 1-0-4., Appendix 'I.' Brooke apologized for the breakdown in communications between them and suggested that McNaughton attend all COS meetings. McNaughton demurred and suggested a liaison officer.

38 Memorandum of a Discussion, General McNaughton – General Sir Alan Brooke (CIGS, War Office), at 1700 hrs, 23 April [28 April] 1943, MP, vol. 132, WD, PA 1-14-1.

39 Ibid. McNaughton would, however, give the divisional commander, Major-General Guy Simonds, the 'right of reference' in 'extreme cases.' Memorandum of Conversation, General McNaughton–Maj-General G.G. Simonds, Comd 1 Cdn Div at 0800 hrs, 30 April 1943, MP, vol. 249, WD, Appendix 'NNN.'

40 Alanbrooke, 'Notes for My Memoirs,' diary entry for 23 April 1943.

41 Memorandum of a Discussion, McNaughton – Brooke, 23 April 1943. Brooke said he would write a personal letter to the GOC of the 3rd British Division 'expressing his regret in having to make this last minute change to give way to Cdn requests for active operations.' McNaughton had already identified 1st Canadian Infantry Division as the assault division; the 2nd would be given refresher training in combined operations. Memorandum

of Conversation, General McNaughton–General Paget, C-in-C Home Forces, 1000 hrs, 12 April [15 April] 1943, MP, vol. 131, WD, PA 4-4, Appendix 'T.'
42 Memorandum of Conversation, General McNaughton – Brig. Simpson, War Office, 6 January [7 January] 1943, MP, vol. 249, WD.
43 The most recent restatement of this theme comes from Douglas Delaney. He has suggested that McNaughton 'maintained an almost uncompromising "hands-off" policy concerning British input on the employment of Canadian formations.' 'When Harry Met Monty: Canadian National Politics and the Crerar – Montgomery Relationship,' in Bernd Horn, ed., *The Canadian Way of War: Serving the National Interest* (Toronto: Dundurn, 2006), 215–16.
44 Crerar told Montgomery it would be 'as fit as ever for operations in another couple of months of strenuous training.' Crerar to Montgomery, 27 September 1942, CP, MG30 E157, vol. 7, file 958C.009(D172), LAC.
45 Alanbrooke, 'Notes for My Memoirs,' 5/2/18, 548.
46 Memorandum of a Discussion, Lt.-Gen. McNaughton – Lt.-Gen. Ismay, at the War Cabinet Offices, 1230 hrs, 27 April [29 April] 1943, MP, vol. 249, WD, PA 1-14-1, Appendix 'GGG.' Ismay suggested that McNaughton could visit the Secret Intelligence Centre. 'The Chiefs of Staff consider,' he stated, 'that as a special, indeed unique, case you should be given access to this Centre, if you would care to have it.' Because of Ismay's illness, McNaughton told Brooke that he had not followed up on his letter suggesting closer liaison with the COS. Memorandum of a Discussion, McNaughton–Brooke, 23 April 1943.

6. The Difficulty of Training in 1940

1 Field Marshal Lord Alanbrooke, 'Notes for my Memoirs', Alanbrooke Papers (hereafter AlP), files 5/2/16 and 5/2/20, Liddell Hart Centre for Military Archives (hereafter LHCMA), King's College, London.
2 Charles P. Stacey, *Official History of the Canadian Army in the Second World War, I: Six Years of War: The Army in Canada, Britain, and the Pacific* (Ottawa: Queen's Printer, 1957), 253.
3 John A. English, *The Canadian Army and the Normandy Campaign: A Study of Failure in High Command* (New York: Praeger, 1991), xiv, 136, 310–11; idem, *Lament for an Army: The Decline of Canadian Military Professionalism* (Toronto: Irwin, 1998), 32; Douglas E. Delaney, 'Looking Back on Canadian Generalship in the Second World War,' *Canadian Army Journal* (hereafter *CAJ*) 7, no. 1 (2004): 18.

4 Stacey, *Six Years of War*, 230.
5 William E.J. Hutchinson, 'Test of a Corps Commander: Lieutenant-General Guy Granville Simonds in Normandy, 1944,' MA thesis, University of Victoria, 1982, 91.
6 Comments on draft of Stacey's *Six Years of War*, file 82/983, folder 1, folio 142, Directorate of History and Heritage, Department of National Defence (hereafter DHH), Ottawa.
7 Major Charles P. Stacey, Historical Officer's Report no. 46, 'Situation of the Canadian Forces in the United Kingdom, Summer 1941, V: The Problem of Equipment, 19 September 1941,' DHH; Gerald W.L. Nicholson, *The Gunners of Canada: The History of the Royal Regiment of Canadian Artillery, II: 1919–1967* (Toronto: McClelland and Stewart, 1967), 100–1. The housing of Canadian troops was another major problem that affected training. See Stacey, Historical Officer's Report no. 38, 'Situation of the Canadian Forces in the United Kingdom, Summer 1941, III: The Problem of Accommodation,' DHH.
8 Alex Danchev and Daniel Todman, eds., *War Diaries, 1939–1945: Field Marshal Lord Alanbrooke* (London: Weidenfeld and Nicolson, 2001), 90. Canadian-made vehicles began arriving in May 1940, and by 1 June the division had taken over on an emergency basis some seven hundred vehicles of all types from British stocks. Nevertheless, well into the summer the principal method of towing guns and moving men and supplies was civilian vans with civilian drivers. Nicholson, *The Gunners of Canada*, II:56.
9 McNaughton made this observation to Major-General Price J. Montague at CMHQ on 20 August 1940. At a meeting ten days later with Major-General Lawrence Carr, the ACIGS, his apprehension was confirmed. Carr explained that to meet urgent requirements in the Middle East it had been necessary to drastically curtail equipment issued to Home Force units.
10 Memorandum, 31 August 1940, McNaughton Papers (hereafter MP), MG30 E133, vol. 251, file 'Memorandum BGS Vols 1 & 2,' LAC.
11 Quoted in J.L. Granatstein and Norman Hitsman, *Broken Promises: A History of Conscription in Canada* (Toronto: University of Toronto Press, 1977), 148–9; Paul D. Dickson, *A Thoroughly Canadian General: A Biography of General H.D.G. Crerar* (Toronto: University of Toronto Press, 2007), 144.
12 Interview Transcript, 'Andrew McNaughton – Canadian,' as broadcast on 11 September 1966, 14, MG31, vol. 3, box E42, LAC. In his short biography of Simonds, J.S. McMahon suggested that in the face of the severe equipment shortages in 1940, Simonds could actually do little beyond assisting and encouraging COs in basic training. *Professional Soldier: Guy Simonds – A Memoir* (Winnipeg: McMahon Investments, 1985), 26–7.
13 Interview Transcript, 'First and Second World Wars: McNaughton 1965,' session 1, tape 4, p. 12, MG 31, vol. 3, box E42, file 1, LAC.

14 Kent Roberts Greenfield, Robert B. Palmer, and Bell I. Wiley, *The Army Ground Forces: The Organization of Ground Combat Troops* (Washington: Historical Division, Department of the Army, 1947), 11.
15 Marshall to Morris Sheppard, 5 June 1940, in Larry I. Bland, ed., *The Papers of George Catlett Marshall, II: 'We Cannot Delay,' July 1, 1939–December 6, 1941* (Baltimore: Johns Hopkins University Press, 1986), 236.
16 McNaughton to Crerar, 8 August 1940, MP, vol. 251, file 'Memorandum of BGS Vols 1 & 2.' Granatstein made this point abundantly clear in *The Generals: The Canadian Army's Senior Commanders in the Second World War* (Toronto: Stoddart, 1993), 30–1. For the difficulties experienced by the U.S. Army in finding suitable commanders from the National Guard, see Peter R. Mansoor, *The GI Offensive in Europe: The Triumph of American Infantry Divisions, 1941–1945* (Lawrence: University Press of Kansas, 1999), 19.
17 Lieutenant-General Howard Graham, *Citizen and Soldier: The Memoirs of Lieutenant-General Howard Graham* (Toronto: McClelland and Stewart, 1987), 113. Graham gave a concrete example to Reginald Roy. When Pearkes asked him to be General Staff Officer (GSO) 1 during the temporary absence of Chris Vokes, he replied: 'Well, I'd be delighted, I don't know much about the job,' even though he had attended the Militia Staff Course and the Advanced Militia Staff Course in 1937. Interview with Lt.-Gen. H.D. Graham, 24 September 1970, p. 4, PP, box 25.
18 General B.M. Hoffmeister, interview with Brereton Greenhous and William J. McAndrew, 1980, p. 12, William J. McAndrew Collection, Royal Military College of Canada (hereafter RMC).
19 George Kitching, *Mud and Green Fields: The Memoirs of Major-General George Kitching* (Langley: Battleline, 1985), 112; Lieutenant-Colonel John A. Macdonald, 'In Search of Veritable: Training the Canadian Staff Officer, 1899 to 1945,' MA thesis, RMC, 1992, ch. 3 and App. III. Sixty-one officers attended the course from 2 January to 12 April 1941. During 1940 the worst shortage appears to have been in the position of GSO 2. McNaughton to Crerar, 8 August 1940.
20 Macdonald, 'In Search of Veritable,' 98; Geoffrey Hayes, 'Science and the Magic Eye: Innovations in the Selection of Canadian Army Officers, 1939–1945,' *Armed Forces and Society* 22, no. 2 (1995–6): 275.
21 Memorandum on Procedure in Connection with Canadian Business and Relations Between CMHQ. and 7 Corps, 21 July 1940, MP, vol. 251.
22 Stacey, *Six Years of War*, 231–2; Roy interview with Major-General C.B. Price, 12 June 1966, PP, box 25; Granatstein, *The Generals*, 69; Douglas E. Delaney, *The Soldier's General: Bert Hoffmeister at War* (Vancouver: UBC Press, 2005), 23.
23 Interview Transcript, 'First and Second World Wars,' session 3, tape 4, pp. 14–15.

24 Quoted in Macdonald, 'In Search of Veritable,' 106. McNaughton said after the war that after the initial course was set up it was the intention to 'take it back to Canada lock, stock and barrel as a going concern and set it up.' Interview Transcript, 'First and Second World Wars,' session 3, tape 4.
25 Quoted in Macdonald, 'In Search of Veritable,' 100.
26 Stacey, Historical Officer's Report no. 14, 'Visit to Canadian Junior War Staff Course,' DHH; English, *The Canadian Army and the Normandy Campaign*, 99, 101.
27 Interview Transcript, 'First and Second World Wars,' session 3, tape 4, p. 14. In July 1940, McNaughton had established a committee consisting of Brigadiers Sansom and Turner and Colonel Maurice A. Pope to ascertain all available and potential staff officers. Memorandum on Procedure in Connection with Canadian Business and Relations Between CMHQ and 7 Corps, 21 July [25 July] 1940, MP, vol. 251, file Memorandum of BGS, vols. 1 & 2.
28 English, *The Canadian Army and the Normandy Campaign*, 101. Guy Turner, McNaughton's GSO 1, suggested that the establishment of the CJWSC was 'one of the most valuable and important steps taken as regards training during the war.' He told Stacey that it was entirely McNaughton's idea 'and I trust that he will be given full credit.' Turner comments on draft of Stacey's *Six Years of War*, vol. I, file 83/983, folder 1, folio 60, DHH.
29 Reginald H. Roy, *Sherwood Lett: His Life and Times* (Vancouver: UBC Alumni Association, 1991), 101; Roy interview with Major-General C.B. Price, 12 June 1966, PP, box 25.
30 General Sir William Morgan, 'With Alexander to Dunkirk: Reminiscences of a GSO 1,' *Army Quarterly and Defence Journal* (hereafter *AQDJ*) 102, no. 3 (1972): 357, 359. One observer has questioned whether Alexander was a complete master of his profession like Montgomery or Slim. Major-General J.D. Hunt, 'Alex,' *AQDJ* 103 (July 1973): 440–1.
31 Roy interview with Price, 12 June 1966. It has been the author's experience, at least at the armoured squadron level, that the officer most directly involved with training subalterns to perform their duties is not the OC, but the Battle Captain (BC), the 'operations' officer of the squadron.
32 Charles P. Stacey, *A Date With History: Memoirs of a Canadian Historian* (Ottawa: Deneau, 1983), 124; Dominick Graham, *The Price of Command: A Biography of General Guy Simonds* (Toronto: Stoddart, 1993), 15; J.L. Granatstein interview with M.P. Bogert, 8 September 1991, Granatstein Interviews, DHH. Bogert was keenly resentful for not being promoted at the pace of his peers, and this should be remembered when weighing his comments.
33 Stacey, *Six Years of War*, 232; David J. Bercuson, *Maple Leaf against the Axis: Canada's Second World War* (Toronto: Stoddart, 1995), 21; E.W. Sansom to

wife, 17 January 1940, Sansom Papers, MG30 E537, vol. 1, LAC. Sansom had been the DMT at NDHQ prior to joining the division.
34 'Andrew McNaughton – Canadian,' p. 15. For similar criticisms of McNaughton, see the comments of Brigadier G.E. Beament and Brigadier N.E. Rodger in the Granatstein Interviews, 21 and 24 May 1991, DHH; Colonel A.S. Price correctly placed part of the blame on McNaughton's principal staff officers 'for allowing him to get involved' in so many low-level technical problems. 'Andrew McNaughton – Canadian,' p. 15.
35 Raymond E. Lee Diary, 13 October 1940, United States Army Military History Institute (hereafter USAMHI), Carlisle Barracks, Pennsylvania.
36 Memorandum of Discussions with General Paget, C-in-C 21 Army Gp, Col. Ralston, Minister of National Defence, and Lt.-Gen. Stuart, Chief of the General Staff, Canada, 15 November 1943, MP, vol. 250, PA 5-0-3-2, Appendix 'L.'
37 Eedson L.M. Burns to Richard Malone, 26 March 1969, and R.H. Stokes-Rees to Richard Malone, 25 February 1969, Malone Papers (hereafter MaP), Queen's University Archives; Granatstein interview with General Elliot Rodger, Ottawa, 21 May 1991, Granatstein Interviews; Danchev and Todman, *War Diaries*, 98.
38 Heinz Guderian, *Achtung Panzer! The Development of Armoured Forces, Their Tactics, and Operational Potential* (London: Arms and Armour, 1993), 212.
39 David French, *Raising Churchill's Army: The British Army and the War against Germany, 1919–1945* (Oxford: Oxford University Press, 2000), 89.
40 Martin van Creveld, *Command in War* (Cambridge, MA: Harvard University Press, 1985), 189.
41 Major Ferdinand O. Miksche, *Blitzkrieg* (London: Faber and Faber, 1941), 59; Basil H. Liddell Hart, *This Expanding War* (London: Faber and Faber, 1942), 190; Len Deighton, *Blitzkrieg: From the Rise of Hitler to the Fall of Dunkirk* (London: Jonathan Cape, 1979), 178; Bryan Perrett, *A History of Blitzkrieg* (London: Robert Hale, 1983), 66; S.L.A. Marshall, *Blitzkrieg: Its History, Strategy, Economics, and the Challenge to America* (New York: William Morrow, 1940), 128.
42 Paddy Griffith, *Forward into Battle: Fighting Tactics from Waterloo to the Near Future* (Novato: Presidio, 1991), 111, 113.
43 This was part of the Third Army manoeuvres. Martin Blumenson, ed., *The Patton Papers*, 2 vols. (Boston: Houghton Mifflin, 1972 and 1974), I:949.
44 Minutes of Conference Held at H.Q., 1 Cdn. Div. at 1200 Hrs., 9 Jun 40, PP, box 2. This quote is based on Turner's preparation of the Minutes.
45 Massey to External (Containing Message from McNaughton to Skelton), 29 June 1940, Crerar Papers (hereafter CP), MG30 E157, vol. 1, 958C.009(D15), LAC.

46 Stacey, *Six Years of War*, 233.
47 Minutes of Conference held at HQ 1 Cdn Div at 1200hrs, 9 June 1940, MP, vol. 239.
48 Daniel J. Hughes, ed., *Moltke on the Art of War: Selected Writings* (Novato: Presidio, 1993), 35.
49 Greenfield, *The Organization of Ground Combat Troops*, 46.
50 Ibid., 46, 50.

7. The Politics of Training

1 Douglas E. Delaney, *A Soldier's General: Bert Hoffmeister at War* (Vancouver: UBC Press, 2005), 26.
2 Charles P. Stacey, *Official History of the Canadian Army in the Second World War, I: Six Years of War: The Army in Canada, Britain, and the Pacific* (Ottawa: Queen's Printer, 1957), 289.
3 Interview Transcript, 'First and Second World War: McNaughton 1965,' session 3, tape 2, pp. 14–15, MG31, vol. 3, box E42, file 1, LAC.
4 It should also be borne in mind that Montgomery never saw a single tank in France. His understanding of how an armoured division worked was purely theoretical. Nigel Hamilton, *Monty: The Making of a General, 1887–1942* (London: Hamish Hamilton, 1981), 353, 531.
5 Wade H. Haislip, 'Corps Command in World War II,' *Military Review* 70 (May 1990): 22; Harold R. Winton, *Corps Commanders of the Bulge: Six American Generals and Victory in the Ardennes* (Lawrence: University Press of Kansas, 2007), 7–8.
6 Matthew B. Ridgway, *Soldier: The Memoirs of Matthew B. Ridgway* (New York: Harper and Row, 1956), 18.
7 Edward N. Luttwak, 'The Operational Level of War,' *International Security* 5, no. 3 (1980–1): 61–76.
8 Stephen Ashley Hart, *Montgomery and 'Colossal Cracks': The 21st Army Group in Northwest Europe, 1944-45* (Westport: Praeger, 2000), 12–13. For a good discussion of the imprecise nature of the term 'operational,' see Richard Simpkin, *Race to the Swift: Thoughts on Twenty-First Century Warfare* (London: Brassey's, 1985), 23–4; Robert M. Citino, *Quest for Decisive Victory: From Stalemate to Blitzkrieg in Europe, 1899–1940* (Lawrence: University Press of Kansas, 2002), xii–xiii.
9 Karl-Heinz Frieser, *The Blitzkrieg Legend: The 1940 Campaign in the West* (Annapolis: United States Naval Institute Press, 2005), 6.
10 General Sir Brian Horrocks, *Corps Commander* (New York: Scribner, 1977), 30–1; John A. English, *The Canadian Army and the Normandy Campaign: A Study of Failure in High Command* (New York: Praeger, 1991), xiii.

Notes to pages 98–103   291

11 Larry I. Bland, ed., *The Papers of George Catlett Marshall, II: 'We Cannot Delay,' July 1, 1939–December 6, 1941* (Baltimore: Johns Hopkins University Press, 1986), 632.
12 Memorandum on Procedure in Connection with Canadian Business and Relations Between CMHQ and 7 Corps, 21 July 1940, and Memorandum of a Staff Conference held at HQ 7 Corps, 22 July 1940, MP, MG30 E133, vol. 251, file 'Memorandum of BGS Vols. 1 & 2,' LAC.
13 Interview Transcript, 'First and Second World War,' session 3, tape 3, p. 2; Major Charles P. Stacey, Historical Officer's Report no. 36, 'Situation of the Canadian Forces in the United Kingdom, Summer, 1941: II, Recent Changes in Commands and Staffs, 7 July 1941,' DHH. 'It would seem,' Stacey declared, 'the British authorities (following honourable precedents of the last war) assigned very able officers to serve with the Canadians.' Murison was Canadian by birth.
14 Stacey, *Six Years of War*, 289.
15 Interview Transcript, 'First and Second World War,' session 3, tape 4, p. 11.
16 Military Training Pamphlet (hereafter MTP) no. 47, *Movement by Road*, War Office, September 1941, 26.
17 Reginald H. Roy, *For Most Conspicuous Bravery: A Biography of Maj.-Gen. George R. Pearkes, V.C., through Two World Wars* (Vancouver: UBC Press, 1977), 163.
18 Peter Chaddick-Adams, 'General Sir Miles Christopher Dempsey (1896–1969): "Not a Popular Leader,"' *RUSI Journal* 150, no. 5 (October 2005): 68.
19 Interview transcript, session 3, tape 5, p. 9.
20 Chris Ellis and Peter Chamberlain, eds., *Handbook on the British Army 1943* (London: Arms and Armour, 1976), 13–15. At brigade the Brigade Major (BM) was responsible for training and was assisted by Staff Captains (hereafter SCs) for operations, supply, and transport.
21 Interview transcript, session 3, tape 4, p. 9.
22 Reginald H. Roy, interview with Lt.-Gen. H.D. Graham, 24 September 1970, PP, box 25, file 74-1, University of Victoria Archives, MacPherson Library, Special Collections; *Movement by Road*, 26. Pearkes's solution – to move in an echelon system with the vehicles most needed at the time up front – was reasonable and pragmatic.
23 Major Charles P. Stacey, Canadian Historical Report no. 11, 'Canadian Corps Exercise FOX, 22 February 1941,' DHH; 'South Eastern Command Exercise BEAVER II, Final Conference – 2 January 1942, Remarks by the Army Commander,' file 171.009(D4), DHH.
24 Pearson Diary, 5 to 20 April 1940, LLP, MG 226 N8, vol. 3, LAC.
25 Stacey, Report no. 13, 'Canadian Corps Exercise DOG, 26–28 February 1941,' DHH.

26 Interview Transcript, 'First and Second World War,' session 3, tape 4, 2. Bill Rawling incorrectly places this statement in the context of early 1942. 'The Generalship of Andrew McNaughton: A Study in Failure,' in Bernd Horn and Stephen Harris, eds., *Warrior Chiefs: Perspectives on Senior Canadian Military Leaders* (Toronto: Dundurn, 2001), 78. McNaughton added: 'We had to be allergic to any lane that we couldn't get down at speed, you know, like those sunken roads and things. They're bloody awful.'

27 Ralston, Substance of a Conversation with Brooke and Stuart, 15 November 1943, 'Overseas Trip in 1943,' Ralston Papers (hereafter RP), MG27 III BII, vol. 59, LAC.

28 Alex Danchev and Daniel Todman, eds., *War Diaries, 1939–1945: Field Marshal Lord Alanbrooke* (London: Weidenfeld and Nicolson, 2001), 151. Armand Smith was injured in an automobile accident in the autumn of 1940 and was officially replaced by Roberts on 22 February 1941.

29 Reginald H. Roy, interview with Maj.-Gen. A.E. Potts, 15 October 1971, 14, PP, box 25, file 74-1.

30 Crerar to McNaughton, 20 April 1942, Crerar Papers (hereafter CP), MG30 E157, vol. 2, file 958C.009(D21), LAC; Chris Vokes to Reginald H. Roy, 5 March 1971, PP, box 25, file 71-1.

31 Canadian Corps, Exercise HARE, 9-10-11 Apr 1941, file 171.009(D4), DHH.

32 Danchev and Todman, *War Diaries*, 58; Brooke to McNaughton, 11 September 1941, McNaughton Papers (hereafter MP), vol. 193.

33 Carlo D'Este, *Bitter Victory: The Battle for Sicily, 1943* (New York: Dutton, 1991), 36, 51; Sir Bernard Law Montgomery, *The Memoirs of Field Marshal Montgomery of Alamein* (London: Collins, 1958), 59. Brooke certainly had the aura of command. As General Sir William Jackson and Field Marshal Lord Bramall observed, Brooke's 'determined face, rapid staccato voice, decisiveness and imperturbability impressed everyone who met him.' *The Chiefs: The Story of the United Kingdom Chiefs of Staff* (London: Brassey's, 1992), 186. For less flattering views of Brooke, see John Grigg, *1943: The Victory That Never Was* (London: Eyre Methuen, 1980), 4; and Sir John Colville, *The Churchillians* (London: Weidenfeld and Nicolson, 1981), 145.

34 Hamilton, *Monty*, 377. Sir John Smyth suggested that even though Brooke had limited battle experience, he was a 'fine commander of troops in battle – as well as being a great staff officer.' *Leadership in War 1939–1945: The Generals in Victory and Defeat* (London: David and Charles, 1974), 48.

35 Danchev and Todman, *War Diaries*, 33–4, 68.

36 Ibid., 95, 97; David French, 'Colonel Blimp and the British Army: British

Divisional Commanders in the War against Germany, 1939–1945,' *English Historical Review* 111, no. 444 (1996): 1197. Ironside relieved eleven commanders.
37 Interview transcript, 'First and Second World War,' session 3, tape 4, p. 10. On 20 August, Brooke informed Evans of the reason for his relief and recorded that it 'resulted in an unpleasant interview.' Danchev and Todman, *War Diaries*, 101.
38 Brooke obviously considered Evans's performance in the BEF unacceptable, but it is clear that Evans grasped the requirement for more infantry in the armoured division much more quickly than his successor, Major-General C.W. Norrie. David French, *Raising Churchill's Army: The British Army and the War against Germany, 1919–1945* (London: Oxford University Press, 2000), 191–2.
39 McNaughton faced Major-General J.E. Utterson-Kelso, commander of the 47th (London) Division, who also had the 55th Division, the 25th Army Tank Brigade, the 24th Guards Brigade, and the 168th Infantry Brigade.
40 Stacey, Historical Section Report no. 34, 'South-Eastern Command Exercise WATERLOO, 14–16 June 1941,' DHH.
41 Quoted in Terry Copp, *The Brigade: The Fifth Canadian Infantry Brigade, 1939–1945* (Stoney Creek: Fortress, 1992), 24; George G. Blackburn, *Where the Hell Are the Guns: A Soldier's Eye View of the Anxious Years, 1939–1944* (Toronto: McClelland and Stewart, 1997), 121.
42 Stacey, Historical Section Report no. 34.
43 Major-General Eedson L.M. Burns, *General Mud: Memoirs of Two World Wars* (Toronto: Clarke, Irwin, 1970), 105.
44 Less than six weeks later Brooke visited Odlum's 2nd Division and was 'agreeably surprised at the state of efficiency ... They have come on a great deal lately and are I think better than the 1st Can Div now.' Danchev and Todman, *War Diaries*, 164, 175. The improvement was no doubt effected by Simonds, the GSO 1, rather than by Odlum.
45 Roger A. Beaumont, 'Command Method: A Gap in the Historiography,' *Naval War College Review* 31, no. 3 (1979): 61, 64, 68.
46 Alanbrooke, 'Notes for My Memoirs,' Alanbrooke Papers (hereafter AbP), file 5/2/17, 341, CIGS Papers, Liddell Hart Centre for Military Archives (hereafter LHCMA), King's College, London. It seems ironic that Brooke also declared, 'I was genuinely attached to him.'
47 Carr had the VIII Corps (1st and 46th Divisions) and the XI Corps (43rd and 53rd Divisions), as well as the 6th and 9th Armoured Divisions.
48 Christopher H.N. Hull, 'A Case Study of Professionalism in the Canadian

Army in the 1930s and 1940s: Lieutenant-General G.G. Simonds,' MA thesis, Purdue University, 1989, 41; William E. Hutchinson, 'Test of a Corps Commander: Lieutenant-General Guy Granville Simonds in Normandy, 1944,' MA thesis, University of Victoria, 1982, 102. Lieutenant-Colonel C.R.S. Stein was Acting/AQMG.

49 Major Charles P. Stacey, Report no. 49, 'Home Forces Exercise BUMPER (Army Manoeuvres, 29 September–3 October 1941),' DHH; idem, *Six Years of War*, 240.

50 Brooke, 'Exercise BUMPER, 27 September–3 October 1941: Comments by C-in-C Home Forces,' War Office, 199/727, Public Records Office (hereafter PRO), Kew.

51 Danchev and Todman, *War Diaries*, 133, 143.

52 Alanbrooke, 'Notes for my Memoirs,' file 5/2/16. Many officers felt that Brooke's criticism of Alexander was in fact unjustified, but Alexander chose not to defend himself, fearing he might undermine the broader lessons Brooke was trying to highlight. William G.F. Jackson, *Alexander of Tunis as Military Commander* (London: Batsford, 1971), 113; John North, ed., *The Alexander Memoirs, 1940–1945* (London: Cassell, 1962), 81.

53 Timothy Harrison Place, *Military Training in the British Army, 1940–1944: From Dunkirk to D-Day* (London: Frank Cass, 2000), 36; George Kitching, *Mud and Green Fields: The Memoirs of General George Kitching* (Langley: Battleline, 1985), 112; Reginald H. Roy, interview with His Honourable Major-General G.R. Pearkes, 15 July 1966, PP, box 2, file 74-1. Not until March 1942 did Home Forces have enough room on the South Downs for an entire brigade to practise with live ammunition and air support while dispensing with all unnecessary safety precautions. French, *Raising Churchill's Army*, 199.

54 Crerar to McNaughton, 26 June 1941, CP, vol. 1, file 958C.009 'Personal Correspondence Lt.Gen. A. McNaughton,' LAC; Paul D. Dickson, 'The Politics of Army Expansion: General H.D.G. Crerar and the Creation of First Canadian Army, 1940–1941,' *Journal of Military History* (hereafter *JMH*) 60, no. 2 (1996): 288.

55 Crerar to Montague, 9 June 1941, CP, vol. 1, file 958C.009(D7), 'Personal Correspondence with P.J. Montague.'

56 Major Charles P. Stacey, Report no. 40, '55th Division Exercise ALBERT (Divisional phase), 31 July–2 August 1941,' DHH. The troops quickly got bored with such 'make-believe exercises.' Burns to Crerar, 7 May 1941, CP, vol. 19, 958C.009(D333). See also Strome Galloway, *The General Who Never Was* (Belleville: Mika, 1981), 72; Chris Vokes, *Vokes: My Story* (Ottawa: Gallery, 1985), 69–71; Howard Graham, *Citizen and Soldier: The Memoirs of Lieutenant-General Howard Graham* (Toronto: McClelland and Stewart, 1987), 128.

57 Place, *Military Training in the British Army*, 39. It was also a fundamental fact that major exercises such as WATERLOO and BUMPER did not facilitate the practice of minor tactics at the unit level.
58 Timothy T. Luper, *The Dynamics of Doctrine: The Changes in German Tactical Doctrine during the First World War* (Fort Leavenworth: Combat Studies Institute, 1981), vii; Robert A. Doughty, *The Seeds of Disaster: The Development of French Army Doctrine, 1919–1939* (Hamden: Archon, 1985), x–xi.
59 William McAndrew, 'Fire or Movement? Canadian Tactical Doctrine, Sicily – 1943,' *Military Affairs* 51, no. 3 (1987): 140.
60 Lieutenant-General Sir William Bartholomew was the GOC, Northern Command. French, *Raising Churchill's Army*, 191.
61 Correlli Barnett, *The Desert Generals*, rev. ed. (Bloomington: Indiana University Press, 1982), 101. There was little chance of corrective measures being identified by the training army in Britain because even during BUMPER no accepted technique for employing armoured divisions had been confirmed. Giffard Le Q. Martel, *An Outspoken Soldier: His Views and Memoirs* (London: Sifton Praed, 1949), 181. The Americans had the same problem in Tunisia, where divisions were repeatedly broken up and scattered about the battlefield. Rick Atkinson, *An Army at Dawn: The War in North Africa, 1942–1943* (New York: Henry Holt, 2002), 333, 390.
62 English, *The Canadian Army and the Normandy Campaign*, 160; David A. Wilson, 'The Development of Tank-Infantry Co-operation in the Canadian Army for the Normandy Campaign of 1944,' MA thesis, University of New Brunswick, 1992, ii. The War Office disseminated doctrine through a system of officially sanctioned publications. The MTPs took a long time to prepare. To spread the latest tactical lessons more quickly it published Army Training Instructions (ATIs). These filled the void until new MTPs could be published. The Canadian Army published Canadian Army Training Memoranda (CATMs), which reproduced doctrine from a wide range of British doctrinal publications. For a full discussion of the varied types of doctrinal publications available at the time, see Place, *Military Training in the British Army*, 10–14.
63 Memorandum, 31 August 1940, MP, vol. 251, file 'Memorandum BGS Vols. 1 & 2.' On the decentralization of British artillery, see Major-General Jonathan B.A. Bailey, *Field Artillery and Firepower* (Annapolis: Naval Institute Press, 2004), 286–96.
64 Ronald Lewin, *Montgomery as Military Commander* (New York: Stein and Day, 1971), 56.
65 Shelford Bidwell and Dominick Graham, *Firepower: British Army Weapons and Theories of War, 1904–1945* (London: Allen Unwin, 1982), 226; S.E. Army:

Exercise TIGER, Final Conference – 4 June, 1942, Remarks of Army Commander, file 171.009(D4), DHH.
66 Williamson Murray, 'British Military Effectiveness in the Second World War,' in *Military Effectiveness, III: The Second World War*, ed. Williamson Murray and Allan R. Millett (Boston: Allen Unwin, 1988), 112, 125; French, *Raising Churchill's Army*, 184, 210; Terry Copp, *The Brigade*, 27.

8. Enter Montgomery

1 J.L. Granatstein, *The Generals: The Canadian Army's Senior Commanders in the Second World War* (Toronto: Stoddart, 1993), 68; John A. English, *The Canadian Army and the Normandy Campaign: A Study of Failure in High Command* (New York: Praeger, 1991), 126; John Swettenham, *McNaughton, II: 1939–1943* (Toronto: Ryerson, 1969), 187; Sansom to wife, 29 February and 3 March 1940, Sansom Papers (hereafter SaP), MG30 E537, vol. 1, LAC; Crerar to McNaughton, 19 May and 11 August 1941, Crerar Papers (hereafter CP), MG30 E157, vol. 1, file 958C.009, 'Personal Correspondence – Lt.Gen. A. McNaughton,' LAC; Eedson L.M. Burns, *General Mud: Memoirs of Two World Wars* (Toronto: Clarke, Irwin, 1970), 105. See also Crerar to Burns, 24 May 1941, and Burns to Crerar, 16 July 1941, CP, vol. 19, file 958C.009(D338); Paul D. Dickson, 'The Limits of Professionalism: H.D.G. Crerar and the Canadian Army, 1914–1944,' PhD diss., University of Guelph, 1993, 355.
2 John W. Pickersgill, ed., *The Mackenzie King Record, I: 1939–1944* (Toronto: University of Toronto Press, 1960), 258; Crerar War Diary (hereafter WD), 14 October 1941, CP, vol. 15, file CGS. The age limitations were as follows: major-generals appointed at age 53 and retired at age 57; brigadiers appointed at 49 and retired at 54; and lieutenant-colonels appointed at 45 and retired at 51. Charles P. Stacey, *Official History of the Canadian Army in the Second World War, I: Six Years of War, the Army in Canada, Britain, and the Pacific* (Ottawa: Queen's Printer, 1957), 418.
3 Brooke wanted to relieve fifty-six-year-old Major-General D.G. Johnson, GOC of the 4th Division in France, because of his age, but decided against it and had no reason to regret his decision. French observed that the fault of the British pre-war selection process for senior positions was not that some men were too old but that it 'failed to test their ability to withstand the physical and mental strains of the active service.' The average age of Gort's division commanders in 1940 was fifty-three. David French, *Raising Churchill's Army: The British Army and the War against Germany, 1919–1945* (London: Oxford University Press, 2000), 182; Eisenhower to Leonard T.

Gerow, 16 July 1942, in *The Papers of Dwight David Eisenhower: The War Years*, 5 vols., ed. Alfred D. Chandler, Jr (Baltimore: Johns Hopkins University Press, 1970), I:385. The U.S. Army had 197 division commanders during the Second World War from Pearl Harbor to the end of hostilities. Based on a sample of twenty-five commanders, the average age in 1941 was forty-eight. One study suggests: 'They were probably average division commanders, competent to accomplish the missions given to them.' Lieutenant-Colonel Gary L. Wade, *World War II Division Commanders*, Report no. 7 (Fort Leavenworth: Combat Studies Institute, 1983).
4 Crerar to McNaughton, 16 July 1940, McNaughton Papers (hereafter MP), MG30 E133, vol. 227, file CC7/Crerar/6, LAC; McNaughton to Ralston, 11 November 1941, Cipher no. 2081, CP, vol. 22, file 958C.009 (D335). See also Crerar to Dill, 5 July 1940, 958C.009(D335), CP, vol. 22, and Crerar to McNaughton, 11 August 1941, CP, vol. 1, file 958C.009 Pers Correspondence Lt.Gen. McNaughton. Five days later Brooke and Crerar had dinner together and had 'a long and useful discussion ... about the future of the Canadian Corps. I do hope that they give him the 2nd Canadian Division.' Alex Danchev and Daniel Todman, eds., *War Diaries, 1939–1945: Field Marshal Lord Alanbrooke* (London: Weidenfeld and Nicolson, 2001), 191, 193.
5 English suggested that McNaughton wanted Pearkes to assume command of the corps on a permanent basis, but it is apparent that this was not so. English's source was Roy's biography of Pearkes. According to Pearkes, McNaughton told him: 'I hope this will be a permanent appointment for you.' *For Most Conspicuous Bravery: A Biography of Major-General George R. Pearkes, V.C., through Two World Wars* (Vancouver: UBC Press, 1977), 166; English, *The Canadian Army and the Normandy Campaign*, 127.
6 McNaughton to Ralston, 29 November 1941, Cypher no. 2251, CP, vol. 22, file 'Appointments, Transfers, Reports, October 1939–March 1942', 958C.009(D335); Alanbrooke Papers (hereafter AbP), 5/2/16, diary entry for 6 June 1941, Liddell Hart Centre for Military Archives (hereafter LHCMA). Crerar went to lengths to inform Stacey that he did indeed have battle experience. See his comments on draft of *Six Years of War*, file 82/983, folder 5, vol. II, folio 227, Directorate of History and Heritage (hereafter DHH), Department of National Defence.
7 Interview with D.C. Spry, October 1987, Spry Papers (hereafter SP), MG30 E563, vol. 1, LAC.
8 File 11/2, Alanbrooke Biography: Notes by Mrs M.C. Long, Crerar's thoughts on Brooke, Crerar to Otto Lund, n.d., LHCMA.
9 Ralston War Diary, London Trip, 20 October 1941, CP, vol. 15. See also the CWC Minutes for 6 November 1941, RG2 7C, vol. 6, LAC.

10 Paul D. Dickson, 'The Politics of Army Expansion: General H.D.G. Crerar and the Creation of First Canadian Army, 1940–1941,' *JMH* 60, no. 2 (1996): 294; Crerar to Donnie White, 21 November 1941, CP, vol. 19, file 958C.009(D333); interview with Crerar, n.d., AbP, 12/XI/4/61; David Fraser, *Alanbrooke* (London: Collins, 1982), 188; Nigel Hamilton, *Monty: The Making of a General, 1887–1942* (London: Hamish Hamilton, 1981), 479.
11 Colonel Eedson L.M. Burns, 'Command of the Canadian Army in the United Kingdom,' 18 August 1941, file 112.1, DHH.
12 Dickson, 'Limits of Professionalism,' 357. Burns's proposal coincided with Crerar's intimation to McNaughton that an army was a possibility based on manpower estimates.
13 Paul D. Dickson, *A Thoroughly Canadian General: A Biography of General H.D.G. Crerar* (Toronto: University of Toronto Press, 2007), 147.
14 Brigadier-General Maurice Pope, Memorandum, 8 September 1941, file 112.1, DHH.
15 Alanbrooke, 'Notes for My Memoirs,' 5/2/17, 7 January 1942.
16 Brooke to McNaughton, 7 January 1942, MP, vol. 230.
17 Pope comments on draft of Stacey's *Six Years of War*, vol. I, file 82/983, folder 1, folios 67–8, DHH; Hamilton, *Monty*, 506.
18 McNaughton WD, 30 March 1942, MP, vol. 248.
19 Brigadier Penhale, Memorandum for Army Commander, 'Report on Conference with Director of Staff Duties at the War Office, 13 April 1942 [15 April], MP, vol. 248; Brigadier Turner, The Senior Officer, CMHQ, Formation of First Canadian Army Headquarters in the U.K., 5 April 1942, MP, vol. 248; Memorandum of Conference, 27 December 1941, MP, vol. 249, file CC/Crerar/6; Paul D. Dickson, 'The Hand That Wields the Dagger: Harry Crerar, First Canadian Army Command, and National Autonomy,' *War and Society* 13, no. 2 (1995): 121.
20 Montgomery to Trumball Warren, June 1942, Montgomery Papers, Ancillary Collection 20, Lieutenant Colonel Trumball Warren, Imperial War Museum (hereafter IWM).
21 Dickson, 'The Limits of Professionalism,' 362–8.
22 Simonds to Reginald H. Roy, 29 December 1972. According to George Kitching, Simonds was not impressed with Crerar's habit of searching for lessons learned from a mass of operations logs. *Mud and Green Fields: The Memoirs of Major-General George Kitching* (Langley: Battleline, 1985), 118.
23 Ronald Lewin, *Montgomery as Military Commander* (New York: Stein and Day, 1971), 32; Alun Chalfont, *Montgomery of Alamein* (New York: Atheneum, 1976), 113; Robert Hodgins-Vermass, '"A Bit of 'Binge'": Montgomery's Command of the 3rd Division,' MA thesis, University of Calgary, 1997; Hamilton, *Monty*, 348.

24 Alex Graeme-Evans, 'Field-Marshal Bernard Montgomery: A Critical Assessment,' *Army Quarterly and Defence Journal* (hereafter *AQDJ*) 104, no. 4 (1974): 418; Shelford Bidwell, *Gunners at War: A Tactical Study of the Royal Artillery in the Twentieth Century* (London: Arrow, 1972), 137; Robert W. Tooley, 'Montgomery as Military Trainer: Preparation for Alamein,' MA thesis, University of New Brunswick, 1984; Williamson Murray, 'British Military Effectiveness in the Second World War,' in *Military Effectiveness, III: The Second World War*, ed. Williamson Murray and Alan R. Millett, (Boston: Allen Unwin, 1988), 126.

25 Sir Bernard Law Montgomery, *The Memoirs of Field Marshal Montgomery of Alamein* (London: Collins, 1958), 71, 73; Stephen E. Ambrose, *The Supreme Commander: The War Years of General Dwight D. Eisenhower* (Garden City: Doubleday, 1970), 176.

26 Christopher H.N. Hull, 'A Case Study in Professionalism in the Canadian Army in the 1930s and 1940s: Lieutenant-General G.G. Simonds,' MA thesis, Purdue University, 1989, 41. Hamilton has argued that Simonds's 'extreme professionalism' was the real reason why Montgomery's doctrine was disseminated throughout the corps. *Monty*, 507–8. On 10 March 1942, Montgomery praised Crerar for his corps training memorandum, declaring it 'excellent': 'There is no doubt that the training in the Corps is now beginning to move on sound lines.' Montgomery to Crerar, 10 March 1942, CP, vol. 2, file 958C.009(D182).

27 21 Army Group, *Some Notes on the Conduct of War and the Infantry Division in Battle*, Belgium, November 1944, 15–16, 30–1.

28 Montgomery, Exercise CONQUEROR, The initiative in Battle, 14 April 1942, CP, vol. 2, 958C.009(D182). See also his Notes on Exercise FLIP, dated the same day. Two days later Montgomery wrote to Crerar: 'I am sure you will "thump the table" in no uncertain voice on these matters. There is a great deal at stake, and we shall make no progress if we degenerate into a mutual congratulation society.' Montgomery to Crerar, 16 April 1942, CP, vol. 2, file 958C.009(D182).

29 Montgomery, Notes on the Command Element in Canadian Corps, 6 March 1942, MP, vol. 200; Reginald H. Roy interview with Major-General C.B. Price, 12 June 1966, PP, box 25, file 74-1.

30 South Eastern Command Skeleton Exercise 'Lancelot,' Final Conference – 14 March 42, Remarks of Army Commander, file 171.009(D4), DHH.

31 Montgomery, 'BEAVER III, Notes on Commanders,' 25 April 1942, CP, vol. 2, 958C.009(D182); Roy, *For Most Conspicuous Bravery*, 168.

32 Montgomery, 'Some Notes on the Broader Aspects of BEAVER III,' 23 April 1942, and Montgomery, 'BEAVER III Notes on Commanders,' 25 April 1942, CP, vol. 2, 958C.009(D182). Montgomery considered R.A. Wyman

teachable as well; the problem was that he had 'no Divisional Commander to train him and bring him on, as he is Corps Troops.' As a remedy Montgomery recommended that Corps Headquarters, meaning Crerar, see to it that Wyman received the proper guidance and supervision, but Crerar was hardly qualified to provide such mentorship. Montgomery, 1 Canadian Army Tank Brigade, 7 May 1942, CP, vol. 2, file 958C.009(D182).

33 Crerar, 'Notes regarding Enclosed Correspondence with C-in-C South-Eastern Command,' CP, vol. 2. English suggests that there is 'little doubt' that Montgomery's judgements were 'generally free from all but professional bias.' *The Canadian Army and the Normandy Campaign*, 128; David French, 'Colonel Blimp and the British Army: British Divisional Commanders in the War against Germany, 1939–1945,' *English Historical Review* 111, no. 444 (1996): 1196. Terry Copp, however, disagrees. *The Brigade: The Fifth Canadian Infantry Brigade, 1939–1945* (Stoney Creek: Fortress, 1992), 28.

34 Crerar to Montgomery, 25 February 1942, CP, vol. 1, file 958C.009. Crerar had already given this matter some thought, for in January he wrote to his subordinate commanders: 'I set great importance on each commander, down to the junior officers, training his own subordinates and troops.' Crerar to All Commanders and Commanding Officers, Canadian Corps, 14 January 1942, CP, vol. 1, file 958C.009.

35 Crerar to Montgomery, 6 January 1942, CP, vol. 1, file 958C.009.

36 Montgomery, 'Notes on BEAVER IV,' 13 May 1942, MP, vol. 200, and 'Notes on Commander 3 Div,' 13 May 1942, CP, vol. 2, file 958C.009(D182). According to Crerar, Price never improved for another exercise in July. See Crerar, Notes on Exercise HAROLD, 25–31 July 1942, 6 August 1942, file 171.009, DHH.

37 McNaughton to Montgomery, 26 May 1942, MP, vol. 200; Montgomery to Trumball Warren, 27 June 1942.

38 Montgomery to Crerar, 30 May 1942, CP, vol. 2, file 958C.009(D182). Montgomery concluded: 'I am very well satisfied with the Corps; if we can now put the polish on, there will be no other Corps to touch it. I hope we shall have some good battles together on the other side.'

39 Major Charles P. Stacey, Historical Officer's Report no. 73, South Eastern Command Exercise TIGER, 19–30 May 42, DHH.

40 Quoted in Stacey, *Six Years of War*, 244.

41 Montgomery, Report on Lt. Gen. Bucknall, 2 August 1944, BLM 119/14, IWM.

42 French, 'Colonel Blimp and the British Army,' 1190.

43 Crerar to McNaughton, 10 August 1942, CP, vol. 1, file 958C.009, 'Confidential Report on Officers'; Crerar to McNaughton, 5 March 1941, CP, vol. 1, file 958C.009, 'Personal Correspondence with Lt.-Gen. McNaughton.'

44 Neither Crerar nor Simonds questioned Pearkes's energy or drive, but they did highlight his inability to get along with others. Crerar to McNaughton, 20 April 1942, CP, vol. 2, file 958C.009 (D21); Simonds to Reginald H. Roy, 29 December 1972. When Pearkes was relieved, Vokes suggested that the army was 'deprived of its best general officer.' Chris Vokes to Reginald H. Roy, 5 March 1971. Potts called Pearkes 'a highly proficient professional soldier ... He left us full of confidence in his abilities to command.' Major-General A.E. Potts interview with Reginald H. Roy, 15 October 1971, PP.
45 Swettenham, *McNaughton*, II:172.

9. Exercise SPARTAN

1 Memorandum on Discussion – Army Commander and A/P.R.O., 24 December [28 December] 1942, McNaughton Papers (hereafter MP), MG30 E133, vol. 249, War Diary (hereafter WD), Appendix 'P,' LAC; Memorandum, Minutes of Meeting in McNaughton's Office, 29 December 1942, MP, vol. 249; Memorandum of a Conversation by General McNaughton with General Paget, C.-in-C., Home Forces, at G.H.Q., Home Forces at 1745 hrs, 4 November [5 November] 1942, MP, vol. 249.
2 Quoted in Charles P. Stacey, *Official History of the Canadian Army in the Second World War, I: Six Years of War: The Army in Canada, Britain, and the Pacific* (Ottawa: Queen's Printer, 1957), 250.
3 Montgomery to Crerar, 9 January 1943, Crerar Papers (hereafter CP), MG30 E157, vol. 7, file 958C.009(D172), LAC.
4 The Director of Military Training at the War Office, Major-General J.A.C. Whitaker, informed McNaughton of an Army Council decision that 1st Division could not miss any more combined training; therefore, 'I should be much obliged if you would inform the Commander 1 Canadian Corps that 1 Canadian Division will not be available for exercise SPARTAN.' McNaughton conceded to Whitaker's demand the next day. Whitaker to McNaughton, 30 January 1943, and McNaughton to Whitaker, 31 January 1943, MP, vol. 136. McNaughton had originally intended to use Salmon's division instead of Roberts's. Minutes of a Conference held at H.Q. First Cdn Army at 1130 hrs, 10 January 1943, MP, vol. 249, WD, PA 5-0-3, Appendix 'M.' Combined operations training for 1st Division commenced in late November 1941, behind schedule. Major Charles P. Stacey, Historical Officer's Report no. 93, 'Combined Operations Training, 1st Canadian Corps, 29 April 1943,' Directorate of History and Heritage (hereafter DHH), Department of National Defence, Ottawa.
5 Major Charles P. Stacey, Report no. 91, 'Press Conference Concerning Or-

ganization of First Canadian Army and Arrangements for Press Representatives,' 25 February 1943, DHH. McNaughton was fundamentally opposed to the weakening of the infantry division. Memorandum of a Discussion at HQ, First Canadian Army, 4 June 1942, on British Army Organization, MP, vol. 248, WD, Appendix 'D'; Memorandum of a Discussion with Canadian Observers on Return from the Middle East, 16 June 1942, MP, vol. 248, WD, Appendix 'E.'

6 In March 1941, Gammell was a major-general commanding the 3rd Infantry Division in Montgomery's V Corps. During V Corps Exercise no. 5, Gammell confounded the attack of the opposing 4th Division to such an extent that Montgomery terminated the exercise prematurely. Timothy Harrison Place, *Military Training in the British Army, 1940–1944: From Dunkirk to D-Day* (London: Frank Cass, 2000), 36; Nigel Hamilton, *Monty: The Making of a General, 1887–1942* (London: Hamish Hamilton, 1981), 466–7.

7 Crerar, 'SPARTAN Appreciation,' 16 February 1943, RG24, vol. 10832, file 229C1.8(D7), LAC.

8 GHQ Exercise SPARTAN: Narrative of Events, March 1943, 4–5, file 545.033(D1), DHH.

9 General Sir Bernard Paget, 'GHQ Exercise SPARTAN: Comments by C-in-C Home Forces, March 1943,' 6, file 545.033(D1), DHH.

10 'SPARTAN: Narrative of Events,' 6.

11 Paget, 'Comments by C-in-C Home Forces,' 9; Charles P. Stacey, Report no. 94, GHQ Exercise 'Spartan,' March 1943, 12 May 1943, DHH.

12 Ibid., 15.

13 'SPARTAN: Narrative of Events,' 8. In the event this happened, McNaughton directed Crerar to move northwest on Oxford to give XII British Corps room to deploy.

14 1 Cdn Corps Intelligence Summary no. 9, 1800 hours, 7 March, RG24, vol. 10832, LAC. The attack never materialized until late in the day on 9 March and was far less than predicted. 1 Cdn Corps Operation Order no. 9, 9 March 1943, RG24, vol. 10832, file 229C1.8(D7), LAC.

15 Sir James Grigg interview with Marian C. Long, n.d., Alanbrooke Papers (hereafter AbP), file 11/82, Liddell Hart Centre for Military Archives (hereafter LHCMA), King's College, London.

16 Sir Bernard Law Montgomery, *The Path to Leadership* (New York: Putnam, 1961), 135. Pownall once told Mountbatten that 'when P.J. doesn't like someone, he does not neglect an opportunity to stick in a knife.' Philip Ziegler, *Mountbatten: A Biography* (New York: Alfred A. Knopf, 1985), 298.

17 Charles P. Stacey, Report no. 40, '55th Division Exercise ALBERT (Divisional Phase), 31 Jul–2 Aug 41,' DHH.

18 Charles P. Stacey, Exercise SPARTAN Notes, file 545.003(D1), DHH.

19 Paget, 'Comments by C-in-C Home Forces,' 11.
20 'SPARTAN: Narrative of Events,' 22.
21 Quoted in Stacey, Exercise SPARTAN NOTES.
22 'SPARTAN: Narrative of Events,' 13.
23 McNaughton WD, 7 March 1943, MP, vol. 249. Swettenham stated: 'On the 7th he went forward to the bridging sites to imbue the engineers with a sense of urgency. At 4 P.M., it was reported, the bridge situation had improved.' *McNaughton, II: 1939–1943* (Toronto: Ryerson, 1969), 227. Swettenham's source was the SPARTAN narrative, but it made no declaration that McNaughton went forward on the 7th. All it stated was that 'by midday General McNaughton realized that the bridge situation was unsatisfactory.' 'SPARTAN: Narrative of Events,' 13.
24 Paget, 'Comments by C-in-C Home Forces,' 14. Eisenhower told Patton much the same thing when the latter assumed command of II Corps after Kasserine,' Stephen E. Ambrose, *The Supreme Commander: The War Years of General Dwight D. Eisenhower* (Garden City: Doubleday, 1970), 179–80.
25 Alanbrooke, 'Notes on My Life,' AbP, file 5/2/20, LHCMA; Alex Danchev and Daniel Todman, eds., *War Diaries 1939–1945: Field Marshal Lord Alanbrooke* (London: Weidenfeld and Nicolson, 2001), 637.
26 Charles P. Stacey, *Arms, Men, and Governments: The War Policies of Canada, 1939–1945* (Ottawa: Queen's Printer, 1970), 235. Stacey added only that McNaughton's diary 'establishes that General Brooke visited him on the morning of 7 March, between the issuance of the warning order and its cancellation.'
27 'SPARTAN: Narrative of Events,' 14.
28 Paget, 'Comments by C-in-C Home Forces,' 9. This was the fundamental premise of Eisenhower's broad front strategy.
29 'SPARTAN: Narrative of Events,' 14; McNaughton WD, 7 March 1943.
30 'SPARTAN: Narrative of Events,' 15.
31 Ibid., 18. Obviously, signals were extremely important. British armoured doctrine made it clear that 'since the fighting efficiency of an armoured division and its component units is largely dependent on the reliability of its communications, it is essential that not only communications personnel, but all officers, should have adequate training in this subject.' Military Training Pamphlet no. 41, *The Tactical Handling of an Armoured Division and Its Components, Part I*, War Office, July 1943, 26.
32 McNaughton WD, 8 March 1943. I Canadian Corps estimated that the bridges at Wallingford and Abingdon would not be completed until 'well into the night' of 7 March. 1 Cdn Corps Operation Order no. 6, 7 March 1943, RG24, vol. 10832, file 229C1.8(D7), LAC.
33 Paget, 'Comments by C-in-C Home Forces,' 33. The armoured forces did

not stick to the roads, but when they did, valuable lessons were learned, such as the fact that no more than forty tanks should occupy one mile of road during a night move. See Douglas How, *The 8th Hussars: A History of the Regiment* (Sussex: Maritime, 1964), 178.
34 McNaughton WD, 9 March 1943. Problems with the wireless sets on the exercise were widespread. See Will R. Bird, *North Shore (New Brunswick) Regiment* (Fredericton: Brunswick, 1963), 165.
35 'SPARTAN: Narrative of Events,' 20.
36 Stacey, Report no. 94, 21.
37 Ibid., 21.
38 Ibid., 22.
39 Ibid., 24.

10. The Long Shadow of SPARTAN

1 John Swettenham, *McNaughton, II: 1939–1943* (Toronto: Ryerson, 1969), 283; John A. English, *The Canadian Army and the Normandy Campaign: A Study of Failure in High Command* (New York: Praeger, 1991), 310.
2 McNaughton to Lieutenant-General F.E. Morgan, CGS, GHQ Home Forces, file 212C1(D30), Directorate of History and Heritage (hereafter DHH), Department of National Defence, Ottawa.
3 Swettenham, *McNaughton*, II:266.
4 Comments by Major-General H.A. Young on draft of *Six Years of War*, file 82/983, folder 4, DHH.
5 English, *The Canadian Army and the Normandy Campaign*, 144; General Sir Bernard Paget, 'GHQ Exercise SPARTAN: Comments by the C-in-C Home Forces, March 1943,' foreword, file 545.033(D1), DHH.
6 Minutes of a Conference held at H.Q. First Cdn Army at 1130 hrs, 10 January 1943, McNaughton Papers (hereafter MP), MG30 E133, vol. 249, War Diary (hereafter WD), PA 5-0-3, Appendix 'M,' LAC.
7 Ibid.
8 In late February 1943 Crerar announced: 'The formations and units of 1 Cdn Corps are generally regarded, and quite logically so, as the best trained troops in the United Kingdom.' 'Exercise SPARTAN,' file 229C1.9(C2), DHH.
9 Charles P. Stacey, *Official History of the Canadian Army in the Second World War, I: Six Years of War: The Army in Canada, Britain, and the Pacific* (Ottawa: Queen's Printer, 1957), 249.
10 Charles P. Stacey, *A Date with History: Memoirs of a Canadian Historian* (Ottawa: Deneau, 1983), 116. Swettenham argued that if McNaughton had been

'solely concerned with his own reputation' he would have 'made sure that Crerar's well-trained corps was left as the only major Canadian formation in the exercise.' *McNaughton*, II:274.

11 Lieutenant H.D. Martin, Report no. 43, Historical Section (G.S.), Army Headquarters, 'Training of the 4th and 5th Canadian Armoured Divisions in the United Kingdom, October 1941–July 1944,' DHH; Memorandum of a Discussion with Lt.-Gen. F.E. Morgan, C.G.S., Home Forces, held in General McNaughton's [HQ] at 1245 hrs, 13 November [14 November] 1942, MP, vol. 249, WD, PA 3-6, Appendix 'K.' McNaughton had been told in early September 1942 that the prospect of getting tanks from British sources was 'bad and the future was most unpromising.' Memorandum of a Discussion, Lt.-Gen. A.G.L. McNaughton and Lt.-Gen. Martel, 8 September [11 September] 1942, MP, vol. 248, WD, PA 9-4-15, Appendix 'C.' Similarly, the equipment situation in Worthington's 4th Armoured Division was so bad that crew drills and basic manoeuvres were practised using fence rails. Lawrence J. Zaporzan, 'Rad's War: A Biographical Study of Sydney Valpy Radley-Walters from Mobilization to the End of the Normandy Campaign 1944,' MA thesis, University of New Brunswick, 2001, 107. The situation was difficult for the Guards Armoured Division as well. Re-rolled from infantry, its headquarters was formed on 19 June 1941, but tanks arrived slowly and it did not conduct its first full division exercise, CHEDDAR, until July 1942. Roland Ryder, *Oliver Leese* (London: Hamish Hamilton, 1987), 91.

12 Stein took the Staff College Preparatory Cource at RMC as a captain and was Assistant Director of Engineering Services in the 1930s. He was posted to a British armoured division in 1941 to learn how to handle tanks. J.L. Granatstein interview with W.F.R. Stein, 1 March 1992, Granatstein Interviews, DHH. Prior to taking command of the 5th CAD, Stein had been the BGS of First Canadian Army.

13 Memorandum of a Conversation between Lt.-Gen. Martel and McNaughton, 20 June [22 June] 1942, MP, vol. 248, WD, PA 5-0, Appendix 'J'; Memorandum of a Discussion, McNaughton with C-in-C Home Forces on 8 July [11 July] 1942, MP, vol. 248, WD, PA 3-6, Appendix 'H.' Martel expressed 'certain opinions' on various Canadian commanders but no record was made in these memoranda. Kenneth Macksey, *A History of the Royal Armoured Corps, 1914–1975* (Beaminster: Newton, 1983), 82, 86. Martel was one of the British armoured pioneers to whom Heinz Guderian gave credit for stimulating his thoughts on armoured warfare in the interwar period. *Panzer Leader* (New York: Dutton, 1952), 20. Martel's principal theoretical work was *In the Wake of the Tank* (London: Sifton Praed, 1931). He commanded the

50th Division in Brooke's II Corps in France and took part in the only British tank counter-attack against the Germans.
14 Stacey, *Six Years of War*, 250; Michael P. Cessford, 'Warriors for the Working Day: The 5th Canadian Armoured Division in Italy, 1943–1945,' MA thesis, University of New Brunswick, 1989, 13–14. When the division entered combat in May 1944 in the Liri Valley, Hoffmeister, its new commander, did not know the first thing about how to manoeuvre it and relied heavily on Brigadier J. Desmond Smith, commander of the 5th Canadian Armoured Brigade. Apparently, when Hoffmeister assumed command he told Smith: 'Des, I know sweet bugger all about armour and I am going to depend on you.' J.L. Granatstein interview with Bertram Hoffmeister, 14 September 1991, Granatstein Interviews, DHH.
15 English, *The Canadian Army and the Normandy Campaign*, 144. It is not clear who was to command the British armoured corps, identified by McNaughton as the 'Guards Armoured Corps' in the original order of battle, or whether or not he had armoured experience. II Canadian Corps was based on I Corps but modified to handle both infantry and armoured divisions. Decisions Reached at Series of Discussions Between The Minister, General McNaughton and General Stuart, March 1942, MP, vol. 248, WD, Appendix 'C.'
16 Montgomery, 'Southeast Army: Exercise TIGER, Final Conference, 4 June 1942,' file 171.009(D4), DHH. It is true, however, that he saw no need for special armoured corps headquarters. See Giffard Le Q. Martel, *An Outspoken Soldier: His Views and Memoirs* (London: Sifton Praed, 1949), 179. Oddly enough, when Crerar was CGS in late 1941, he suggested an armoured corps of two divisions for the Canadian Army Programme for 1941–42. Stacey, *Six Years of War*, 95.
17 Nigel Hamilton, *Monty: Master of the Battlefield, 1942–1944* (London: Hamish Hamilton, 1983), 150. Martel stated that as early as July 1941 Brooke and Montgomery were strongly opposed to armoured corps headquarters. *An Outspoken Soldier*, 179.
18 Crerar consistently kept Wyman's 1st Canadian Army Tank Brigade in corps reserve. See I Canadian Corps Operations Orders nos 6–12, Exercise SPARTAN Orders and Instructions, RG24, vol. 10832, file 229C1.8(D7), LAC.
19 Military Training Pamphlet (MTP) no. 41, *The Tactical Handling of an Armoured Division and Its Components, Part 2: The Armoured Regiment*, War Office, February 1943, p. 34.
20 Timothy Harrison Place, *Military Training in the British Army, 1940–1944: From Dunkirk to D-Day* (London: Frank Cass, 2000), 94.
21 MTP no. 41, p. 31.

22 Paget, 'Comments by C-in-C Home Forces,' 11.
23 Extracts from Copy no. 20 of Exercise SPARTAN: Report by Chief Umpire, 22 March 1943, RG24, vol. 10414, file 200G1.013(D3), LAC.
24 Paget, 'Comments by C-in-C Home Forces,' 17.
25 SPARTAN R.E. Final Conference, Summary of Remarks by Chief Engineer Home Forces, RG24, vol. 10436, file 212C1(D31), LAC.
26 McNaughton to Comds – 12 Corps, 1 Cdn Corps & 2 Cdn Corps, 6 March 1943, MP, vol. 248, PA 4-2-1, Appendix 'D.'
27 Paget, 'Comments by C-in-C Home Forces,' 30, 54.
28 Extracts from Copy no. 20 of Exercise SPARTAN: Report by Chief Umpire.
29 The division would conduct a road move exercise, HARLEQUIN, in early September, which suggests that basic skills were still lagging.
30 Larry Worthington, *Worthy: A Biography of Major-General F.F. Worthington* (Toronto: Macmillan, 1961), 189.
31 Paget, 'Comments by C-in-C Home Forces,' 10.
32 James L. Ralston, 'Overseas Trips in 1943: Record of Diary Excerpts and Communications Regarding Change in Canadian Army Command, 27th December 1943,' entry for 3 August 1943, Ralston Papers (hereafter RP), MG27 III BII, vol. 59, LAC.
33 Paget, 'Comments by C-in-C Home Forces,' 13. According to Roman Jarymowycz, both Crerar and Simonds exhibited this deficiency in Normandy. 'Canadian Armour in Normandy: Operation "Totalize" and the Quest for Operational Maneuver,' *Canadian Military History* (hereafter *CMH*) 7, no. 2 (1998): 35.
34 Extracts from Copy no. 20 of Exercise SPARTAN: Report by Chief Umpire, 22 March 1943, RG24, vol. 10414, file 200G1.013(D3), DHH.
35 Frank J. Price, *Troy H. Middleton: A Biography* (Baton Rouge: Louisiana State University Press, 1974), 186.
36 John J.T. Sweet, *Mounting the Threat: The Battle of Bourguebus Ridge, 18–23 July 1944* (San Rafael: Presidio, 1977), 77, 89.
37 Comments by Henry Latham on draft of *Six Years of War*, Comments Received from Abroad, file 82/983, folder 3, DHH. Latham added that 'the severest criticism I heard applied to one or two [Canadian] commanders was "He doesn't know much, but he's a bloody good scout and I bet he'll fight."'
38 Basil Liddell Hart, Memorandum on McNaughton, 16 August 1966, file LH 2/M, Liddell Hart Papers (LHP), Liddell Hart Centre for Military Archives (hereafter LHCMA). Perhaps there is also significant truth to Brigadier Denis Whitaker's assertion that big exercises like SPARTAN were always chaotic because of the umpires. J.L. Granatstein interview with Brigadier Denis Whitaker, 19 March 1991, Granatstein Interviews, DHH.

39 Quoted in Swettenham, *McNaughton*, II:284n.
40 Quoted in English, *The Canadian Army and the Normandy Campaign*, 146.
41 Simonds to Trumbull Warren, 1969, quoted in Dominick Graham, *The Price of Command: A Biography of General Guy Simonds* (Toronto: Stoddart, 1993), 59.
42 Ibid.
43 Quoted in Stacey, *Six Years of War*, 251; Charles P. Stacey, Report no. 94, 'GHQ Exercise SPARTAN, March 1943,' 12 May 1943, DHH.
44 Comments by Crerar on draft of *Six Years of War*, file 82/983, folder 4, DHH; Crerar to Stacey, 7 June 1952, CP, MG30 E157, vol. 21, file 958C.009(D329), LAC.
45 SPARTAN was discussed, but apparently only with regard to the performance of the 'Z' Composite Group. RAF Memorandum of Conversation with Lt-General Sir Archibald Nye, 16 March [18 March] 1943, MP, vol. 249, WD, Appendix 'J.'
46 Draft of *Missing from the Record*, RP, vol. 58; Richard S. Malone, *Missing from the Record* (Toronto: Collins, 1946), 139. The official report in fact made such bad reading that the War Office only ventured to send an abridged edition to Canada.
47 Paget to McNaughton, 15 March 1943, MP, vol. 248, PA 6-9-2.
48 Extracts from Copy no. 20 of Exercise SPARTAN: Report by Chief Umpire; Notes on Differences between 'First Draft' and Printed Versions of: GHQ Exercise SPARTAN: March 1943: Comments by Commander-in-Chief Home Forces, RG24, vol. 10414, file 200G1.033(D2), DHH.
49 Swettenham, *McNaughton*, II:283.
50 Memorandum, Conversation General McNaughton with General Sir Alan Brooke, CIGS, War Office, at 1530 hrs, 5 April [8 April] 1943, MP, vol. 249, WD, PA 1-0-4, Appendix 'I.'
51 Crerar to McNaughton, 9 April 1943, CP, vol. 1, file 958C.009, 'Confidential Report on Officers.'
52 Crerar, 'Memorandum on Action of 2 Canadian Division Exercise SPARTAN 4-6 March,' 9 May 1943, CP, vol. 1, file 958C.009, 'Confidential Report of Officers.' Bullen-Smith would be relieved from command in Normandy in July 1944.
53 Crerar to McNaughton, 9 April 1943. In early May Crerar prepared a further memorandum of his evaluations 'should it become necessary.' Crerar, Memorandum on Action of 2 Canadian Division. One of the British staff officers assigned to the Dieppe planning staff declared after the war that Roberts did not possess 'either the intellectual ability nor the powers of command which would have fitted him for the terrible task which was entrusted to him.'

Goronwy Rees, 'Operation Jubilee,' *Sunday Times*, 28 July 1963. Brigadier Denis Whitaker felt that Roberts was the scapegoat for Dieppe. *Dieppe: Tragedy to Triumph* (Toronto: McGraw-Hill Ryerson, 1993), 290.
54 Crerar to McNaughton, 9 April 1943. In fact, Montgomery also identified this weakness in Roberts. Montgomery, Some Notes on the Broader Aspects of BEAVER III, 23 April 1942, CP, vol. 2, file 958C.009(D182).
55 Crerar to McNaughton, 5 March 1943, CP, vol. 1, file 958.009(D23), GOC file 6-1-7, 'Confidential Report on Officers.' This date simply cannot be right for it was the second day of the exercise.
56 Memorandum of Conversation with Paget, 29 July 1943, MP, vol. 202, PA 6-9-5-2.
57 McNaughton WD, 19 April 1943, MP, vol. 248. Sansom's officers were using the term 'forward body' rather than 'advance guard.' McNaughton returned to Sansom's headquarters four days later, but no observations were recorded. At this time, Colonel Alec H. Gatehouse was the RAC adviser to II Canadian Corps.
58 Memorandum of a Conference with General Sir Alan Brooke, CIGS, at the War Office, 1530 hrs, 14 September [18 September] 1943, MP, vol. 250, WD, PA 1-18-1, Appendix 'G.' During LINK, II Canadian Corps controlled the 1st Polish Armoured Division and the 61st Infantry Division.
59 R.W. Thompson, *Generalissimo Churchill* (London: Hodder and Stoughton, 1973), 93.
60 Basil Liddell Hart, Main Impressions of the SPARTAN Exercise, March 1–12, 1943, file 11/1943/9, LHP, LHCMA; J.L. Granatstein interview with Major-General M.P. Bogert, 8 September 1991, Granatstein Interviews, DHH.
61 Charles P. Stacey, *Arms, Men, and Governments: The War Policies of Canada, 1939–1945* (Ottawa: Queen's Printer, 1970), 232.
62 Montague to Stuart, 18 October 1943, RG24, vol. 10033, file 9/Senior Appointments/1.
63 On 5 January 1944, Stuart (by this time Acting Army Commander) told Crerar in Italy that Sansom 'should be replaced earliest.' Two days later Stuart cabled Ottawa that Brooke and Montgomery 'are equally anxious that Sansom should be replaced earliest.' Stuart to Ralston and Murchie, 7 January 1944, DND, RG24, vol. 10033, file 9/Senior Appointments/1/2.
64 Hamilton, *Monty: Master of the Battlefield*, 174; David Fraser, *Alanbrooke* (London: Collins, 1982), 86.
65 Alanbrooke, 'Notes for My Memoirs,' AbP, file 5/2/20, p. 654, LHCMA.
66 English, *The Canadian Army and the Normandy Campaign*, 145.
67 Alanbrooke, 'Notes for My Memoirs,' file 5/2/20, 654.

11. The Sicily Incident

1 In early June 1943, McNaughton, Crerar, Sansom, Keller, Burns, Worthington, and Stein all watched Simonds and Wyman take their formations through combined training in Scotland. McNaughton War Diary (hereafter WD), 7 June 1943, McNaughton Papers (hereafter MP), MG30 E133, vol. 249, LAC.
2 Memorandum of a Conversation, General McNaughton – Lt.-Gen. Sir Archibald Nye (VCIGS) at the War Office, 1600 hrs, 27 May [28 May] 1943, MP, vol. 131, WD, Appendix 'WW'; John Swettenham, *McNaughton, II: 1939–1943* (Toronto: Ryerson, 1969), 297.
3 Memorandum of a Conversation, McNaughton–Nye, 27 May 1943.
4 Memorandum of a Conversation, General McNaughton–General Sir Bernard Montgomery, 31 May 1943 [1 June], MP, vol. 249, WD, Appendix 'DDD.' Montgomery actually went to Malta on 3 July. Daniel Dancocks stated that McNaughton 'learned via the War Office that Montgomery had suggested instead that McNaughton visit during operation HUSKY rather than before,' but this was simply incorrect. *The D-Day Dodgers: The Canadians in Italy, 1943–1945* (Toronto: McClelland and Stewart, 1991), 48.
5 Memorandum of a Conversation, General McNaughton–Lt.-Gen. Sir Archibald Nye, (VCIGS), at the War Office, 1500 hrs, 12 June 1943 [14 June], MP, vol. 250, WD, Appendix 'R.'
6 Gerald W.L. Nicholson, *Official History of the Canadian Army in the Second World War, II: The Canadians in Italy, 1943–1945* (Ottawa: Queen's Printer, 1957), 178.
7 Telegram, War Office to Freedom (Eisenhower), Algiers, 19 June 1943, quoted in Charles P. Stacey, *Arms, Men, and Governments: The War Policies of Canada, 1939–1945* (Ottawa: Queen's Printer, 1970), 225. Stacey added that McNaughton probably never saw this cable as he made no mention of it in his detailed report prepared after returning from Sicily in July.
8 Memorandum of a Discussion, General McNaughton–Maj.-Gen. Kennedy, (DMO), at the War Office, 1200 hrs, 19 June [20 June] 1943, MP, vol. 250, WD, Appendix 'Z.' Brooke's cable to Eisenhower may be the one originating from the War Office stating that Stuart and one staff officer would arrive on 7 July. McNaughton, Report of Visit to North Africa 6-20 July [22 July] 1943, MP, vol. 249; Swettenham, *McNaughton*, II:298.
9 Stacey, *Arms, Men, and Governments*, 225.
10 Memorandum, 8 May 1943, MP, vol. 136, WD, Appendix 'U'; Swettenham, *McNaughton*, II:297. Stuart's comment was on a message to Montgomery about the Fourth Canadian Victory Loan.

11 The other staff officers included Brigadier Charles Foulkes, BGS of First Canadian Army, Brigadier J.E. Genet, Chief Signal Officer, First Canadian Army, Lieutenant-Colonel D.C. Spry and Major T. Martin, Personal Assistants to McNaughton, and Stuart respectively.
12 Alfred D. Chandler, Jr, ed., *The Papers of Dwight David Eisenhower: The War Years*, 5 vols. (Baltimore: Johns Hopkins University Press, 1970), II:1254. Chandler was unable to locate Bedell Smith's letter to Eisenhower indicating that McNaughton had been denied entry. The letter was designated 114SF. Eisenhower actually disembarked near the Canadian sector and went looking for a Canadian officer to greet and welcome to his command. Carlo D'Este, *Eisenhower: A Soldier's Life* (New York: Henry Holt, 2002), 434.
13 Lieutenant-Colonels Henderson, Lockhart, Sparling, and Love.
14 McNaughton, Report of Visit to North Africa; Comments by Charles Foulkes, October 1968, quoted in Stacey, *Arms, Men, and Governments*, 225.
15 Richard S. Malone, *A Portrait of War, 1939–1943* (Toronto: Collins, 1983), 170. It is a fact that 1st Canadian Infantry Division lost many of its echelon vehicles at sea during the invasion. Verifying Montgomery's threat to have McNaughton arrested is problematic. Malone is not a trustworthy source. However, Richard Lamb does indicate that Montgomery issued the threat. *Montgomery in Europe 1943–1945: Success or Failure?* (London: Buchan and Enright, 1983), 29. Moreover, it does sound like something Montgomery would have said.
16 Nigel Nicolson, *Alex: The Life of Field Marshal Earl Alexander of Tunis* (London: Weidenfeld and Nicolson, 1973), 120; Brian Holden Reid in *Churchill's Generals*, ed. John Keegan (London: Grove Weidenfeld, 1991), 111. William G.F. Jackson stated that Alexander 'tolerated Stillwell, making allowances for his idiosyncrasies. The fact that the two never fell out is to Alexander's credit rather than Stillwell's.' *Alexander of Tunis as Military Commander* (London: Batsford, 1971), 123; McNaughton, Report of Visit to North Africa; Swettenham, *McNaughton*, II:300.
17 Stuart returned to England on 25 July and stayed with the McNaughtons as their house guest for the night.
18 Francis de Guingand, *Operation Victory* (London: Hodder and Stoughton, 1947), 184; Carlo D'Este, *Bitter Victory: The Battle for Sicily, 1943* (New York: Dutton, 1988), 616. De Guingand had apparently authorized, without Montgomery's knowledge, visits by three observers. A gunner named Colonel Godfrey Jeans actually stayed in Sicily for three weeks before Montgomery discovered him and sent him packing. An unnamed infantry officer never got farther than Malta. Lamb, *Montgomery in Europe*, 30.

19 Philip Ziegler, *Mountbatten: A Biography* (New York: Alfred A. Knopf, 1985), 204; Dancocks, *The D-Day Dodgers*, 48.
20 Stacey, *Arms, Men, and Governments*, 227.
21 Sir Bernard Law Montgomery, *The Memoirs of Field Marshal Montgomery of Alamein* (London: Collins, 1958), 184–5.
22 Nigel Hamilton, *Monty: Master of the Battlefield, 1942–1944* (London: Hamish Hamilton, 1983), 332; Dominick Graham, *The Price of Command: A Biography of General Guy Simonds* (Toronto: Stoddart, 1993), 182; Howard Graham, *Citizen and Soldier: The Memoirs of Lieutenant-General Howard Graham* (Toronto: McClelland and Stewart, 1987), 182; Lamb, *Montgomery in Europe*, 30; Charles P. Stacey, *A Date with History: Memoirs of a Canadian Historian* (Ottawa: Deneau, 1983), 234; Terry Copp, 'Examining a General's Dismissal,' *Legion Magazine* (May–June 1997).
23 McNaughton hoped that Malone 'might be able to help our relationship with the British HQ' and 'protect our Canadian interests a bit.' *A Portrait of War*, 171.
24 Colonel Dick Malone, *Missing From the Record* (Toronto: Collins, 1946), 62; *A Portrait of War*, 170. Malone never mentioned Montgomery asking Simonds for his opinion in his later book and actually ran into McNaughton at the Canadian transport pool at Syracuse later in August.
25 David Bercuson, *Maple Leaf against the Axis: Canada's Second World War* (Toronto: Stoddart, 1995), 159; Bill McAndrew, *Canadians and the Italian Campaign, 1943–1945* (Montreal: Art Global, 1996), 56; Swettenham, *McNaughton*, II:300. Dancocks stated: 'This was a clear violation of his right as senior Canadian army officer overseas, to keep tabs on Canadian troops in the field.' *The D-Day Dodgers*, 49.
26 De Guingand, *Operation Victory*, 184. 'A warning order regarding the visit,' he suggested, 'would no doubt have led to a request for a postponement until the situation was sufficiently assured.'
27 Alun Chalfont, *Montgomery of Alamein* (New York: Atheneum, 1976), 210; Hamilton, *Monty: Master of the Battlefield*, 332.
28 Stacey, *Arms, Men, and Governments*, 227. During the 31 May meeting with McNaughton at the War Office, Montgomery proposed that he deal only with Simonds 'on purely domestic Cdn matters.' Indeed, Montgomery accurately viewed Simonds as the Senior Canadian Officer in HUSKY who could 'use his special authority to deal directly with General McNaughton if the situation so demanded.' McNaughton agreed to the proposal with the understanding that it did not overburden Simonds. Memorandum of Conversation, McNaughton–Montgomery, 31 May 1943.
29 Alexander to Montgomery, 16 July 1943, Alexander Papers (hereafter AP), WO 215/55, file 27, PRO.

30 Swettenham, *McNaughton*, II:301; Harry C. Butcher, *My Three Years with Eisenhower* (New York: Simon and Schuster, 1946), 367. Butcher considered the whole affair a 'gross insult to Canada,' that Eisenhower would get the blame and that 'the affront to Canada would be picked up by the anti-British press in America and there would be hell to pay.'
31 Chandler, *The Papers of Dwight David Eisenhower*, I:179; Robert H. Ferrell, ed., *The Eisenhower Diaries* (New York: Norton, 1981), 50. Lieutenant-General Brehon B. Somervell was the U.S. Army's chief logistician.
32 Immediately before the invasion Ottawa had expressed concern that the War Office would refuse Canada the right to have its forces fully identified in the initial press releases. When Eisenhower released his first communiqué the Canadians were not mentioned. However, the omission was corrected quickly when Washington released its own statement giving full credit to Canadian forces. Lester B. Pearson, *Mike: The Memoirs of the Right Honourable Lester B. Pearson, I: 1897–1948* (Toronto: University of Toronto Press, 1972), 241.
33 Norman Hilsman, *Alexander of Tunis* (London: W.H. Allen, 1952), 190.
34 Butcher, *My Three Years with Eisenhower*, 367–8.
35 Alexander to Brooke, 18 July 1943, AP, WO 215/55, file 27. After the war Alexander told Richard Malone: 'It was unfortunate that General McNaughton's visit came when the Canadian division was deeply involved in its first battle against a formidable foe … a visit at this time by such an important person would have caused extra work and some embarrassment.' *A Portrait of War*, 170–1.
36 Brooke to Alexander, 19 July 1943, AP, WO 215/55, file 27.
37 Quoted in Stacey, *Arms, Men, and Governments*, 255. McNaughton also wrote to Paget that facilities afforded the Canadian officers to study the Sicilian campaign were 'both inadequate and embarrassing.' McNaughton to GHQ, Home Forces, 26 July 1943, MP, vol. 131, WD, PA 1-0-3.
38 Memorandum of a Conversation with General Sir Alan Brooke at the War Office, Wednesday, 21 July [23 July] 1943, MP, WD, vol. 250, Appendix 'M'; Alanbrooke Diary, 21 July 1943, 'Notes for My Memoirs,' 5/2/21, 743, Alanbrooke Papers (hereafter AbP), Liddell Hart Centre for Military Archives (hereafter LHCMA), King's College, London; Alex Danchev and Daniel Todman, eds., *War Diaries, 1939–1945: Field Marshal Lord Alanbrooke* (London: Weidenfeld and Nicolson, 2001), 431; Brooke to Montgomery, 29 September 1943, Montgomery Papers, Imperial War Museum (hereafter IWM).
39 Swettenham, *McNaughton*, II:234; Mel Thistle, ed., *The Mackenzie–McNaughton Wartime Letters* (Toronto: University of Toronto Press, 1975), 116.
40 Stacey, *A Date with History*, 123–4.
41 Memorandum of a Conversation with General Sir Alan Brooke, 21 July 1943.

42 McNaughton, Report of Visit to North Africa; Nicholson, *The Canadians in Italy*, 177. Bill McAndrew cited this as the main reason but implied that Alexander told McNaughton that Montgomery did not want any distractions for Simonds. This, however, was not the case. *Canadians and the Italian Campaign*, 56.
43 Telegram, quoted in Captain J.H. Hitsman, Historical Report no. 182, Historical Section, Canadian Military Headquarters, 23 May 1949, 'The Strategic Role of First Canadian Army, 1942-1944,' DHH. Exactly what Stuart was referring to is unclear.
44 Memorandum of a Conversation with General Sir Alan Brooke, 21 July 1943.
45 Swettenham, *McNaughton*, II:305.
46 Hamilton, *Monty: Master of the Battlefield*, 333.
47 Memorandum of a Conversation with General Sir Alan Brooke, 21 July 1943.
48 Alanbrooke Diary, 21 July 1943, 5/2/21; Memorandum of a Conversation with General Sir Alan Brooke, 21 July 1943.
49 Swettenham, *McNaughton*, II:305.
50 Alanbrooke, 'Notes for My Memoirs,' 743-4. Chalfont believed that Alexander 'had no alternative' but to support Montgomery. *Montgomery of Alamein*, 210.
51 Brooke to Alexander, 22 July 1943, AP, WO 215/55, file 27.
52 Alexander to Brooke, 22 July 1943, AP, WO 215/55, file 27. Over in the Seventh Army area Patton had threatened to place Churchill's science adviser, Professor Solly Zuckerman, and his party under close arrest for being in his area of operations. Solly Zuckerman, *From Apes to Warlords* (New York: Harper and Row, 1978), 200-1.
53 Montgomery to Crerar, 23 July 1943, CP, MG30 E157, vol. 7, file 958C.009(D172), LAC.
54 Montgomery to McNaughton, 26 July 1943, MP, vol. 200. Montgomery stated: 'I had to intervene. Simmonds [*sic*] sacked Graham in a very hasty manner, having I think lost his temper; I imagine there were faults on both sides. Anyhow I stepped in, smoothed the thing out, and Graham went back to his Bde.'; Montgomery to Alexander, 6 August 1943, AP, WO 215/55, file 27.
55 John W. Pickersgill, ed., *The Mackenzie King Record, I: 1939-1943* (Toronto: University of Toronto Press, 1960), 523. In fact, it had not been the War Office at all. Eisenhower had raised every objection against mention of Canada in the communiqué. Charles P. Stacey, *Canada and the Age of Conflict: A History of Canadian External Policies, II: 1921-1948* (Toronto: Macmillan,

1981), 339; W.A.B. Douglas and Brereton Greenhous, *Out of the Shadows: Canada and the Second World War*, rev. ed. (Toronto: Dundurn, 1995), 134.
56 Memorandum of a Discussion, Hon. J.L. Ralston–Lt.-Gen. McNaughton–Lt.-Gen. K. Stuart at HQ First Cdn Army, 29 July [2 August] 1943, MP, WD, PA 5-3-1, Appendix 'X.' Though greatly annoyed with Alexander and Montgomery, McNaughton wished to repay Eisenhower's hospitality by addressing a congratulatory message to him with regard to the fine progress of the Sicilian campaign. He deemed it essential to keep the sympathy and goodwill of the Americans in North Africa.
57 Montgomery to Crerar, 25 August 1943, CP, vol. 7, file 958C.009(D172).
58 Memorandum of a Discussion, Ralston–McNaughton–Stuart, 29 July 1943.
59 McNaughton to Montgomery, 15 August 1943, MP, vol. 249; Swettenham, *McNaughton*, II:318.
60 Chalfont, *Montgomery of Alamein*, 210; Malone, *Missing from the Record*, 63; Dancocks, *The D-Day Dodgers*, 112.
61 McNaughton, Visit to General Sir Bernard Montgomery, H.Q. Eighth Army, 24–25 Aug [1 September] 1943, MP, vol. 250, WD, PA 1-3-14-2.
62 Swettenham, *McNaughton*, II:320.
63 Montgomery Diary, 24 September 1943, BLM/45, LHCMA.
64 Montgomery, *Memoirs*, 184–85.

12. Broken Dagger: A Corps in Italy

1 Lieutenant-Colonel D.C. Spry, untitled memorandum, 6 May 1943, McNaughton Papers (hereafter MP), MG30 E133, vol. 136, PA 1-14-1, LAC. Brigadier Elliot N. Rodger called Spry, McNaughton's personal assistant, and informed him that the Director of Staff Duties, Major-General Alexander Galloway, had asked for the state of readiness on behalf of Nye. Rodger made it clear that the War Office did not intend to form a corps.
2 McNaughton to Stuart, 6 May 1943, MP, vol. 238, War Diary (hereafter WD), Appendices 'M,' 'N,' and 'R.' In late July 1943, Paget told McNaughton that the British were short 88,000 for the newly raised 21st Army Group while at the same time the British manpower commitment to the Canadian Army was 90,000. Moreover, as far as McNaughton was concerned, British plans to keep their divisions up to strength 'do not rest on a realistic basis' – a conviction shared by Morgan. Memorandum of a Discussion, Hon. J.L. Ralston – Lt.-Gen. McNaughton – Lt.-Gen. K. Stuart at H.Q. First Cdn Army, 29 July [2 August], MP, vol. 250, WD, Appendix 'X'; Memorandum of a Conference, General McNaughton – General Morgan, Chief of Staff to the

Supreme Allied Commander (hereafter COSSAC) at Norfolk House, 1515 Hrs, 17 May [20 May] 1943, MP, vol. 249, WD, PA 1-0-6, Appendix 'EE.'

3 Of the eleven British divisions, one was an airborne division and another was a Royal Marine division.

4 McNaughton to Stuart, 6 May 1943. BRIMSTONE seems to have already been put on the shelf at this point, as far as the British were concerned.

5 Stuart to McNaughton, 8 May 1943, and McNaughton to Stuart, 9 May 1943, MP, vol. 136, WD, PA 1-14-1, Appendix 'U.' Just how one Canadian division in Sicily and one in Sardinia were to be united, and where, was something Stuart never addressed.

6 Memorandum of a Conversation, General McNaughton – General Paget (Commander-in-Chief, Home Forces), at GHQ, Home Forces, at 1600 hrs, 19 May [28 May] 1943, MP, vol. 131, Appendix 'JJ.' Paget was also concerned that the army's efforts were somewhat divided between training for the assault role and training for the follow-up role.

7 Memorandum of a Conversation, General McNaughton – Lt.-Gen. Sir Archibald Nye, Vice Chief of the Imperial General Staff, at the War Office, 1500 hrs, 12 June [14 June] 1943, MP, vol. 250, Appendix 'R'; Memorandum of a Conversation, Gen. McNaughton – Gen. Paget, 1130 hrs, 17 June [18 June] 1943, MP, vol. 131, WD, PA 5-0-3, Appendix 'V.' On 15 July COSSAC submitted its outline plan for OVERLORD to the COS.

8 Memorandum of Conversation, Lt.-Gen. A.G.L. McNaughton – Lt.-Gen. K. Stuart – Lt.-Gen. F.E. Morgan, at Norfolk House, 1500 hrs, 27 July [31 July] 1943, MP, vol. 250, WD, Appendix 'U.' The military disadvantage of occupying Italy was also discussed, but specific views were not recorded. Stuart may have felt emboldened, for he told Morgan that COSSAC planning seemed to be driven more by politics than by military concerns. According to McNaughton, Morgan made no reply.

9 Memorandum of Conversation, McNaughton – Stuart – Morgan, 27 July 1943.

10 Alex Danchev and Daniel Todman, eds., *War Diaries, 1939–1945: Field Marshal Lord Alanbrooke* (London: Weidenfeld and Nicolson, 2001), 426, 434, diary entries for 5 and 27 July 1943.

11 Ibid., 432–3, diary entries for 21 and 24 July 1943.

12 For specifics, see First Cdn Army Op Instr Number 1, RANKIN Case 'C,' 30 December 1943, 8-3-14/9, DHH, Department of National Defence, Ottawa; Lieutenant-General E.W. Sansom, Notes of Commander-in-Chief's Conference of Army and Corps Commanders on 18 November [19 November] 1943 at HQ 21 Army Group, DHH.

13 Memorandum of a Discussion, Hon. J.L. Ralston – Lt.-Gen. McNaughton

- Lt.-Gen. K. Stuart, at H.Q. First Cdn Army, 29 July [2 Aug] 1943, MP, vol. 250, WD, Appendix 'X.'
14 John R. Campbell, 'James Layton Ralston and Manpower for the Canadian Army,' MA thesis, Wilfrid Laurier University, Waterloo, 1986, 82. 'Ralston's memos after 1943,' observed David A. Wilson, 'reflect exactly those same concerns with shortages of recruits as they had when the Army was still fighting to build a field army in 1941.' 'Close and Continuous Attention: Human Resources Management in Canada during the Second World War,' PhD diss, University of New Brunswick, 1997, 272, 274.
15 Memorandum of a Discussion, Ralston – McNaughton – Stuart, 29 July 1943.
16 James L. Ralston, 'Overseas Trips in 1943,' Record of Diary Excerpts and Communications Regarding Change in Canadian Army Command, 27th December 1943, entry for 29 July, Ralston Papers (hereafter RP), MG27 III BII, vol. 59, LAC.
17 Memorandum of a Discussion, Ralston – McNaughton – Stuart, 29 July 1943. McNaughton felt that a corps in the Mediterranean 'might have opportunities for making a useful contribution and he valued the battle experience which would be gained.'
18 Memorandum of a Discussion, Ralston – McNaughton – Stuart, 29 July 1943.
19 J.L. Granatstein, *The Politics of the Mackenzie King Government, 1939–1945* (Toronto: Oxford University Press, 1975), 265; John W. Pickersgill, ed., *The Mackenzie King Record, I: 1939–1944* (Toronto: University of Toronto Press, 1960), 568–70.
20 Ralston, 'Overseas Trips in 1943,' entry for 29 July 1943. Charles P. Stacey concluded that King's 'contemporary record seems to leave very little doubt that by July Ralston was determined that McNaughton should go.' *Arms, Men, and Governments: The War Policies of Canada, 1939–1945* (Ottawa: Queen's Printer, 1970), 232. Malone informed Ralston in writing that he intended to make it clear that it was 'the military chiefs who desired the replacement, not yourself.' Malone to Ralston, 15 May 1946, RP, vol. 58.
21 Memorandum of Conversation, Ralston – McNaughton – Stuart, 29 July 1943.
22 Ralston, 'Overseas Trips in 1943,' entry for 29 July 1943.
23 Conversation General McNaughton – Hon. J.L. Ralston, Minister of National Defence – General Stuart, Chief of the General Staff (hereafter CGS), 2 Aug [6 Aug] 1943, MP, vol. 250, WD, Appendix 'C'; Ralston, 'Overseas Trips in 1943,' entry for 2 August 1943.
24 Ralston, 'Overseas Trips in 1943,' entry for 3 August 1943.

25 Danchev and Todman, *War Diaries*, 435, entry for 3 August 1943; Vincent Massey, *What's Past Is Prologue: The Memoirs of the Right Honourable Vincent Massey* (Toronto: Macmillan, 1963), 381.
26 Conversation, General McNaughton – Hon. J.L. Ralston, Minister of National Defence – Lt.-Gen. K. Stuart, CGS, 5 August [9 August] 1943, MP, vol. 250, WD, Appendix 'F.' Ralston explained that because of changed circumstances, Brooke had revised his view of the need to maintain the army.
27 Conversation, McNaughton – Ralston – Stuart, 5 August 1943; Ralston, 'Overseas Trips in 1943,' entry for 5 August 1943.
28 Conversation, McNaughton – Ralston – Stuart, 5 August 1943.
29 Ibid. For Ralston, this was 'the cardinal point,' and he noted that he would 'not want to quote that as his [McNaughton's] considered opinion in view of what had been said at our conference on Thursday, July 29th, re battle experience and morale.' Ralston, 'Overseas Trips in 1943,' entry for 5 August 1943.
30 Massey, *What's Past Is Prologue*, 382, entry for 5 August 1943. Massey noted that Ralston 'put me fully into the picture' regarding the fate of the army, 'telling me that the only other person who knew the full story was Stuart.'
31 Ralston, 'Overseas Trips in 1943,' entry for 5 August 1943.
32 Memorandum, 6 August 1943, MP, vol. 250, WD, Appendix 'H.'
33 Ibid.
34 Ibid.
35 Memorandum of Discussion, General McNaughton – Hon. J.L. Ralston, Minister of National Defence – Lt.-Gen. K. Stuart, CGS, at HQ First Cdn Army, 7 August [9 August] 1943, MP, vol. 250, WD, Appendix 'J'; Ralston, 'Overseas Trips in 1943,' entry for 7 August 1943.
36 War Cabinet Chiefs of Staff Committee, Minutes of Meeting held in Hotel Frontenac, 11 August 1943, file 314.009(D332), DHH. The other Canadian chiefs included Vice Admiral P.W. Nelles and Air Marshal L.S. Breadner. The British chiefs included Admiral of the Fleet Sir Dudley Pound, Air Chief Marshal Sir Charles F.A. Portal, Vice Admiral the Lord Louis Mountbatten, and Lieutenant-General Sir Hastings L. Ismay.
37 The division came under Dempsey's XIII Corps, which began to plan for invading Italy on 13 August.
38 Pickersgill, *The Mackenzie King Record*, I:504, 545. As James Eayrs stated: 'In the urgency of the need to follow up the successes in Sicily with a strike across the Straits of Messina ... the precise role of the Canadian forces had been neglected.' *The Art of the Possible: Government and Foreign Policy in Canada* (Toronto: University of Toronto Press, 1961), 81. Donald G. Creighton has suggested that the experience in Sicily had a somewhat intoxicating

effect on Ottawa. *The Forked Road: Canada, 1939–1957* (Toronto: McClelland and Stewart, 1976), 86.
39 Memorandum of Conversation, McNaughton – Paget, 17 June, 1943. Granatstein has argued that 'whatever General McNaughton wanted, it made no military sense to send the division back to England after just five weeks in action.' *Canada's Army: Waging War and Keeping the Peace* (Toronto: University of Toronto Press, 2002), 227.
40 Quoted in Charles P. Stacey, *A Very Double Life: The Private World of Mackenzie King* (Toronto Macmillan, 1976), 199.
41 Gerald W.L. Nicholson, *Official History of the Canadian Army in the Second World War, II: The Canadians in Italy, 1943–1945* (Ottawa: Queen's Printer, 1957), 188.
42 Memorandum of Discussion, Lt.-Gen. McNaughton and Maj.-Gen. Kennedy, Director of Military Operations, War Office, 16 August [18 August] 1943, MP, vol. 138, WD, PA 1-14-1, Appendix 'P.' That same day the War Office sent a message to Allied Forces Headquarters (hereafter AFHQ) 15th Army Group that the Canadians were authorized to participate. War Office to AFHQ 15AG, 17 August 1943, MP, vol. 138, PA 1-17-1. To cover himself, McNaughton sought out Simonds and asked for an assurance that the plan was feasible. McNaughton noted that Simonds 'gave me an unqualified assurance that he and all Comds under his orders were satisfied and confident of success.' McNaughton, Memorandum of Conversation with Maj-Gen. Simonds, 25 August [31 August] 1943, MP, vol. 250, WD, PA 1-3-14-2, Appendix 'V.'
43 Albert N. Garland and Howard McGraw Smyth, *Sicily and the Surrender of Italy* (Washington: Center of Military History, 1986), 440.
44 'We entered into a major war on the continent of EUROPE,' Montgomery exclaimed, 'without a very clear plan as to how we would fight the campaign.' Nigel Hamilton, *Monty: Master of the Battlefield, 1942–1944* (London: Hamish Hamilton, 1983), 387, 431; Sir Bernard Law Montgomery, *El Alamein to the River Sangro* (London: Hutchinson, 1952), 99–100.
45 Memorandum of a Conversation, McNaughton – Kennnedy, 16 August 1943; Pickersgill, *The Mackenzie King Record*, I:558; Nicholson, *The Canadians in Italy*, 341.
46 Memorandum of a Conversation, Lt.-Gen. A.G.L. McNaughton – Lt.-Gen. Sir Archibald Nye,VCIGS, at the War Office, 2 September [6 September] 1943, MP, vol. 250, WD, Appendix 'A.'
47 Ibid.
48 Memorandum of a Conference with General Sir Alan Brooke, Chief of the Imperial General Staff, at the War Office, 1530 hrs, 14 September [18 Sep-

tember] 1943, MP, vol. 139, PA 1-18-1, Appendix 'G.' When McNaughton suggested that political considerations might override military ones, Brooke replied that the arrangements for BOLERO 'were regarded by the US as firm and must be implemented to the target date; he himself felt there should be more elasticity in these sorts of agreements, but the US regarded them as in the nature of a contract.'

49 Danchev and Todman, *War Diaries*, 452, entry for 13 September 1943.
50 Memorandum of a Conference, McNaughton – Brooke, 14 September 1943. Brooke referred to 1 May as the date for the invasion as premature.
51 Alanbrooke Diary, 15 September 1943, 'Notes for My Memoirs,' 5/2/22, Alanbrooke Papers, Liddell Hart Centre for Military Archives (hereafter LHCMA), King's College, London.
52 Quoted in Nicholson, *The Canadians in Italy*, 342.
53 Alanbrooke Diary, 15 September 1943, Garland and Smyth, *Sicily and the Surrender of Italy*, 436. Brooke noted on 1 November that his Mediterranean strategy had been 'emasculated' by the siphoning off of strength at American insistence for what he incredibly termed a 'nebulous 2nd Front.' He actually believed that had his 'soft underbelly' strategy prevailed, 'the war might have been finished in 1943!' Danchev and Todman, *War Diaries*, 465.
54 Memo to the Minister [Crerar to Ralston], 29 November 1943, RP, vol. 59, GOC 6-9. He argued that the issue of keeping First Canadian Army headquarters 'is a matter of high Government policy' and that it required consideration of both political and military factors derived from 'more authoritative knowledge of United Nations strategy and intentions than have been placed at my disposal now, or in the last two years.' It was for this reason that Crerar pressed McNaughton for operational experience two days later, for 'if the sequence of forthcoming events follows a normal course, those on HQ of a Cdn Corps will be engaged in operational responsibilities before HQ Cdn Army.' McNaughton denied his request three days later, declaring that 'it simply is not practicable to send any Senior Officer to the Mediterranean.' Crerar to McNaughton, 17 September 1943, and McNaughton to Crerar, 20 September 1943, Crerar Papers (hereafter CP), MG30 E157, vol. 2, file 958.009(D21), LAC.
55 This message from King to Churchill was relayed in a letter from Massey to Churchill on 4 October 1943. Quoted in Captain J.T. Hitsman, Historical Report no. 182, 'The Strategic Role of First Canadian Army, 1942–1944,' 23 May 1949, 176, DHH.
56 Massey, *What's Past Is Prologue*, 384.
57 Robert Speaight, *Vanier: Soldier, Diplomat, and Governor General* (Toronto: Collins, 1970), 260.

58 Memorandum of a Discussion with General Sir Alan Brooke, CIGS, War Office, Thursday, 7 October [14 October] 1943, MP, vol. 250, WD, Appendix 'E.' Brooke said that the plan now was to bring back the personnel of the 50th, 51st, 1st Airborne, and 7th Armoured as well as XXX Corps and leave their equipment behind.
59 Memorandum of a Discussion, McNaughton – Brooke, 7 October 1943.
60 Ralston, 'Overseas Trips in 1943,' entry for 5 November 1943; Memorandum of Discussions with Col. Ralston (Minister of National Defence) and Lt.-Gen. K. Stuart (CGS), 5 November [10 November] 1943, MP, vol. 250, WD, PA 5-0-3-2, Appendix 'F.'
61 Charles P. Stacey, *Official History of the Canadian Army in the Second World War, III: The Victory Campaign: The Operations in Northwest Europe, 1944–1945* (Ottawa: Queen's Printer 1960), 31. Paget proposed that British XII Corps come under McNaughton's First Canadian Army but that the 3rd Canadian Infantry Division and the 2nd Army Tank Brigade should be placed under I British Corps, which in turn was to serve under an American army for the invasion. McNaughton saw no alternative, telling Stuart on 28 October that there was no choice 'until the progress of the war makes possible a re-collection of Canadian units and formations in a homogenous Canadian Army ... Profound changes in our system of organization will be necessary ... I would appreciate early instructions for general guidance as to policy to be followed.' Minutes of a Discussion held with Lt.-Gen. Morgan CGS 21 Army Gp and General Paget C-in-C, 4 November 1943, MP, vol. 250, WD, Appendix 'C'; McNaughton to Stuart, 28 October 1943, General Service (hereafter GS) 2671, quoted in Historical Report no. 182, 'The Strategic Role of First Canadian Army,' 186–7; John Swettenham, *McNaughton, II: 1939–1943* (Toronto: Ryerson, 1969), 328–9.
62 Quoted in Nicholson, *The Canadians in Italy*, 341; Montgomery to Alexander, 28 October 1943, Alexander Papers (hereafter AP), WO 214/55, file # 27, PRO.
63 Alexander to Brooke, 16 October 1943, and Brooke to Alexander, 19 October 1943, AP, WO 215/55, file 27. According to Maurice Pope's diary for 3 August 1943, during Stuart's visit to Sicily in July both Alexander and Eisenhower seem to have responded favourably to Stuart's suggestion that an additional Canadian division be exchanged with a British division. Historical Report No. 182, 'The Strategic Role of First Canadian Army,' 166.
64 Alexander to Whiteley, 14 and 16 October 1943, and Alexander to Eisenhower, 15 October 1943, AP, WO 214/55, file 27. On this day the War Office relayed to Alexander a message from the COS stating that in accordance with TRIDENT decisions, the 50th, 51st, and 1st Airborne were coming

back, but that this allocation 'has the disadvantage that it leaves us deficient of battle experienced armoured div for OVERLORD. We have therefore examined the possibility of exchanging a British Armd Div ... for a Canadian Armd Div ... This would have the added advantage that we should meet the Canadian desire to form a Canadian Corps in the Mediterranean. We therefore intend to exchange the personnel of a Canadian and British Armoured Division and to despatch a Canadian Corps HQ ... The Prime Minister of Canada has agreed to this move of the Canadians.' FREEDOM signed C-in-C to 15th Army Group, 15 October 1943, AP, WO 215/55, file 27.
65 Whiteley to Alexander, 15 October 1943, AP, WO 214/55, file 27.
66 Alfred D. Chandler, Jr, ed., *The Papers of Dwight David Eisenhower: The War Years* (Baltimore: Johns Hopkins University Press, 1970), III:1516-17. It should also be noted that McNaughton had told Eisenhower in August, 'provided it was decided that the Mediterranean theatre would remain active we would welcome developing Cdn participation to a Corps level.' There is no evidence, however, that Eisenhower welcomed the idea. Memorandum of a Conversation with General Eisenhower at Allied Forces Headquarters (hereafter AFHQ), 20 August [31 August] 1943, MP, vol. 250, Appendix 'G.'
67 Massey, *What's Past Is Prologue,* 383; Historical Report no. 182, 'The Strategic Role of First Canadian Army,' 181.
68 Alexander to McNaughton, 9 November 1943, MP, vol. 138, PA 1-14-1-18.
69 See George Kitching, *Mud and Green Fields: The Memoirs of Major-General George Kitching* (Langley: Battleline, 1985), 162; Daniel G. Dancocks, *The D-Day Dodgers: The Canadians in Italy, 1943-1945* (Toronto: McClelland and Stewart, 1991), 208; Montgomery to Alexander, 28 October 1943, AP, WO 214/55, file 27; Crerar to McNaughton, 11 November 1943, CP, vol. 7, file 958C.009(D140).
70 Ralston, 'Overseas Trips in 1943,' entry for 27 November 1943.
71 Nicholson, *The Canadians in Italy,* 362-3. In early January 1944 the 11th, CIB of the 5th CAD relieved the 3rd CIB in the line along the Arielli River north of Ortona. It attacked on 17 January but quickly floundered and was withdrawn. The armoured division and the 1st Division were at the time in separate British corps, and Crerar had difficulty persuading Leese to combine them. They did not come together until 9 February 1944. William J. McAndrew, 5th Canadian Armoured Division: The Background, file 84/68, folder 3, DHH; Nicholson, *The Canadians in Italy,* 379.
72 J.L. Granatstein, *The Generals: The Canadian Army's Senior Commanders in the Second World War* (Toronto: Stoddart, 1993), 107.

73 D.C. Spry interview, October 1987, Spry Papers, MG30 E563, vol. 1, LAC; John Swettenham, *McNaughton, III: 1944–1966* (Toronto: Ryerson, 1969), 20; Desmond Morton, *A Military History of Canada* (Edmonton: Hurtig, 1990), 218. However, Burns pointed out after the war that the move of I Canadian Corps to join First Canadian Army only saved perhaps 2,000 men. *Manpower in the Canadian Army, 1939–1945* (Toronto: Clarke, Irwin, 1956), 24–43. In early August, McNaughton rationalized that if sufficient reinforcements were not forthcoming he would have to keep the 4th CAD out of the order of battle and perhaps even recommend its disbandment. Ralston admitted even here that manpower was a problem, while Stuart declared: 'This was no time in the conduct of the war to be disbanding formations.' Conversation, General McNaughton – Hon. J.L. Ralston, Minister of National Defence – General Stuart, CGS, 2 August [6 August] 1943, MP, vol. 250, WD, Appendix 'C.' See also D.C. Spry to The Senior Officer, Canadian Military Headquarters, 27 May 1943, MP, vol. 249, WD, Appendix 'TT.'
74 Stacey, *Arms, Men, and Governments*, 247; Bill McAndrew, *Canadians and the Italian Campaign, 1943-1945* (Montreal: Art Global, 1996), 156.
75 Ralston, 'Overseas Trips in 1943,' entry for 6 November 1943; John W. Pickersgill and Donald F. Forster, eds., *The Mackenzie King Record, II: 1944–1945* (Toronto: University of Toronto Press, 1968), 128. That the Canadian situation was not a unique one is proven by the debate in New Zealand over whether to leave the 2nd New Zealand Division in the Mediterranean at the end of the North African campaign or bring it home to join the other New Zealand divisions fighting in the South Pacific. The New Zealand official historian declared that the ultimate decision to send the 2nd Division to Italy 'issued from an interplay of complex and often conflicting forces.' Nothing described the Canadian situation better. Neville Crompton Phillips, *Official History of New Zealand in the Second World War, 1939–45, Italy, I: The Sangro to Cassino* (Wellington: War History Branch, Department of Internal Affairs, 1957), 25.
76 There had been considerable press publicity about the Canadian Army doing so. See COSSAC Memorandum no. 6, 4 May 1943, MP, vol. 249, WD.

13. The Final Months of McNaughton's Command

1 John W. Pickersgill, ed., *The Mackenzie King Record, I: 1939–1944* (Toronto: University of Toronto Press, 1960), 498. At the end of March, Crerar and Brooke discussed 'the best method of having McNaughton recalled back to Canada to avert his commanding a Canadian Army which he is totally inca-

pable of doing.' Alanbrooke Diary, 31 March 1943, 'Notes for My Memoirs,' 5/2/20, 661, Alanbrooke Papers (hereafter AbP), Liddell Hart Centre for Military Archives (hereafter LHCMA), King's College, London.

2 J.L. Granatstein, *The Generals: The Canadian Army's Senior Commanders in the Second World War* (Toronto: Stoddart, 1993), 218-219; McNaughton to Brooke, 21 November 1941, McNaughton Papers (hereafter MP), MG30 E133, vol. 197, PA 6-9-C-4, LAC. Stuart had no command experience since commanding the 7th Battalion, C.E., in 1918. 'Record of Service,' Stuart Papers (hereafter StP), MG30 E520, vol. 1, LAC.

3 Quoted in Paul D. Dickson, 'The Limits of Professionalism: General H.D.G. Crerar and the Canadian Army, 1914–1944,' PhD diss., University of Guelph, 1993, 430.

4 Ibid., 437; Charles P. Stacey, *A Date with History: Memoirs of a Canadian Historian* (Ottawa: Deneau, 1983), 126. The assessment document Dickson referenced is undated, but there seems to be no reason to dispute his suggestion of May.

5 Stuart to Ralston, 12 November 1943, Tel 2844, Massey to External, Ralston Papers (hereafter RP), MG27 III BII, vol. 59, LAC; J.L. Ralston, 'Overseas Trips in 1943,' Record of Diary Excerpts and Communications Regarding Change in Canadian Army Command, 27th December 1943, RP, vol. 59. Stuart indicated that the original discussion took place in June but he meant May. It also appears that Brooke first spoke to Maurice Pope about McNaughton's performance on SPARTAN prior to talking to Ralston and Stuart. John Swettenham, *McNaughton, II: 1939–1943* (Toronto: Ryerson, 1969), 285–6. Pope made no mention of this in his memoirs.

6 Dick Malone, *Missing From the Record* (Toronto: Collins, 1946), 140.

7 Stuart could not recall whether he or Brooke first raised the issue, but stated: 'In any event I cannot see that it makes the slightest difference. If the Chief of the Imperial General Staff (hereafter CIGS) had not elaborated his remarks made in Washington, I should have asked him to do so. Otherwise I should have failed in my duty to the Canadian Army and to the Canadian Government.' Memorandum, Substance of Conversations between General Sir Alan Brooke, CIGS, and Lieut-General K. Stuart, Chief of the General Staff (hereafter CGS) Canada, in connection with the fitness of Lieut-General A.G.L. McNaughton to comd in the field, 13 November 1943, RP, vol. 59.

8 Ibid. When Brooke read this memorandum he signed and dated it, adding: 'The above is a fair and accurate statement of my conversations with General Stuart.'

9 Charles P. Stacey, *Arms, Men, and Governments: The War Policies of Canada, 1939–1945* (Ottawa: Queen's Printer, 1970), 232.

10 Ralston, undated memorandum, quoted in Stacey, *Arms, Men, and Governments*, 233; Stuart to Ralston, 12 November 1943, Tel 2844, RP, vol. 59.
11 Conversation, General McNaughton – Hon. J.L. Ralston, Minister of National Defence – General Stuart, CGS, 5 August [9 August] 1943, MP, vol. 250, PA 5-3-1, Appendix 'F'; Memorandum of a Conversation General McNaughton – Lt.-Gen. Sir Archibald Nye, Vice Chief of the Imperial General Staff (hereafter VCIGS) at the War Office, 1600 hrs, 27 May [28 May], 1943, MP, vol. 131, PA 1-0-4, Appendix 'WW.' McNaughton pointed out to Ralston that 'very great value had been gained by the attachment of Cdn personnel to British First Army.' See also Chief of Staff to the Supreme Allied Commander (hereafter COSSAC) Memorandum no. 9, 23 May 1943, MP, vol. 249.
12 Stopford had commanded XII Corps on SPARTAN. Minutes of Discussion held with Lt.-Gen. Morgan CGS 21 Army Gp and General Paget, C-in-C, 4 November [10 November] 1943, MP, vol. 250, War Diary (hereafter WD), Appendix 'C'; Ralston, 'Overseas Trips in 1943,' entry for 23 November and 13 December 1943; Charles P. Stacey, 'Canadian Leaders of the Second World War,' *Canadian Historical Review* (hereafter *CHR*) 66, no. 1 (1985): 67.
13 Bill Rawling, 'The Generalship of Andrew McNaughton: A Study in Failure,' in *Warrior Chiefs: Perspectives on Senior Canadian Military Leaders*, ed. Bernd Horn and Stephen J. Harris (Toronto: Dundurn, 2001), 88.
14 Leese to Kennedy, 20 January 1944, John Noble Kennedy Papers (hereafter KP), box 5, WWII Manuscript Diaries, 1939–1943, Imperial War Museum (hereafter IWM).
15 Conversation, McNaughton – Ralston – Stuart, 5 August 1943; Ralston, 'Overseas Trips in 1943,' entry for 5 August 1943.
16 Pickersgill, *The Mackenzie King Record*, I:606–7.
17 Copy of Memorandum in General Stuart's Handwriting, Substance of a Conversation between the Minister, General McNaughton and General Stuart, held at Army HQ on 5 November 1943, RP, vol. 58, 'War Notes, Ralston'; Ralston, 'Overseas Trips in 1943,' entry for 5 November 1943; Memorandum of Discussion with Col. Ralston (Minister of National Defence) and Lt.-Gen. K. Stuart (CGS), 5 November [10 November] 1943, MP, vol. 250, WD, PA 5-0-3-2, Appendix 'F.' This source contains memoranda for 5, 8, 9, and 10 November.
18 Ralston, 'Overseas Trips in 1943,' entry for 8 November 1943.
19 Memorandum of Discussions with Col. Ralston (Minister of National Defence) and Lt.-Gen. K. Stuart (CGS), 8 November [10 November] 1943, MP, vol. 250, WD, Appendix 'F'; Copy of Memorandum in General Stuart's Handwriting, The Minister, General McNaughton and General Stuart at HQ, First Cdn Army on Monday, 8 November 1943, RP, vol. 58. Ralston

and Stuart had visited Paget earlier in the day and recorded that Paget said McNaughton was ill-suited healthwise and temperamentally to command. It is obvious that Paget did not look forward to the confrontation, for he recommended waiting until a definite replacement had been found before informing him officially. Copy of Memorandum in General Stuart's Handwriting, Substance of a Conversation between the Minister, General Paget and General Stuart on Monday, 8 November 1943, RP, vol. 58, 'War Notes, Ralston.'

20 Memorandum of Discussions with Ralston – Stuart, 8 November 1943.
21 Memorandum of Discussions with Ralston – Stuart, 9 November 1943.
22 Memorandum of Discussions with Ralston – Stuart, 10 November 1943. McNaughton and Paget had always had a good relationship, but not everyone liked Paget. Morgan considered him 'exceedingly difficult to work with' and 'narrow minded.' Memorandum of a Conference, General McNaughton – General Morgan (COSSAC), at Norfolk House 1515 hrs, 17 May [20 May] 1943, MP, vol. 249, PA 1-0-6, Appendix 'EE.'
23 Ralston, 'Overseas Trips in 1943,' entry for 9 November 1943. Ralston reiterated this in a cable to King. Massey to External, For Prime Minister from Ralston, G.S. 2819, 11 November 1943, RP, vol. 59.
24 McNaughton to King, G.S. 2826, 10 November 1943, RP, vol. 59.
25 Massey to External, From Ralston for Prime Minister, G.S. 2820, 11 November 1943, RP, vol. 59. When Paget read McNaughton's telegram to King he told Stuart that he considered it 'unbalanced.' Paget to Stuart, 11 November 1943, RP, vol. 59. When Brooke saw the telegram he concluded: 'It rather looks to me as if McNaughton is going off his head!' and 'He gave every indication and behaved as if he was approaching a bad nervous breakdown.' Alex Danchev and Daniel Todman, eds., *War Diaries, 1939–1945: Field Marshal Lord Alanbrooke* (London: Weidenfeld and Nicolson, 2001), 470, diary entry for 12 November 1943; Alanbrooke Diary, 12 November 1943, 818.
26 Crerar actually ordered Simonds to report for a mental and physical examination. Dominick Graham, *The Price of Command: A Biography of General Guy Simonds* (Toronto: Stoddart, 1993), 116–19. When McNaughton decided to openly challenge the cause of his relief in early 1944, Ralston noted: 'My own feeling is that ... this latest episode and some of the earlier seemingly inexplicable contradictions might be explained by increasing worsening of physical and nervous conditions brought on by strain.' Ralston to Stuart, 4 February 1944, RP, vol. 58, 'War Notes, Ralston.'
27 Paget to McNaughton, 11 November 1943, RP, vol. 59. Massey sent this letter off to King the next day. Massey to External, 12 November 1943, RP, vol. 59.

28 Massey to External, For Prime Minister from McNaughton, G.S. 2837, 12 November 1943, RP, vol. 59.
29 Massey to External, For Prime Minister from Ralston, G.S. 2831, 11 November 1943, RP, vol. 59; Stacey, *Arms, Men, and Governments*, 243.
30 Ralston, 'Overseas Trips in 1943,' entry for 12 November 1943.
31 Memorandum, Substance of conversation between General Paget, Lieut-General McNaughton and Lieut-General Stuart, at General Paget's Headquarters at 1600 hrs, 13 November 1943, RP, vol. 59. McNaughton's memorandum of the meeting lists Ralston as being present, but this appears to be an error.
32 Memorandum of Discussions with General Paget, C-in-C 21 Army Gp, Col. Ralston, Minister of National Defence, and Lt.-Gen. Stuart, CGS, Canada, 13 November [15 November] 1943, MP, vol. 250, WD, Appendix 'L.' Circumstantial proof that this passage does not make sense lay in the fact that some previous researcher scribbled two large question marks beside it.
33 Memorandum of Discussions, Paget – Ralston – Stuart, 13 November 1943; Ralston, 'Overseas Trips in 1943,' entry for 14 November 1943.
34 Nye offered three options. The position of army commander could be left vacant for an 'indefinite period,' Crerar could be brought back from Italy to take command immediately, or a British officer could be placed in command temporarily until Crerar could return. According to Nye, Paget favoured the second option but 'Crerar would lack any practical experience.' Nye to Brooke, 13 December 1943, RP, vol. 59.

Epilogue

1 Ralston, 'Overseas Trips in 1943,' entry for 30 November 1943, Ralston Papers (hereafter RP), MG27 III BII, vol. 59, LAC. Indeed, considerable tension existed between Crerar and Stuart. For Crerar's part it was because Stuart was his junior. Ralston noted that things got 'rather heated' when they were together.
2 McNaughton to Ralston, 25 December 1943, Cypher no. 3261, RP, vol. 59.
3 'Fighting Fit,' *Time*, 14 February 1944. J.L. Granatstein suggests that he returned to Canada 'a loose cannon, one posing great potential danger to the ship of state and the war effort' because of his refusal to stay quiet about his relief. King, however, felt that Ralston was 'prepared to precipitate the most appalling domestic situation and also to help destroy the possibility of the furtherance of a world order and actually assisting the enemy by showing division in the Dominion.' *Canada's War: The Politics of the Mackenzie King Government, 1939–1945* (Toronto: Oxford University Press, 1975), 352;

John W. Pickersgill and Donald F. Forster, eds., *The Mackenzie King Record, II: 1944–1945* (Toronto: University of Toronto Press, 1968), 157.
4 'McNaughton Talks,' *Time*, 9 October 1944.
5 R. McGregor Dawson argued that McNaughton returned to Canada with a 'smouldering grievance' against both Ralston and Stuart. *The Conscription Crisis of 1944* (Toronto: University of Toronto Press, 1961), 57; idem, 'The Revolt of the Generals,' *Weekend Magazine* 10, no. 44 (1960): 21–49; Brian Nolan, *King's War: Mackenzie King and the Politics of War, 1939–1945* (Toronto: Random House, 1988), 156.
6 'The Chief Steps Down,' *Time*, 3 October 1944.

Conclusion

1 The standard interpretation is well ingrained. For example, see Mark M. Boatner III's biographical sketch of McNaughton in *The Biographical Dictionary of World War II* (Novato: Presidio, 1996), 353–4.
2 Memorandum of a Discussion with Lt.-Gen F.E. Morgan, at Largiebeg, Sunday, 12 December 1943, McNaughton Papers (hereafter MP), MG30 E133, vol. 250, War Diary (hereafter WD), Appendix 'D,' LAC. Martha Ann Hooker has argued that McNaughton's efforts to 'entrench Canadian control of her armed forces practically cost him his military career,' but it is clear that this was only one aspect of his larger difficulties. 'In Defence of Unity: Canada's Military Policies, 1935–1944,' MA thesis, Carleton University, 1985, 236; Desmond Morton, *A Military History of Canada*, rev. ed. (Edmonton: Hurtig, 1990), 123.
3 J.L. Granatstein, *Canada's Army: Waging War and Keeping the Peace* (Toronto: University of Toronto Press, 2002), 214.
4 McNaughton quoted in C.J.V. Murphy, 'The First Canadian Army,' *Fortune* (January 1944): 164.
5 Interview with D.C. Spry, 16 May 1985. It is also apparent that both Pope and Pearson tried to deflect some of the negative criticism of McNaughton's performance on SPARTAN. Pope Diary, 16 June 1943, Pope Papers, MG27 IIIF4, vol. 1, LAC.
6 Alanbrooke Diary, 19 August 1944, 998; Alanbrooke Papers (hereafter AbP), LHCMA, Kings College, London. J.L. Ralston, 'Overseas Trips in 1943,' Record of Diary Excerpts and Communications Regarding Change in Canadian Army Command, 27th December 1943, entry for 11 December 1943, Ralston Papers (hereafter RP), MG27 III BII, vol. 59, LAC.
7 Nigel Hamilton, *Monty: Master of the Battlefield, 1942–1944* (London: Hamish Hamilton, 1983), 465.

8 Montgomery to Ellis, 19 March 1960, Some Comments on Major Ellis's Volume of the British Official History Victory in the West, Vol. 1, The Battle of Normandy, CAB 140/106, Public Records Office (PRO), Kew; Jeffrey Williams, *The Long Left Flank: The Hard Fought Way to the Reich, 1944–1945* (London: Leo Cooper, 1988), 23; Nigel Hamilton, *Monty: The Making of a General, 1887–1942* (London: Hamish Hamilton, 1981), 507.
9 Interview Transcript, 'First and Second World War: McNaughton 1965,' session 3, tape 5, MG31, vol. 3, box E42, file 1, LAC. Roman Jarymowycz has suggested that an adaption of the Soviet system to overcome the layered German defence 'may have been the answer to Allied frustrations in Normandy.' *Tank Tactics: From Normandy to Lorraine* (Boulder: Lynne Rienner, 2001), 304. Jarymowycz's argument, however, suffers from serious flaws.
10 Brereton Greenhous, Stephen J. Harris, William C. Johnston, and William G.P. Rawling, *The Official History of the Royal Canadian Air Force, III: The Crucible of War, 1939–1945* (Toronto: University of Toronto Press, 1994), 244.
11 Interview Transcript, 'Andrew McNaughton – Canadian,' as broadcast on 11 September 1966, MG 31, vol. 3, box E42, LAC.
12 Douglas Fisher, 'Between Ourselves,' *Legion Magazine* (May 1983): 4; Chris Vokes, *Vokes: My Story* (Ottawa: Gallery, 1985), 152–3; George Kitching, *Mud and Green Fields: The Memoirs of Major General George Kitching* (Langley: Battleline, 1985), 176; James Alan Roberts, *The Canadian Summer: The Memoirs of James Alan Roberts* (Toronto: Oxford University Press, 1981), 108; Charles P. Stacey, *A Date With History: Memoirs of a Canadian Historian* (Ottawa: Deneau, 1983), 126.
13 Farley Mowat, *The Regiment* (Toronto: McClelland and Stewart, 1974), 246.
14 'Andrew McNaughton – Canadian,' 16.
15 Ralston, 'Overseas Trips in 1943,' entry for 3 August 1943; Massey to External, 12 November 1943, For Prime Minister from Ralston, RP, vol. 59.
16 Alanbrooke Diary, 12 November 1943, 818.
17 Forrest C. Pogue, *George C. Marshall: Organizer of Victory, 1943–1945* (New York: Viking, 1973), 6.
18 Ralston, 'Overseas Trips in 1943,' entry for 8 December 1943.
19 C.P. Stacey, *Arms, Men, and Governments: The War Policies of Canada, 1939–1945* (Ottawa: Queen's Printer, 1970), 233–4. As a result of Stuart's approach to the entire problem, several individuals became aware of the War Office's negative views of McNaughton when they did not need to know, including Gibson, Foulkes, Pope, and Murchie.
20 Quoted in John Swettenham, *McNaughton, II: 1939–1943* (Toronto: Ryerson, 1969), 347–8.
21 J.L. Granatstein, *The Generals: The Canadian Army's Senior Commanders in the*

Second World War (Toronto: Stoddart, 1993), 226; Stacey, *Arms, Men, and Governments*, 234. There is considerable evidence that Stuart worked in an unethical manner regarding the manpower estimates. In early August, Stuart told the CWC that there was no need to be concerned about reinforcements even though he had a letter from Crerar indicating concern that very month. In October, Stuart admitted to King that he 'had made a mistake'; King was particularly incensed that Crerar's letter had not been brought to the attention of either Ralston or the CWC until 24 October. King considered the situation 'strange' and was tremendously annoyed with Stuart. John W. Pickersgill and Donald F. Forster, eds., *The Mackenzie King Record, II: 1944-1945* (Toronto: University of Toronto Press, 1968), 145; Stacey, *Arms, Men, and Governments*, 426; Hooker, 'In Defence of Unity,' 206–9.

22 Stacey, *Arms, Men and Governments*, 247; Dominick Graham, *The Price of Command: A Biography of General Guy Simonds* (Toronto: Stoddart, 1993), 107; David Bercuson, *Maple Leaf against the Axis: Canada's Second World War* (Toronto: Stoddart, 1995), 178; Morton, *A Military History of Canada*, 212.

23 McNaughton, Memorandum of Discussions with General Paget, C-in-C 21 Army Gp, Col Ralston, Minister of National Defence, and Lt-Gen Stuart, CGS, Canada, 13 November [15 November] 1943, MP, vol. 250, PA 5-0-3-2.

24 Hamilton, *Monty: Master of the Battlefield*, 506–7; Paul D. Dickson, 'The Limits of Professionalism: General H.D.G. Crerar and the Canadian Army, 1914–1944,' PhD diss., University of Guelph, 1993, 366; Montgomery to Trumbull Warren, 1 January 1969, Richard Malone Papers, Queen's University Archives, Kingston.

25 Crerar, Memorandum of Conversation with Lieut.-General B.L. Montgomery, Commanding S.E. Army on 4 July 42 Commencing 1800 hrs, Crerar Papers (hereafter CP), MG30 E157, vol. 2, file 958.009(D21), LAC. According to Crerar, Montgomery seemed to take this quite well and invited both McNaughton and Crerar to RAF headquarters. Montgomery was not overly impressed when Crerar replaced Pearkes as corps commander, calling it 'musical chairs.'

26 Crerar to Stuart, 13 May 1944, file 312.009(D59), Directorate of History and Heritage (hereafter DHH), Department of National Defence; Montgomery to Brooke, 20 May 1944, AlP, 6/2/25 CIGS Papers.

27 Quoted in Terry Copp and Robert Vogel, 'No Lack of Rational Speed: 1st Canadian Army Operations, September 1944,' *Journal of Canadian Studies* 16, nos. 3 and 4 (1981): 154n2.

28 Nigel Hamilton, *Monty: The Field Marshal, 1944–1976* (London: Hamish Hamilton, 1986), 34; Williams, *The Long Left Flank*, 36.

29 Peter Simonds, *Maple Leaf Up, Maple Leaf Down* (New York: Ireland, 1946), 101; Brooke to Crerar, 16 February 1945, CP, vol. 3, file 958C.009(D171), DHH.
30 Dickson has cited Brooke's comment of March 1944 – 'I am afraid I have lost a very good friend' – as evidence of McNaughton's exaggerated antipathy to Brooke. Arthur Bryant, ed., *Triumph in the West, 1943–1945* (London: Collins, 1958), 175; Dickson, 'The Limits of Professionalism,' 435.
31 Alanbrooke Diary, 12 November 1943, 818.
32 Memorandum of Discussions, 15 November 1943.
33 W.A.B. Douglas, 'Marching to Different Drums: Canadian Military History,' *Journal of Military History* 56, no. 2 (1992): 258.
34 His press conference of 7 January 1945, where he greatly insulted the Americans for their handling of the Bulge, is but one case in point. See David Belchem, *All in the Day's March* (London: Collins, 1978), 240.
35 Burns comments on official history, Collation of Comments, vol. II (4th Draft), file 82/985, folder 6, DHH.

Bibliography

I. PRIMARY SOURCES

Archives

Directorate of History and Heritage, Department of National Defence, Ottawa
Drafts of the official histories file, 82/985, Folders 1–9, and 82/983, Folders 1–7
Historical narratives by the army historians
Interviews conducted by J.L. Granatstein for *The Generals: The Canadian Army's Senior Commanders in the Second World War*
Kardex Collection: H.D.G. Crerar; C.C. Foulkes; George Beament
Morrison, Major W. Alexander, 'Major-General A.G.L. McNaughton, The Conference of Defence Associations, and the 1936 N.P.A.M. Re-organization: A Master Military Bureaucratic Politician at Work,' File 82/470.

Imperial War Museum, London
David Belchem Papers
Field Marshal Sir Bernard L. Montgomery Papers
Trumball Warren Papers
Lieutenant-General Oliver Leese Papers
Major-General John Noble Kennedy Papers

Liddell Hart Centre for Military Archives, King's College, London
Field Marshal Lord Alanbrooke Papers
General Richard Dewing Papers
Lord Hastings Ismay Papers

Library and Archives Canada, Ottawa
E.L.M. Burns Papers   MG31 G6

H.D.G. Crerar Papers	MG30 E157
J. Hughes-Hallett Papers	MG30 E463
A.G.L. McNaughton Papers	MG30 E133
A.G.L. McNaughton Papers	MG31
V. Odlum Papers	MG30 E300
Lester Pearson Papers	MG26 N8
M.H.S. Penhale Papers	MG31 G21
J.L. Ralston Papers	MG27 III BII
E.W. Sansom Papers	MG30 E537
D.C. Spry Papers	MG30 E563
K. Stuart Papers	MG30 E520

Public Records Office, Kew
Field Marshal Sir Harold Alexander Papers, WO214

Royal Military College of Canada, Kingston
William J. McAndrew Collection

United States Military History Institute, Carlisle Barracks, Pennsylvania
General R.E. Lee Papers
Forrest C. Pogue World War II Interviews

University of Victoria Archives, MacPherson Library, Special Collections
Major-General George R. Pearkes Papers

Queen's University Archives, Kingston
Norman Rogers Papers
Grant Dexter Papers
Richard Malone Papers

Training Documents

Army Training Instruction No. 1: Notes on Tactics as Affected by The Reorganization of the Infantry Division, 1941, The War Office, 21 January 1941.
Army Training Instruction No. 2: The Employment of Army Tanks in Co-operation with Infantry, The War Office, March 1941.
Army Training Instruction No. 3: Handling of an Armoured Division, The War Office, 19 May 1941.
Canadian Army Training Memorandum, Special Supplement to CATM Number 28.
Current Reports from Overseas.

Military Training Pamphlet no. 23: Operations: General Principles, Fighting Troops and Their Characteristics, War Office, September 1939.
Military Training Pamphlet no. 25: Operations: Part 9: The Infantry Division in the Attack, War Office, 1941.
Military Training Pamphlet no. 41: The Tactical Handling of the Armoured Division and Its Components, Part 2: The Armoured Regiment, War Office, February 1943.
Military Training Pamphlet no. 47: Movement by Road, War Office, September 1941.
Notes from Theatres of War, War Office, 1942–43.
Tanks and Their Employment in Co-operation with Other Arms, August 1918.

II: PRIMARY SOURCES, PUBLISHED

Papers

Blake, Robert, ed. *The Private Papers of Douglas Haig, 1914–1919*. London: Eyre and Spottiswoode, 1952.
Bland, Larry I., ed. *The Papers of George Catlett Marshall, II: 'We Cannot Delay,' July 1, 1939– December 6, 1941*. Baltimore: Johns Hopkins University Press, 1986.
– *The Papers of George Catlett Marshall, III: 'The Right Man for the Job,' December 7, 1941–May 31, 1943*. Baltimore: Johns Hopkins University Press, 1991.
Blumenson, Martin, ed. *The Patton Papers*. 2 vols. Boston: Houghton Mifflin, 1972 and 1974.
Chandler, Jr, Alfred D., ed. *The Papers of Dwight D. Eisenhower: The War Years*. 5 vols. Baltimore: Johns Hopkins University Press, 1970.
Debates of the Senate of the Dominion of Canada, 1943–1944, Official Report. Ottawa: Edmund Cloutier, 1944.
Documents Relating to New Zealand's Participation in the Second World War, 1939–45, vols. I (1949) and II (1951). Wellington: War History Branch, Department of Internal Affairs.
Ferrell, Robert H., ed. *The Eisenhower Diaries*. New York: Norton, 1981.
Gilbert, Martin, ed. *The Churchill War Papers, I: At the Admiralty, September 1939– May 1940*. London: Heinemann, 1993.
– *The Churchill War Papers, II: Never Surrender, May 1940–December 1940*. London: Heinemann, 1994.
Gibson, Frederick W., and Barbara Robertson, eds. *Ottawa at War: The Grant Dexter Memoranda, 1939–1945*. Winnipeg: Manitoba Record Society, 1994.
Heiber, Helmuth, and David M. Glantz, eds. *Hitler and His Generals*. New York: Enigma, 2002.

Loewenheim, Francis L., Harold D. Langley, and Manfred Jonas, eds. *Roosevelt and Churchill: Their Secret Wartime Correspondence.* New York: Saturday Review Press/Dutton, 1975.

Munro, John A., ed. *Documents on Canadian External Relations,* vol. VI, *1936–1939.* Ottawa: Department of Foreign Affairs, 1972.

Memoirs and Diaries

Adair, Allan H.S. *A Guard's General: The Memoirs of Major General Sir Allan Adair.* London: Hamish Hamilton, 1986.

Belchem, David. *All in the Day's March.* London: Collins, 1978.

Blackburn, George G. *Where the Hell Are the Guns: A Soldier's Eye View of the Anxious Years, 1939–1944.* Toronto: McClelland and Stewart, 1997.

Bradley, Omar. *A General's Life.* New York: Simon and Schuster, 1983.

Bryant, Arthur, ed. *Triumph in the West, 1943–1945.* London: Collins, 1958.

– *The Turn of the Tide, 1939–1943.* London: Collins, 1957.

Burns, Eedson L.M. *General Mud: Memoirs of Two World Wars.* Toronto: Clarke, Irwin, 1970.

Butcher, Harry C. *My Three Years with Eisenhower.* New York: Simon and Schuster, 1946.

Carton de Wiart, Sir Adrian. *Happy Odyssey.* London: Jonathan Cape, 1950.

Churchill, Winston S. *Their Finest Hour.* London: Cassell, 1949.

– *The Gathering Storm.* London: Cassell, 1948.

– *The Grand Alliance.* London: Cassell, 1950.

– *Great Contemporaries.* London: Thornton Butterworth, 1937.

– *The Hinge of Fate.* London: Cassell, 1951.

Danchev, Alex, and Daniel Todman, eds. *War Diaries, 1939–1945: Field Marshal Lord Alanbrooke.* London: Weidenfeld and Nicolson, 2001.

de Guingand, Francis. *Operation Victory.* London: Hodder and Stoughton, 1947.

Douglas, Sholto. *Years of Command.* London: Collins, 1966.

Earl of Bessborough. *Return to the Forest.* London: Weidenfeld and Nicolson, 1962.

Eden, Anthony. *The Memoirs of Sir Anthony Eden: The Reckoning.* London: Cassell, 1965.

Eisenhower, Dwight D. *Crusade in Europe.* New York: Doubleday, 1948.

Galloway, Strome. *The General Who Never Was.* Belleville: Mika, 1981.

Graham, Howard. *Citizen and Soldier: The Memoirs of Lieutenant-General Howard Graham.* Toronto: McClelland and Stewart, 1987.

Guderian, Heinz. *Achtung Panzer! The Development of Armoured Forces, Their Tactics and Operational Potential.* London: Arms and Armour Press, 1993.

– *Panzer Leader*. New York: Dutton, 1952.
Grigg, P.J. *Prejudice and Judgment*. London: Jonathan Cape, 1948.
Horrocks, Sir Brian. *Corps Commander*. New York: Scribner, 1977.
– *A Full Life*. London: Collins, 1960.
Ismay, Lord. *The Memoirs of General the Lord Ismay*. London: Heinemann, 1960.
Kennedy, Sir John N. *The Business of War: The War Narrative of Major-General Sir John Noble Kennedy*. London: Hutchinson, 1957.
Kitching, George. *Mud and Green Fields: The Memoirs of Major-General George Kitching*. Langley: Battleline, 1985.
Leasor, James. *War at the Top: The Experiences of General Sir Leslie Hollis*. London: Michael Joseph, 1959.
Leutze, James, ed. *The London Journal of General Raymond E. Lee, 1940–41*. Boston: Little, Brown, 1971.
Lloyd George, David. *War Memoirs of David Lloyd George*, 6 vols. London: Nicholson and Watson, 1936.
MacArthur, Douglas. *Reminiscences*. New York: McGraw-Hill, 1964.
Macleod, Colonel Roderick, and Kelly Davis, eds. *Time Unguarded: The Ironside Diaries, 1937-1940*. New York: David McKay, 1962.
Maisky, Ivan. *Memoirs of a Soviet Ambassador: The War, 1939–1945*. New York: Scribner, 1968.
Malone, Dick. *Missing from the Record*. Toronto: Collins, 1946.
Malone, Richard S. *A Portrait of War, 1939–1943*. Toronto: Collins, 1983.
– *A World in Flames, 1944–1945: A Portrait of War, Part Two*. Toronto: Collins, 1984.
Martel, Giffard Le Q. *An Outspoken Soldier: His Views and Memoirs*. London: Sifton Praed, 1949.
Massey, Vincent. *What's Past Is Prologue: The Memoirs of the Right Honourable Vincent Massey*. Toronto: Macmillan, 1963.
Montgomery, Sir Bernard Law. *The Memoirs of Field Marshal Montgomery of Alamein*. London: Collins, 1958.
Montgomery, Brian. *A Field Marshal in the Family*. London: Constable, 1973.
Moran, Lord. *Winston Churchill: The Struggle for Survival, 1940–1965*. London: Constable, 1966.
Morgan, Sir Frederick. *Overture to Overlord*. London: Hodder and Stoughton, 1950.
Mowat, Farley. *The Regiment*. Toronto: McClelland and Stewart, 1974.
North, John, ed. *The Alexander Memoirs, 1940–1945*. London: Cassell, 1962.
Patton, Jr., George S. *War As I Knew It*. New York: Bantam, 1979 [1947].
Pawle, Gerald. *The War and Colonel Warden: Based on the Recollections of Commander C.R. Thompson, Personal Assistant to the Prime Minister, 1940–45*. London: George Harrap, 1963.

Pearson, Lester B. *Mike: The Memoirs of the Right Honourable Lester B. Pearson, I: 1897–1948.* Toronto: University of Toronto Press, 1972.
Pershing, John J. *My Experiences in the World War.* 2 vols. New York: Stokes, 1931.
Pickersgill, John W., ed. *The Mackenzie King Record, I: 1939–1944.* Toronto: University of Toronto Press, 1960.
Pickersgill, John W., and Donald F. Forster, eds. *The Mackenzie King Record, II: 1944–1945.* Toronto: University of Toronto Press, 1968.
Pile, Sir Frederick. *Ack-Ack.* London: Harrap, 1949.
Pope, Maurice A. *Soldiers and Politicians: The Memoirs of Lt.-Gen. Maurice A. Pope.* Toronto: University of Toronto Press, 1962.
Rees, Goronwy. *A Bundle of Sensations: Sketches in Autobiography.* London: Chatto and Windus, 1960.
Ridgway, Matthew B. *Soldier: The Memoirs of Matthew B. Ridgway.* New York: Harper and Row, 1956.
Roberts, James Alan. *The Canadian Summer.* Toronto: Oxford University Press, 1981.
Spears, Sir Edward. *Assignment to Catastrophe.* New York: Wyn, 1954.
Stacey, Charles P. *A Date With History: Memoirs of a Canadian Historian.* Ottawa: Deneau, 1983.
Truscott, Lucian K. *Command Missions: A Personal Story.* Novato: Presidio, 1990.
Vokes, Chris. *Vokes: My Story.* Ottawa: Gallery, 1985.
Ward, Norman, ed. *A Party Politician: The Memoirs of Chubby Power.* Toronto: Macmillan, 1966.
Wedemeyer, Albert C. *Wedemeyer Reports!* New York: Henry Holt, 1958.
Wheeler-Bennett, Sir John, ed. *Action This Day: Working With Churchill.* London: Macmillan, 1968.
Zuckerman, Solly. *From Apes to Warlords.* New York: Harper and Row, 1978.

III: SECONDARY SOURCES

Official Histories

Bean, C.E.W. *Official History of Australia in the War of 1914–1918: The Story of ANZAC, I: From the Outbreak of the War to the End of the First Phase of the Gallipoli Campaign, May 4, 1915.* (Sydney: Angus and Robertson, 1921).
Butler, James Ramsay Montagu. *Grand Strategy, II: September 1939–June 1941.* London: HMSO, 1957.
Cline, Ray S. *Washington Command Post: The Operations Division.* Washington: Office of the Chief of Military History, 1951.

Collier, Basil. *The Defence of the United Kingdom.* London: HMSO, 1957.

Duguid, A. Fortescue. *Official History of the Canadian Forces in the Great War, 1914–1919,* 2 vols. Ottawa: Patenaude, 1938.

Dziuban, Stanley W. *Military Relations between the United States and Canada, 1939–1945.* Washington: Center of Military History, 1990.

Garland, Albert N., and Howard McGraw Smyth. *Sicily and the Surrender of Italy.* Washington: Center of Military History, 1986.

Greenfield, Kent Roberts, Robert B. Palmer, and Bell I. Wiley. *The Army Ground Forces: The Organization of Ground Combat Troops.* Washington: Historical Division, Department of the Army, 1947.

Greenhous, Brereton, Stephen J. Harris, William C. Johnston, and William P. Rawling. *The Official History of the Royal Canadian Air Force, III: The Crucible of War, 1939–1945.* Toronto: University of Toronto Press, 1994.

Harrison, Gordon A. *Cross-Channel Attack.* Washington: Office of the Chief of Military History, 1951.

Hinsley, Francis H. *British Intelligence in the Second World War: Its Influence on Strategy and Operations, I.* London: HMSO, 1979.

Howard, Michael. *British Intelligence in the Second World War, V: Strategic Deception.* London: HMSO, 1990.

Martin, Lieutenant-General H.J., and Colonel Neil D. Opren, *South African Forces World War II, Vol. VII: South Africa at War: Military and Industrial Organization and Operations in Connection with the Conduct of the War, 1939–1945.* Cape Town: Purnell, 1979.

Molony, C.J.C. *The Mediterranean and the Middle East, V: The Campaigns in Sicily and Italy 3rd September 1943 to 31st March 1944.* London: HMSO, 1973.

Nicholson, Gerald W.L. *Canadian Expeditionary Force, 1914–1919: Official History of the Canadian Army in the First World War.* Ottawa: Queen's Printer, 1962.

– *Official History of the Canadian Army in the Second World War, II: The Canadians in Italy, 1943–1945.* Ottawa: Queen's Printer, 1957.

Phillips, Neville Crompton. *Official History of New Zealand in the Second World War, 1939–45, Italy, I: The Sangro to Cassino.* Wellington: War History Branch, Department of Internal Affairs, 1957.

Pogue, Forrest C. *The Supreme Command.* Washington: Center of Military History, 1989.

Roskill, Captain Stephen W. *Official History of the Second World War, The War at Sea, II: The Period of Balance.* London: HMSO, 1956.

Stacey, Charles P. *Official History of the Canadian Army in the Second World War, I: Six Years of War: The Army in Canada, Britain, and the Pacific.* Ottawa: Queen's Printer, 1957.

– *Official History of the Canadian Army in the Second World War, III: The Victory*

Campaign: The Operations in Northwest Europe, 1944–1945. Ottawa: Queen's Printer, 1960.
– *Arms, Men, and Governments: The War Policies of Canada, 1939–1945*. Ottawa: Queen's Printer, 1970.

Theses

Campbell, John R. 'James Layton Ralston and Manpower for the Canadian Army.' MA thesis, Wilfrid Laurier University, 1996.
Cessford, Michael P. 'Hard in the Attack: The Canadian Army in Sicily and Italy, July 1943–June 1944.' PhD diss., Carleton University, 1996.
– 'Warriors for the Working Day: The 5th Canadian Armoured Division in Italy, 1943–1945.' MA thesis, University of New Brunswick, 1989.
Coleman, R.C. 'General A.G.L. McNaughton and the Command and Control of the Canadian Army 1939–1943.' MA thesis, Royal Military College, 1967.
Dickson, Paul D. 'The Limits of Professionalism: General H.D.G. Crerar and the Canadian Army, 1914–1944.' PhD diss., University of Guelph, 1993.
Gimblett, Richard H. 'Buster Brown: The Man and His Clash with "Andy" McNaughton.' BA thesis, Royal Military College, 1979.
Hayes, Geoffrey. 'The Development of the Canadian Army Officer Corps, 1939–1945.' PhD diss., Duke University, 1979.
Hisdal, Howard V. 'Lieutenant-General Guy Granville Simonds and the Battle of the Scheldt: A Study in Canadian Generalship.' MA thesis, Carleton University, 1986.
Hodgins-Vermass, Robert. '"A Bit of 'Binge'": Montgomery's Command of the 3rd Division.' MA thesis, University of Calgary, 1997.
Hooker, Martha Ann. 'In Defence of Unity: Canada's Military Policies, 1935–1944.' MA thesis, Carleton University, 1985.
Hutchinson, William E. 'Test of a Corps Commander: Lieutenant-General Guy Granville Simonds in Normandy, 1944.' MA thesis, University of Victoria, 1982.
Hull, Christopher H.N. 'A Case Study in Professionalism in the Canadian Army in the 1930s and 1940s: Lieutenant-General G.G. Simonds.' MA thesis, Purdue University, 1989.
Macdonald, Lieutenant-Colonel John A. 'In Search of Veritable: Training the Canadian Army Staff Officer, 1899 to 1945.' MA thesis, Royal Military College, 1992.
Stewart, William A. 'Attack Doctrine in the Canadian Corps, 1916–1918.' MA thesis, University of New Brunswick, 1980.
Tooley, Robert W. 'Montgomery as Military Trainer: Preparation for Alamein.' MA thesis, University of New Brunswick, 1984.

Wilson, David A. 'Close and Continuous Attention: Human Resources Management in Canada during the Second World War.' PhD diss., University of New Brunswick, 1997.
- 'The Development of Tank-Infantry Co-operation Doctrine in the Canadian Army for the Normandy Campaign of 1944.' MA thesis, University of New Brunswick, 1992.
Zaporzan, Lawrence J. 'Rad's War: A Biographical Study of Sydney Valpy Radley-Walters from Mobilization to the End of the Normandy Campaign 1944.' MA thesis, University of New Brunswick, 2001.

Books

Ambrose, Stephen E. *The Supreme Commander: The War Years of General Dwight D. Eisenhower.* New York: Simon and Schuster, 1970.
Atkin, Ronald. *Dieppe 1942: The Jubilee Disaster.* London: Macmillan, 1980.
Atkinson, Rick. *An Army at Dawn: The War in North Africa, 1942–1943.* New York: Henry Holt, 2002.
Bailey, Major-General Jonathan B.A. *Field Artillery and Firepower.* Annapolis: Naval Institute Press, 2004.
Barnett, Correlli. *Britain and Her Army, 1509–1970: A Military, Political, and Social Survey.* London: Allen Lane, 1970.
- *The Desert Generals*, rev. ed. Bloomington: Indiana University Press, 1982.
Bennett, Ralph. *Ultra and Mediterranean Strategy.* New York: William Morrow, 1989.
Bercuson, David. *Maple Leaf against the Axis: Canada's Second World War.* Toronto: Stoddart, 1995.
Berger, Carl. *The Sense of Power: Studies in the Ideas of Canadian Imperialism, 1867–1914.* Toronto: University of Toronto Press, 1970.
Berton, Pierre. *Vimy.* Toronto: McClelland and Stewart, 1986.
Bidwell, Shelford. *Gunners at War: A Tactical Study of the Royal Artillery in the Twentieth Century.* London: Arrow, 1972.
Bidwell, Shelford, and Dominick Graham. *Firepower: British Army Weapons and Theories of War, 1904–1945.* London: Allen Unwin, 1982.
Bird, Will R. *North Shore (New Brunswick) Regiment.* Fredericton: Brunswick, 1963.
Bissell, Claude. *The Imperial Canadian: Vincent Massey in Office.* Toronto: University of Toronto Press, 1986.
Black, Ernest G. *I Want One Volunteer.* Toronto: Ryerson, 1965.
Blaxland, Gregory. *The Plain Cook and the Great Showman: The First and Eighth Armies in North Africa.* London: William Kimber, 1977.

Boatner, Mark M. *The Biographical Dictionary of World War II.* Novato: Presidio, 1996.
Bond, Brian. *The Victorian Army and the Staff College, 1854–1914.* London: Eyre Methuen, 1972.
Brown, Anthony Cave. *Bodyguard of Lies.* New York: Harper and Row, 1975.
– *'C': The Secret Life of Sir Stewart Graham Menzies, Spymaster to Winston Churchill.* New York: Macmillan, 1987.
Bryden, John. *Best-Kept Secret: Canadian Secret Intelligence in the Second World War* Toronto: Lester, 1993.
Buchan, John. *Augustus.* London: Hodder and Stoughton, 1937.
Burns, Eedson L.M. *Manpower in the Canadian Army, 1939–1945.* Toronto: Clarke, Irwin, 1956.
Campbell, John P. *Dieppe Revisited: A Documentary Investigation.* London: Frank Cass, 1993.
Carver, Field Marshal Sir Michael, ed. *The War Lords: Military Commanders of the Twentieth Century.* New York: Little, Brown, 1976.
Carver, Field Marshal Lord. *The Apostles of Mobility: The Theory and Practice of Armoured Warfare.* London: Weidenfeld and Nicolson, 1979.
Chalfont, Alun. *Montgomery of Alamein.* New York: Atheneum, 1976.
Citino, Robert M. *Quest for Decisive Victory: From Stalemate to Blitzkrieg in Europe, 1899–1940.* Lawrence: University Press of Kansas, 2002.
Cole, Major D.H. *Imperial Military Geography: General Characteristics of the Empire in Relation to Defence.* London: Sifton Praed, 1935.
Colville, Sir John. *The Churchillians.* London: Weidenfeld and Nicolson, 1981.
Condell, Bruce, and David T. Zabecki, eds. *On the German Art of War: Truppenführung.* Boulder: Lynne Rienner, 2001.
Connell, John. *Wavell: Scholar and Soldier.* London: Collins, 1964.
Cook, Tim. *Clio's Warriors: Canadian Historians and the Writing of the World Wars.* Vancouver: UBC Press, 2006.
Cooke, Owen. *The Canadian Military Experience 1867–1997: A Bibliography.* Ottawa: Department of National Defence, 1997.
Copp, Terry. *The Brigade: The Fifth Canadian Infantry Brigade, 1939–1945.* Stoney Creek: Fortress, 1992.
– *Fields of Fire: The Canadians in Normandy.* Toronto: University of Toronto Press, 2003.
Corrigan, Gordon. *Blood, Sweat, and Arrogance and the Myths of Churchill's War.* London: Weidenfeld and Nicolson, 2006.
Creighton, Donald G. *Canada's First Century, 1867–1967.* Toronto: Macmillan, 1970.

- *The Forked Road: Canada, 1939–1957*. Toronto: McClelland and Stewart, 1976.
Creveld, Martin van. *Command in War*. Cambridge, MA: Harvard University Press, 1985.
Cruickshank, Charles. *Deception in World War II*. Oxford: Oxford University Press, 1979.
Dancocks, Daniel G. *The D-Day Dodgers: The Canadians in Italy, 1939–1945*. Toronto: McClelland and Stewart, 1991.
- *Legacy of Valour: The Canadians at Passchendaele*. Edmonton: Hurtig, 1986.
- *Spearhead to Victory: Canada and the Great War*. Edmonton: Hurtig, 1987.
Dawson, R. MacGregor. *The Conscription Crisis of 1944*. Toronto: University of Toronto Press, 1961.
Deighton, Len. *Blitzkrieg: From the Rise of Hitler to the Fall of Dunkirk*. London: Jonathan Cape, 1979.
Delaney, Douglas E. *A Soldier's General: Bert Hoffmeister at War*. Vancouver: UBC Press, 2005.
D'Este, Carlo. *Bitter Victory: The Battle for Sicily, 1943*. New York: Dutton, 1988.
- *Eisenhower: A Soldier's Life*. New York: Henry Holt, 2002.
Dickson, Paul D. *A Thoroughly Canadian General: A Biography of General H.D.G. Crerar*. Toronto: University of Toronto Press, 2007.
- ed. *1943: The Beginning of the End*. Waterloo: Wilfrid Laurier University Press, 1995.
Douglas, W.A.B. & Brereton Greenhous. *Out of the Shadows: Canada and the Second World War*. Rev. ed. Toronto: Dundurn, 1995.
Doughty, Robert A. *The Seeds of Disaster: The Development of French Army Doctrine, 1919–1939*. Hamden: Archon, 1985.
Dunmore, Spencer. *Wings of Victory: The Remarkable Story of the British Commonwealth Air Training Plan in Canada*. Toronto: McClelland and Stewart, 1994.
Eayrs, James. *The Art of the Possible: Government and Foreign Policy in Canada*. Toronto: University of Toronto Press, 1961.
- *In Defence of Canada: Appeasement and Rearmament*. Toronto: University of Toronto Press, 1965.
- *In Defence of Canada: From the Great War to the Great Depression*. Toronto: University of Toronto Press, 1964.
Eggleston, Wilfrid. *Scientists at War*. London: Oxford University Press, 1950.
Ellis, Chris, and Peter Chamberlain, eds. *Handbook on the British Army 1943*. London: Arms and Armour, 1976.
English, John A. *The Canadian Army and the Normandy Campaign: A Study of Failure in High Command*. New York: Praeger, 1991.

- *Lament for an Army: The Decline of Canadian Military Professionalism.* Contemporary Affairs No. 3. Toronto: Irwin, 1998.
Fergusson, Bernard. *The Watery Maze: The Story of Combined Operations.* London: Collins, 1961.
Fraser, David. *Alanbrooke.* London: Collins, 1982.
- *And We Shall Shock Them: The British Army in the Second World War.* London: Sceptre, 1988.
French, David. *Raising Churchill's Army: The British Army and the War against Germany, 1919–1945.* London: Oxford University Press, 2000.
Frieser, Karl-Heinz. *The Blitzkrieg Legend: The 1940 Campaign in the West.* Annapolis: United States Naval Institute Press, 2005.
Gabel, Christopher R. *The US Army GHQ Maneuvers of 1941.* Washington: Office of the Chief of Military History, 1991.
Galbraith, John Kenneth. *The Anatomy of Power.* Boston: Houghton Mifflin, 1983.
Gilbert, Martin. *Finest Hour: Winston S. Churchill 1939–1941.* London: Heinemann, 1983.
Goodwin-Austen, A.R. *The Staff and the Staff College.* London: Constable, 1977.
Goodspeed, Donald J. *The Road Past Vimy: The Canadian Corps 1914–1918.* Toronto: Macmillan, 1969.
Graham, Dominick. *The Price of Command: A Biography of General Guy Simonds.* Toronto: Stoddart, 1993.
Graham, Dominick, and Shelford Bidwell. *Coalitions, Politicians, and Generals: Some Aspects of Command in Two World Wars.* London: Brassey's, 1993.
- *Tug of War: The Battle for Italy, 1943–1945.* New York: St Martin's, 1986.
Granatstein, J.L. *Canada's Army: Waging War and Keeping the Peace.* Toronto: University of Toronto Press, 2002.
- *Canada's War: The Politics of the Mackenzie King Government, 1939–1945.* Toronto: Oxford University Press, 1975.
- *The Generals: The Canadian Army's Senior Commanders in the Second World War.* Toronto: Stoddart, 1993.
- *The Politics of Survival: The Conservative Party of Canada, 1939–1945.* Toronto: University of Toronto Press, 1967.
Granatstein, J.L., and Norman Hitsman. *Broken Promises: A History of Conscription in Canada.* Toronto: University of Toronto Press, 1977.
Greenhous, Brereton. *Dieppe, Dieppe.* Montreal: Art Global, 1992.
- *Dragoon: The Centennial History of the Royal Canadian Dragoons, 1883–1983.* Belleville: Guild of the Royal Canadian Dragoons, 1983.
Griffith, Paddy. *Forward into Battle: Fighting Tactics from Waterloo to the Near Future.* Novato: Presidio, 1991.

Grigg, John. *1943: The Victory That Never Was*. London: Eyre Methuen, 1980.
Hackett, Sir John. *The Profession of Arms*. London: Sidgwick and Jackson, 1983.
Hamilton, Nigel. *Monty: The Making of a General, 1887–1942*. London: Hamish Hamilton, 1981.
– *Monty: Master of the Battlefield, 1942–1944*. London: Hamish Hamilton, 1983.
Harris, Stephen J. *Canadian Brass: The Making of a Professional Army, 1860–1939*. Toronto: University of Toronto Press, 1988.
Hart, Russell A. *Clash of Arms: How the Allies Won in Normandy*. Boulder: Rienner, 2001.
Hart, Stephen Ashley. *Montgomery and 'Colossal Cracks': The 21st Army Group in Northwest Europe, 1944–45*. Westport: Praeger, 2000.
Haycock, Ronald G. *Sam Hughes: The Public Career of a Controversial Canadian, 1885–1916*. Ottawa: Wilfrid Laurier University Press, 1986.
Higgins, Trumbull. *Soft Underbelly: The Anglo-American Controversy over the Italian Campaign, 1939–1943*. London: Collier-Macmillan, 1968.
Hilsman, Norman. *Alexander of Tunis*. London: W.H. Allen, 1952.
Horn, Bernd. *Bastard Sons: An Examination of Canada's Airborne Experience, 1942–1995*. St Catharines: Vanwell, 2001.
– ed. *The Canadian Way of War: Serving the National Interest*. Toronto: Dundurn, 2006.
Horn, Bernd, and Stephen Harris, eds. *Warrior Chiefs: Perspectives on Senior Canadian Military Leaders*. Toronto: Dundurn, 2001.
Houston, Donald E. *Hell on Wheels: The 2nd Armoured Division*. San Rafael: Presidio, 1977.
How, Douglas. *The 8th Hussars: A History of the Regiment*. Sussex: Maritime, 1964.
Howard, Michael. *Soldiers and Governments: Nine Studies in Civil-Military Relations*. Bloomington: Indiana University Press, 1959.
Huntington, Samuel P. *The Soldier and the State: The Theory and Practice of Civil Military Relations*. Cambridge, MA: Harvard University Press, 1957.
Hyatt, A.M.J. *Sir Arthur Currie: A Military Biography*. Toronto: University of Toronto Press, 1987.
Jackson, Bill, and Dwin Bramall. *The Chiefs: The Story of the United Kingdom Chiefs of Staff*. London: Brassey's, 1992.
Jackson, William G.F. *Alexander of Tunis as Military Commander*. London: Batsford, 1971.
– *Overlord: Normandy 1944*. London: Davis-Poynter, 1978.
James, Lawrence. *The Rise and Fall of the British Empire*. New York: St Martin's, 1994.
Janowitz, Morris. *The Professional Soldier: A Social and Political Portrait*. New York: Free Press, 1960.

Jarymowycz, Roman J. *Tank Tactics: From Normandy to Lorraine.* Boulder: Lynne Rienner, 2001.

Keegan, John, ed. *Churchill's Generals.* New York: Grove Weidenfeld, 1991.

Lamb, Richard. *Churchill as War Leader.* New York: Carroll and Graf, 1991.

– *Montgomery in Europe 1943–1945: Success or Failure?* London: Buchan and Enright, 1983.

Lang, Kurt. *Military Institutions and the Sociology of War: A Review of the Literature with Annotated Bibliography.* Beverly Hills: Sage, 1972.

Lewin, Ronald. *Montgomery as Military Commander.* New York: Stein and Day, 1971.

Liddell Hart, Basil H. *This Expanding War.* London: Faber and Faber, 1942.

– *The Memoirs of Captain Liddell Hart.* 2 vols. London: Cassell, 1965.

Luper, Timothy T. *The Dynamics of Doctrine: The Changes in German Tactical Doctrine during the First World War.* Fort Leavenworth: Combat Studies Institute, 1981.

Luvaas, Jay. *The Education of an Army: British Military Thought 1815–1940.* London: Cassell, 1970 [1964].

Macksey, Kenneth. *A History of the Royal Armoured Corps, 1914–1975.* Beaminster: Newton, 1983.

Manchester, William. *American Caesar: Douglas MacArthur 1880–1964.* New York: Little, Brown, 1978.

Mansoor, Peter R. *The GI Offensive in Europe: The Triumph of American Infantry Divisions, 1941–1945.* Lawrence: University Press of Kansas, 1999.

Marteinson, John, and Michael R. McNorgan. *The Royal Canadian Armoured Corps: An Illustrated History.* Kitchener: Royal Canadian Armoured Corps Association, 2000.

McAndrew, Bill. *Canadians and the Italian Campaign, 1943–1945.* Montreal: Art Global, 1996.

McMahon, J.S. *Professional Soldier: General Guy Simonds – A Memoir.* Winnipeg: McMahon Investments, 1985.

Miksche, Major Ferdinand O. *Blitzkrieg.* London: Faber and Faber, 1941.

Monash, Sir John. *The Australian Victories in France in 1918.* London: Hutchinson, 1920.

Montgomery, Sir Bernard Law. *El Alamein to the River Sangro.* London: Hutchinson, 1952.

– *The Path to Leadership.* New York: Putnam, 1961.

Morton, Desmond. *Canada and War: A Military and Political History.* Toronto: Butterworth, 1981.

– *A Military History of Canada.* Rev. ed. Edmonton: Hurtig, 1990.

- *A Peculiar Kind of Politics: Canada's Overseas Ministry in the First World War.* Toronto: University of Toronto Press, 1982.
Mordal, Jacques. *Dieppe: The Dawn of Decision.* Toronto: Ryerson, 1962.
Morton, William L., ed. *The Shield of Achilles: Aspects of Canada in the Victorian Age.* Toronto: McClelland and Stewart, 1968.
Munro, Ross. *Gauntlet to Overlord: The Story of the Canadian Army.* Toronto: Macmillan, 1946.
Nicholson, Gerald W.L. *The Gunners of Canada: The History of the Royal Regiment of Canadian Artillery.* 2 vols. Toronto: McClelland and Stewart, 1967.
Nicolson, Nigel. *Alex: The Life of Field Marshal Earl Alexander of Tunis.* London: Weidenfeld and Nicolson, 1973.
Nolan, Brian. *King's War: Mackenzie King and the Politics of War, 1939–1945.* Toronto: Random House, 1988.
Perrett, Bryan, *A History of Blitzkrieg.* London: Hale, 1983.
Pile, Sir Frederick. *Ack-Ack.* London: Harrap, 1949.
Pitt, Barrie. *Crucible of War: Western Desert 1941.* London: Jonathan Cape, 1980.
Place, Timothy. *Military Training in the British Army, 1940–1944: From Dunkirk to D-Day.* London: Frank Cass, 2000.
Preston, Richard A. *Canada and Imperial Defence: A Study of the British Commonwealth Defence Organization.* Toronto: University of Toronto Press, 1967.
- *Canada's RMC: A History of the Royal Military College.* Toronto: University of Toronto Press, 1969.
Preston, Richard A., Sydney F. Wise, and Herman O. Werner. *Men in Arms: A History of Warfare and Its Interrelationships with Western Society.* New York: Praeger, 1968.
Price, Frank J. *Troy H. Middleton: A Biography.* Baton Rouge: Louisiana State University Press, 1974.
Rawling, Bill. *Surviving Trench Warfare: Technology and the Canadian Corps, 1914–1918.* Toronto: University of Toronto Press, 1992.
Reader's Digest. *The Canadians at War, 1939–1945.* 2 vols. 1969.
Reid, Brian A. *No Holding Back: Operation Totalize, Normandy, August 1944.* Toronto: Robin Brass Studio, 2005.
Reid, Brian Holden. *The American Civil War.* London: Cassell, 1999.
Reynolds, Quentin. *Dress Rehearsal: The Story of Dieppe.* New York: Random House, 1942.
Robertson, Terence. *The Shame and the Glory: Dieppe.* Toronto: McClelland and Stewart, 1962.
Rosse, Earl of, and Colonel E.R. Hill. *The Story of the Guards Armoured Division.* London: Geoffrey Bles, 1956.

Roy, Reginald. *1944: The Canadians in Normandy.* Canadian War Museum, Publication no. 19. Toronto: Macmillan, 1984.
– *For Most Conspicuous Bravery: A Biography of Maj.-Gen. George R. Pearkes, V.C., through Two World Wars.* Vancouver: UBC Press, 1977.
– *Sherwood Lett: His Life and Times.* Vancouver: UBC Alumni Association, 1991.
Ryder, Roland. *Oliver Leese.* London: Hamish Hamilton, 1987.
Sainsbury, Keith. *The North Africa Landings, 1942: A Strategic Decision.* London: Davis-Poynter, 1976.
Schreiber, Shane B. *Shock Army of the British Empire: The Canadian Corps in the Last 100 Days of the Great War.* Westport: Praeger, 1997.
Simonds, Peter. *Maple Leaf Up, Maple Leaf Down.* New York: Ireland, 1946.
Simpkin, Richard. *Race to the Swift: Thoughts on Twenty-First Century Warfare.* London: Brassey's, 1985.
Sixsmith, Eric K.G. *British Generalship in the Twentieth Century.* London: Arms and Armour Press, 1970.
Smyth, Sir John. *Leadership in War, 1939–1945: The Generals in Victory and Defeat.* London: David and Charles, 1974.
Speaight, Robert. *Vanier: Soldier, Diplomat, and Governor General.* Toronto: Collins, 1970.
Stacey, Charles P. *Canada and the Age of Conflict: A History of Canadian External Policies, I: 1867–1921.* Toronto: Macmillan, 1977.
– *Canada and the Age of Conflict: A History of Canadian External Policies, II: 1921–1948.* Toronto: Macmillan, 1981.
– *A Very Double Life: The Private World of Mackenzie King.* Toronto: Macmillan, 1976.
Stanley, George F.G. *Canada's Soldiers: The Military History of an Unmilitary People.* 3rd ed. Toronto: Macmillan, 1974.
Steele, Richard W. *The First Offensive 1942: Roosevelt, Marshall, and the Making of American Strategy.* Bloomington: Indiana University Press, 1973.
Study Group of the Royal Institute of International Affairs. *Political and Strategic Interests of the United Kingdom: An Outline.* London: Oxford University Press, 1939.
Sweet, John J.T. *Mounting the Threat: The Battle of Bourguebus Ridge, 18–23 July 1944.* San Rafael: Presidio, 1977.
Swettenham, John. *McNaughton.* 3 vols. Toronto: Ryerson, 1968–69.
– *To Seize the Victory: The Canadian Corps in World War I.* Toronto: Ryerson, 1965.
Thistle, Mel, ed. *The Mackenzie–McNaughton Wartime Letters.* Toronto: University of Toronto Press, 1975.
Thompson, R.W. *Generalissimo Churchill.* London: Hodder and Stoughton, 1973.

Urquhart, Hugh M. *Arthur Currie: Biography of a Great Canadian.* Toronto: Dent, 1950.
Van Doorn, Jacques, ed. *Armed Forces and Society: Sociological Essays.* The Hague: Mouton, 1968.
Villa, Brian Loring. *Unauthorized Action: Mountbatten and the Dieppe Raid.* Oxford: Oxford University Press, 1989.
von Mellenthin, Major-General F.W. *German Generals of World War II: As I Saw Them.* Norman: University of Oklahoma Press, 1977.
Wade, Lieutenant-Colonel Gary L. *World War II Division Commanders.* Report no. 7. Fort Leavenworth: Combat Studies Institute, 1983.
Wavell, General Sir Archibald. *Allenby: A Study in Greatness.* London: Harrap, 1940.
West, Nigel. *Unreliable Witness: Espionage Myths of the Second World War.* London: Weidenfeld and Nicolson, 1984.
Whitaker, Dennis. *Dieppe: Tragedy to Triumph.* Toronto: McGraw-Hill Ryerson, 1993.
Williams, Jeffrey. *Byng of Vimy: General and Governor General.* London: Leo Cooper, 1983.
– *The Long Left Flank: The Hard Fought Way to the Reich, 1944–1945.* London: Leo Cooper, 1988.
Wilt, Alan F. *War from the Top: German and British Military Decision Making during World War II.* Bloomington: Indiana University Press, 1990.
Winter, Denis. *Haig's Command: A Reassessment.* New York: Viking, 1991.
Winton, Harold R. *Corps Commanders of the Bulge: Six American Generals and Victory in the Ardennes.* Lawrence: University Press of Kansas, 2007.
Winton, Harold R., and David R. Mets, eds. *The Challenge of Change: Military Institutions and New Realities, 1918–1941.* Lincoln: University of Nebraska Press, 2000.
Worthington, Larry. *Worthy: A Biography of Major-General F.F. Worthington.* Toronto: Macmillan, 1961.
Young, Lieutenant-Colonel F.W., ed. *The Story of the Staff College.* Camberley: Staff College, 1958.
Ziegler, Philip. *Mountbatten: A Biography.* New York: Alfred A. Knopf, 1985.

Articles and Book Contributions

Barnett, Correlli. 'The Education of Military Elites.' *Journal of Contemporary History* 2, no. 3 (1967): 15–35.
Barrett, Captain Roger B. 'General Sir Arthur William Currie: A Common Genius for War.' *ADTB* 2, no. 3 (1999): 53–7.

Baxter, Colin F. 'Winston Churchill: Military Strategist?' *Military Affairs* 47, no. 1 (1983): 7–10.
Beaumont, Roger A. 'Command Method: A Gap in the Historiography.' *Naval War College Review* 31, no. 3 (1979): 61–74.
Ben-Moshe, Tuvia. 'Winston Churchill and the "Second Front": A Reappraisal.' *Journal of Modern History* 62, no. 3 (1990): 503–37.
Berlin, Robert H. 'United States Army World War II Corps Commanders: A Composite Biography.' *JMH* 53, no. 2 (1989): 147–67.
Bolger, Major Daniel P. 'Zero Defects: Command Climate in First U.S. Army, 1944–1945.' *Military Review* 71 (1991): 61–73.
Brennan, Joseph G. 'Ambition and Careerism.' *Naval War College Review* 34, no. 1 (1991): 76–82.
Brooke, Alan F. 'The Evolution of Artillery in the Great War.' *Royal Artillery Journal.* Part I, 51, no. 5 (1924): 250–67; Part II, 51, no. 6 (1925): 359–72; Part III, 52, no. 1 (1925): 37–51; Part IV, 52, no. 3 (1925): 369–87; Part V, 53, no. 1 (1926): 76–93; Part VI, 53, no. 2 (1926): 232–49; Part VII, 53, no. 3 (1926): 320–9; Part VIII, 53, no. 4 (1927): 469–82.
Brown, R. Craig. 'Sir Robert Borden, the Great War, and Anglo-Canadian Relations.' In *Character and Circumstance: Essays in Honour of Donald Grant Creighton*, ed. John S. Moir. 201–24. Toronto: Macmillan, 1970.
Brown, R. Craig, and Desmond Morton. 'The Embarrassing Apotheosis of a Great Canadian: Sir Arthur Currie's Personal Crisis in 1917.' *CHR* 60, no. 1 (1979): 41–63.
Burns, Eedson L.M. 'The Army in India.' *CDQ* 8, no. 1 (1930): 33–41.
– 'The Defence of Canada.' *CDQ* 13, no. 4 (1936): 379–94.
– 'A Division That Can Attack.' *CDQ* 15, no. 3 (1938): 282–98.
– 'The Principles of War: A Criticism of Colonel J.F.C. Fuller's Book "The Foundation of the Science of War."' *CDQ* 4, no. 2 (1927): 168–75.
– 'Protection of the Rearward Services and Headquarters in Modern War.' *CDQ* 10, no. 3 (1933): 295–311.
– 'The Remaking of Modern Armies: A Review.' *CDQ* 5, no. 1 (1927): 115–17.
– 'Speculations on Increased Mobility.' *CDQ* 3, no. 3 (1926): 319–23.
– 'A Step Towards Modernization.' *CDQ* 12, no. 3 (1935): 298–305.
– 'Theory of Military Organization.' *CDQ* 14, no. 3 (1937): 326–31.
– 'Where Do the Tanks Belong?' *CDQ* 16, no. 1 (1938): 28–31.
Caforio, Giuseppe. 'The Military Profession: Theories of Change.' *Armed Forces and Society* 15, no. 1 (1988): 55–69.
Campbell, John P. 'The Ultra Revelations: The Dieppe Raid in a New Light as an Example of Now Inevitable Revisions in Second World War Historiography.' *CDQ* 6, no. 1 (1976): 36–42.
Carravagio, Angelo. 'A Re-Evaluation of Generalship: Lieutenant General Guy

Simonds and Major General George Kitching in Normandy 1944.' *CMH* 11, no. 4 (2002): 5–19.
Cirillo, Roger. Review of Martin Gilbert's *D-Day* in *World War II Quarterly* 5, no. 1 (2008): 73–9.
Collins, General J. Lawton. 'Leadership at Higher Echelons.' *Military Review* (1990): 33–45.
Craig, David. 'Bernard Law Montgomery: A Question of Competence.' *Armor* (1992): 26–31.
Crerar, Henry D.G. 'Empire Interests in the Near and Middle East.' *CDQ* 7, no. 3 (1930): 289–98.
Dawson, R. MacGregor. 'The Revolt of the Generals.' *Weekend Magazine* 10, no. 44 (1960): 21–4, 48–9.
Delaney, Douglas E. 'Looking Back on Canadian Generalship in the Second World War.' *CAJ* 7, no. 1 (2004): 13–22.
Dickson, Paul D. 'The Hand That Wields the Dagger: Harry Crerar, First Canadian Army Command, and National Autonomy.' *War and Society* 13, no. 2 (1995): 113–41.
– 'Harry Crerar and an Army for Strategic Effect.' *CMH* 17, no. 1 (2008): 37–48.
– 'The Politics of Army Expansion: General H.D.G. Crerar and the Creation of First Canadian Army, 1940–1941.' *JMH* 60, no. 2 (1996): 271–98.
Dietrich, Steve E. 'The Professional Reading of General George S. Patton, Jr.' *JMH* 53, no. 4 (1989): 387–418.
Douglas, William A.B. 'Filling Gaps in the Military Past: Recent Developments in Canadian Official History.' *Journal of Canadian Studies* 19, no. 3 (1984): 112–24.
– 'Marching to Different Drums: Canadian Military History.' *JMH* 56, no. 2 (1992): 245–60.
Durflinger, Serge M. 'The *Canadian Defence Quarterly* 1933–1935: Canadian Military Writing of a Bygone Era.' *CDQ* 20, no. 6 (1991): 43–8.
English, John A. 'Lessons From the Great War.' *CMJ* 4, no. 2 (2003): 55–62.
Farrell, Brian P. 'Yes, Prime Minister: Barbarossa, Whipcord, and the Basis of British Grand Strategy, Autumn 1941.' *JMH* 57, no. 4 (1993): 599–625.
'Fighting Fit.' *Time*, 14 February 1944.
Finan, James S., and W.J. Hurley. 'McNaughton and Canadian Operational Research at Vimy.' *Journal of Operational Research Society* 48, no. 1 (1997): 10–14.
French, David. 'Colonel Blimp and the British Army: British Divisional Commanders in the War against Germany, 1939–1945.' *English Historical Review* 111, no. 444 (November 1996): 1182–201.
Froude, Chester J. 'Chief of Staff: The Personality of Major-General McNaughton.' *Saturday Night*, 26 December 1931.

Fuller, J.F.C. 'The Application of Recent Developments in Mechanics and Other Scientific Knowledge to Preparation and Training for Future War on Land.' *Royal United Services Institute Journal* 65 (May 1920): 240–1, 252–6.
Gates, John M. 'The "New" Military Professionalism.' *Armed Forces and Society* 11, no. 3 (1985): 427–36.
Graeme-Evans, Alex. 'Field Marshal Bernard Montgomery: A Critical Assessment.' *AQDJ* 104, no. 4 (1973–1974): 412–26.
Graham, Dominick. 'Stress Lines and Grey Areas: The Utility of the Historical Method to the Military Profession.' In *Military History and the Military Profession*, ed. David A. Charters, Marc J. Milner, and J. Brent Wilson. Westport: Praeger, 1991.
Granatstein, J.L. 'Researching Guy Simonds.' *CMH* 2, no. 2 (1993): 107–8.
Granatstein, J.L., and Robert D. Bothwell. '"A Self-Evident National Duty": Canadian Foreign Policy, 1935–1939.' *Journal of Imperial and Commonwealth History* 3, no. 2 (1975): 212–33.
Granatstein, J.L., and Peter Suedfeld. 'Tommy Burns as a Military Leader: A Case Study Using Integrative Complexity.' *CMH* 3, no. 2 (1994): 63–7.
Haislip, Wade A. 'Corps Command in World War II.' *Military Review* 70 (May 1990): 22–32.
Hammond, Jamie W. 'The Pen before the Sword: Thinking about Mechanization between the Wars.' *CMJ* 1, no. 2 (2000): 95–104.
Harris, Stephen J. 'The Canadian General Staff and the Higher Organization of Defence, 1919–1939: A Problem of Civil–Military Relations.' *War and Society* 3, no. 1 (1985): 83–98.
– 'From Subordinate to Ally: The Canadian Corps and National Autonomy, 1914–1918.' *Revue Internationale d'Histoire Militaire* No. 51 (1982): 109–30.
– 'Or There Would Be Chaos: The Legacy of Sam Hughes and Military Planning in Canada, 1919–1939.' *Military Affairs* 46, no. 3 (1982): 120–6.
Hauser, William L. 'Careerism versus Professionalism in the Military.' *Armed Forces and Society* 10, no. 3 (1984): 449–63.
Hayes, Geoffrey. 'Science and the Magic Eye: Innovations in the Selection of Canadian Army Officers, 1939–1945.' *Armed Forces and Society* 22, no. 2 (1995–6): 275–95.
Henshaw, Peter J. 'The British Chiefs of Staff Committee and the Preparation of the Dieppe Raid, March–August 1942: Did Mountbatten Really Evade the Committee's Authority?' *War in History* 1, no. 2 (1994): 73–90.
– 'The Dieppe Raid: The Product of Misplaced Canadian Nationalism.' *CHR* 77, no. 2 (1996): 250–66.
Hilmer, Norman. 'Defence and Ideology: The Anglo-Canadian Military Alliance in the 1930s.' *International Journal* 33 (Summer 1978): 40–7.

Hunt, Major-General J.D. 'Alex.' *AQDJ* 103 (July 1973): 440–1.
Hutton, Eric. 'A Scientist Commands Canada's First Division.' *Star Weekly*. 9 December 1939.
Hutton, Eric, and William J. McAndrew. 'The Cunning of Restraint: General J.H. MacBrien and the Problems of Peacetime Soldiering.' *CDQ* 7, no. 4 (1979): 40–7.
Hyatt, A.M.J. 'Canadian Generals of the First World War and the Popular View of Military Leadership.' *Social History* 12 (November 1979): 418–30.
Ignatieff, George. 'General A.G.L. McNaughton: A Soldier in Diplomacy.' *International Journal* 22, no. 3 (1967): 402–14.
Jacob, Sir Ian. 'Statesmen and Soldiers in War.' *Foreign Affairs* 38, no. 4 (1960): 656–64.
Jarymowycz, Roman J. 'Canadian Armour in Normandy: Operation "Totalize" and the Quest for Operational Maneuver.' *CMH* 7, no. 2 (1998): 19–40.
– 'Der Geganangriff vor Verrieres: German Counterattacks during Operation Spring, 25–26 July 1944.' *CMH* 2, no. 1 (1993): 74–89.
Leighton, Richard M. 'OVERLORD Revisited: An Interpretation of American Strategy in the European War, 1942–1944.' *American Historical Review* 68, no. 4 (1963): 919–36.
Lunt, Major-General J.D. 'Alex.' *AQDJ* 103, no. 4 (1973): 440–1.
Luttwak, Edward N. 'The Operational Level of War.' *International Security* 5, no. 3 (1980–1): 61–76.
Lutz, James H. 'Canadian Military Thought, 1923–1939: A Profile Drawn from the Pages of the Old *Canadian Defence Quarterly*.' *CDQ* 9, no. 2 (1979): 40–8.
MacBrien, Major-General James H. 'The British Army Manoeuvres September 1925.' *CDQ* 2 (January 1926): 132–50.
Mackenzie King, William Lyon. 'Canada's Foreign Policy.' *CDQ* 15, no. 4 (1938): 380–401.
Mann, Churchill C. 'On the Real Purpose of the Dieppe Raid.' *CDQ* 9, no. 1 (1979): 57.
McAndrew, William J. 'Fire or Movement? Canadian Tactical Doctrine, Sicily –1943.' *Military Affairs* 51, no. 3 (1987): 140–5.
McNaughton, Andrew G.L. 'Canada's Land and Air Forces.' *Empire Club of Canada Speeches 1929*. 189–99. Toronto: Empire Club of Canada, 1930.
– 'The Capture of Valenciennes: A Study in Co-ordination.' *CDQ* 10, no. 3 (1933): 279–94.
– 'The Development of Artillery in the Great War.' *CDQ* 6, no. 2 (1929): 160–71.
– 'The Military Engineer and Canadian Defence.' *CDQ* 7, no. 2 (1930): 150–4.

- 'Military Survey.' *CDQ* 10, no. 1 (1932): 17–22.
- 'The Progress of Air Survey in Canada.' *CDQ* 14, no. 3 (1937): 311–16.
- 'Some Aspects of the Work of the Department of National Defence.' *CDQ* 4, no. 2 (1927): 143–9.

'McNaughton Talks.' *Time*, 9 October 1944.

Milner, Marc. 'Reflections on the State of Canadian Army History in the Two World Wars.' *Acadiensis* 18, no. 2 (1989): 135–50.

Montgomery, Bernard Law. 'The Growth of Modern Infantry Tactics.' Five part series in *The Antelope* (1924–5).
- 'The Human Factor in My Life.' *Listener* 84 (5 November 1970): 625–7.
- 'The Major Tactics of the Encounter Battle.' *Army Quarterly* 36, no. 2 (1938): 268–72.
- 'The Problem of the Encounter Battle as Affected by Modern British War Establishment.' *CDQ* 15 (October 1937): 13–25.

Morgan, General Sir William. 'With Alexander to Dunkirk: Reminiscences of a GSO 1.' *AQDJ* 102, no. 3 (1972): 357–64.

Morton, Desmond. 'Changing Operational Doctrine in the Canadian Corps, 1916–1917.' *ADTB* 2, no. 4 (1999): 35–9.
- '"Junior but Sovereign Allies": The Transformation of the Canadian Expeditionary Force, 1914–1918.' *Journal of Imperial and Commonwealth History* 8, no. 1 (1979): 56–67.

Mountbatten, Louis. 'Operation Jubilee: The Place of the Dieppe Raid in History.' *Journal of the Royal United Services Institute* 119, no. 1 (1974): 25–30.

Murphy, C.J.V. 'The First Canadian Army.' *Fortune* (January 1944): 164.

Murray, Williamson. 'British Military Effectiveness in the Second World War.' In *Military Effectiveness, III: The Second World War*, ed. Alan R. Millet and Williamson Murray. 90–135. Boston: Allen Unwin, 1988.

Nenninger, Timothy K. 'Leavenworth and Its Critics: The U.S. Army Command and General Staff School, 1920–1940.' *JMH* 58, no. 2 (1994): 199–231.

Pearkes, George R. 'The 1914 Campaign in East Prussia.' *CDQ* 8, no. 2 (1932): 248–54.
- 'The Burthen and the Brunt: A Short Description of the Success of the British Regular Army in Canada.' *CDQ* 12, no. 4 (1935): 387–96.
- 'The Evolution and Control of Her Majesty's Canadian Forces.' *CDQ* 10, no. 4 (1933): 465–80.
- 'The Field of Legend.' *CDQ* 9, no. 3 (1932): 383–8.
- 'The Winter March of a Brigade of Guards through New Brunswick, 1862.' *CDQ* 11, no. 1 (1934): 100–10.

Preston, Richard A. 'Canadian Military History: A Reinterpretation Challenge

of the Eighties.' *American Review of Canadian Studies* 19, no. 1 (1989): 95–104.
- 'Military Education, Professionalism, and Doctrine.' *Revue Internationale d'Histoire Militaire* No. 51 (1982): 273–301.

Roy, Reginald H. 'Morale in the Canadian Army in Canada during the Second World War.' *CDQ* 16, no. 2 (1986): 40–5.

Schreiner, Jr, Lt.-Col. Charles W. 'The Dieppe Raid: Its Origins, Aims, and Results.' *Naval War College Review* (May–June 1973): 83–97.

'Shush!' *Time Magazine* (28 February 1944).

Sexton, Donal J. 'Phantoms of the North: British Deceptions in Scandinavia, 1941–1944.' *Military Affairs* 47, no. 3 (1983): 109–14.

Simonds, Guy G. 'An Army That Can Attack – A Division That Can Defend.' *CDQ* 15, no. 4 (1938): 413–17.
- 'The Attack.' *CDQ* 16, no. 4 (1939): 379–90.
- 'What Price Assault without Support?' *CDQ* 16, no. 2 (1939): 142–7.
- 'Where We've Gone Wrong on Defense.' *Maclean's*, 23 June 1956: 28–68.

Skaggs, David Curtis. 'Michael Howard and the Dimensions of Military History.' *Military Affairs* 49, no. 4 (1985): 179–83.

[Special Correspondent]. 'The Army Manoeuvres September 1925.' *The Fighting Forces* 12 (October 1935).

Stacey, Charles P. 'Canadian Defence Policy.' *Canadian Journal of Economics and Political Science* 4, no. 4 (1938): 490–504.
- 'Canadian Leaders of the Second World War.' *CHR* 66, no. 1 (1985): 64–72.
- 'The Life and Hard Times of an Official Historian.' *CHR* 51, no. 1 (1970): 21–47.

Strange, Joseph L. 'The British Rejection of Operation SLEDGEHAMMER, an Alternative Motive.' *Military Affairs* 46, no. 1 (1982): 6–14.

Stuart, Kenneth. Editorial: 'The Anti-Machine Gun Problem.' *CDQ* 8, no. 2 (1931): 147.
- Editorial: 'The Future of Infantry.' *CDQ* 10, no. 1 (1932): 1–4.
- 'The Training of the Non-Permanent Active Militia, Being a Discussion of the Influence of Ground and the Potentialities of the Sand Table.' *CDQ* 7, no. 4 (1930): 482–9.

'The Canadians.' *Time*, 10 August 1942.

'The Chief Steps Down.' *Time*, 3 January 1944.

'The Minister of National Defence.' *CDQ* 4, no. 2 (1927): 134–8.

Travers, Timothy. 'Currie and 1st Canadian Division at Second Ypres, April 1915: Controversy, Criticism, and Official History.' *CMH* 5, no. 2 (1996): 7–15.

Villa, Brian Loring. 'Mountbatten, the British Chiefs of Staff, and the Approval of the Dieppe Raid.' *JMH* 54, no. 2 (1990): 201–26.
Villa, Brian Loring, and Peter Henshaw. 'The Dieppe Raid Debate.' *CHR* 79, no. 2 (1998): 304–15.
Webb, Lt.-Col. Daniel J. 'The Dieppe Raid: An Act of Diplomacy.' *Military Review* 60 (May 1980): 30–7.
Williamson, O.T.G. 'Responsibility for Dieppe.' *Saturday Night* June 12, 1943.
Wilson, Lt.-Col. Sir James. 'Dieppe: Vindication.' *AQDJ* 124, no. 1 (1994): 68–72.

Index

ABERCROMBIE, Operation, 78–9, 220, 240, 276n47
Adair, Maj.-Gen. Alan H.S., 129
Adam, Gen. Sir Ronald, 143
Afrika Korps, 48–9
Alam El Halfa, 112
ALBERT, Exercise, 111
Aldershot, 48, 91, 97
Alexander, Gen. Sir Harold, 14–16, 18, 25, 35–6, 90, 98, 109–11, 172–6, 178–9, 181–3, 185, 187, 198, 204–5
Allenby, Sir Edmund, 5, 20
American Expeditionary Force (First World War), 20
Amiens, 17, 31, 44
Anderson, Gen. W.A.B., 56
Anderson, Maj.-Gen. Hastings, 24
Anderson, Maj.-Gen. Kenneth A.N., 15, 106
Anderson, Maj.-Gen. Thomas V., 28
ANGEL MOVE, Operation, 45, 239
Appreciation of the Situation, 131
ARCADIA Conference, 54
Armies: *American:* Third, 158; *British:* First (First World War), 17; First (Second World War), 62; Second, 200; Eighth, 24, 49, 62, 75, 115, 126, 171, 173–4, 176–7, 179–80, 183, 185, 204, 211; *Canadian:* First, 3–4, 19, 61–2, 66, 77, 118–20, 128–30, 140, 149, 152, 160, 174, 186–7, 189–91, 194, 200, 202, 203, 205–8, 211, 216–19, 222–3, 226, 228–9, 237–8, 243; *French:* Tenth (First World War), 17; *Italian:* Tenth, 49
Armies (Army Groups): 15th, 172–3; 21st, 121, 174–5, 186, 189, 213
Armies (Artillery): *Canadian:* 'B' Battery, 15; 4th Battery / 2nd Brigade, 15; 7th Battery, 15; 21st Battery / 6th Howitzer Brigade, 16; 11th Brigade, 16; Heavy (Canadian Corps), 3, 18, 182
Armour, Christina Mary Ann, 11
Armour, Murray William, 12
Ashton, Maj.-Gen. E.C., 28
Attlee, Clement, 67
Auchinleck, Sir Claude, 25–6, 36, 49, 115
AVALANCHE, Operation, 198, 242

Barker, Lt.-Gen. Michael, 115

Index

Bartholomew Report, 111–12
Battle of Britain, 48, 98
BAYTOWN, Operation, 197–8, 211, 220, 242
BEAVER III, Exercise, 123
BEAVER IV, Exercise, 124
BENITO, Exercise, 108
Bennett, R.B., 27, 33
Berger, Carl, 12
Berlin, Robert H., 20
Bishop's College School, 13–14
Black, Ernest G., 19
Blackader, Brig. Ken, 124
Blackburn, George G., 108
Bogert, M. Pat, 90–1, 99, 159, 163
BOLERO, Operation, 60, 241
Borden, Sir Robert, 21
Brigades: *British:* 8th Infantry, 35; 9th Infantry, 35; 17th Infantry, 24; Nowshera, 35; 31st Tank, 129; 34th Tank, 129; *Canadian:* 1st Canadian Army Tank, 3, 78, 129, 141, 153, 306n18; 1st CIB, 89, 94; 2nd CIB, 63, 236, 240; 5th CIB, 125; 6th CIB, 89; 7th CIB, 124
BRIMSTONE, Operation, 72, 78–9, 188, 220, 242, 316n4
Britain, 12, 15, 19, 24, 28, 34, 42–3, 45–52, 55–6, 58–60, 62, 66–7, 70, 72, 74–6, 78–9, 83–6, 89, 94, 96, 98, 103, 105, 110, 114–18, 126–7, 152–3, 164, 175, 178, 180, 183–4, 187, 190, 192, 198, 200, 203, 205
British Army, 14, 18, 23, 29, 39, 41, 47–8, 59–62, 77, 79, 106, 111, 114, 120, 171, 202, 226
British Empire, 12–13, 178
British Expeditionary Force (BEF), 3, 45, 79, 93, 98, 106, 111, 120, 239, 250n24, 266n34, 293n39

British Joint Planning Staff, 60
Brittany Redoubt, 45–6, 56, 239
Brocas Burrows, Maj.-Gen. Montague, 152
Brooke, Sir Alan F., 5–6, 14, 18–9, 23–6, 32, 35, 45–6, 55–62, 64, 67–74, 77–80, 83, 86, 92, 96–7, 100, 104–6, 109–12, 114–20, 127–8, 140, 143–4, 149–51, 153–4, 157, 161–4, 166–7, 173, 176, 179–83, 186–7, 189–90, 193–4, 196, 200–2, 204, 206–12, 215–16, 219–28
Brown, Col. James Sutherland, 27
Bucknall, Lt.-Gen. Gerald, 126
Bullen-Smith, Maj.-Gen. D.C., 126, 162
BUMPER, Exercise, 109–10, 112, 127–8, 235, 237, 294n50, 51, 295n58, 62
Burns, Eedson L.M., 14, 25, 29, 32, 35–6, 48, 62, 92, 108–9, 116–17, 163, 165, 206
Butcher, Capt. Harry C., 178–9
Byng, Sir Julian, 55

Cabinet War Committee (CWC), 40, 50, 116, 191, 194, 202
Cambrai, 22
Camp Borden, 29, 116–17
Canadian Active Service Force (CASF), 3, 85
Canadian Army, 4–5, 15, 26, 28–9, 36, 39, 46, 51–2, 54, 59–60, 66–8, 70, 73–4, 77–80, 83, 88, 92, 95–6, 113, 115–19, 157, 160, 165, 171, 176–7, 180, 182–3, 189–91, 193, 195, 200–1, 206, 211, 213, 218–19, 220–1
Canadian Army Programme (1942–3), 59
Canadian Chiefs of Staff, 196

Canadian Defence Quarterly (*CDQ*), 29–32
Canadian Expeditionary Force (CEF), 15, 23
'Canadian Force,' 96
Canadian Forces Artillery (CFA), 16
Canadian Junior War Staff Course (CJWSC), 88–9
Canadian Military Headquarters (CMHQ), 46, 86, 111, 152, 218
Canadian Society of Civil Engineers, 13
Canal-du-Nord, 18, 22
Canary Islands, 69–70, 79, 182, 240, 242
Carolina Manoeuvres, 35
Carr, Lt.-Gen. Laurence, 109–11
Carton de Wiart, Maj.-Gen. Sir Adrian, 41
Casablanca Conference (SYMBOL), 73
Chalfont, Alun, 177–8, 185
Chamberlain, Sir Neville, 40
Cherbourg Peninsula, 73
CHESTERFIELD, Operation, 206
Chief of Staff to the Supreme Allied Commander (COSSAC), 189, 202
Chiefs of Staff (COS) Committee, British, 47–9, 52, 56, 60–1, 63–4, 66–7, 69–70, 72–3, 78, 80, 202, 205, 227
Churchill, Winston S., 21, 28, 40–1, 45–50, 52, 54, 56, 59–64, 66–9, 73–4, 76–7, 79, 163, 175, 188, 195, 197–8, 200–2, 220, 225, 239
Combined Chiefs of Staff (CCOS), 26, 54, 73, 195, 204–5
Combined Commanders, 61
Combined Operations Headquarters (COHQ), 63

COMPASS, Operation, 49, 239
CONQUEROR, Exercise, 122, 299n28
conscription, 4, 34, 40, 75–6, 192, 262n2
Copp, Terry, 300n33
Corps: *American:* VIII, 158; XV, 97; *Australian:* Australian–New Zealand Army Corps (ANZAC), 16; *British:* I, 115; II, 45, 105; VII, 3, 47, 51, 55, 97–9, 106, 155; VIII, 130, 135, 144, 146, 158; X, 126; XI, 130, 135, 138, 140; XII, 108, 119, 125, 129, 135, 138, 140, 144, 146, 203, 210; XXX, 98, 126; *Canadian:* Canadian Corps (First World War), 17–18, 20–2, 31, 33, 55, 58, 94, 182, 228; I Canadian, 29, 70, 98, 108, 119–20, 124–7, 129, 131, 133, 138–40, 144, 150–1, 155, 157, 167, 182, 187, 200, 202, 205–6, 208, 218, 220, 228; II Canadian, 118, 129, 138, 140–1, 144–6, 149–51, 153, 157–9, 161, 163–5
Crerar, Henry Duncan Graham, 14, 19, 25, 36, 46–7, 50–2, 59, 63, 70–1, 73–4, 83, 86–9, 105, 111, 114–17, 119–20, 124–9, 133, 137–41, 144, 146, 151, 153, 157–60, 162–7, 171, 183–4, 191, 200–1, 205
CRUSADER, Operation, 49, 112, 185, 240
Cunningham, Lt.-Gen. Sir Alan, 25, 36, 49, 106
Currie, Sir Arthur, 4, 17, 19–24, 31, 33, 40, 51, 55, 182, 217, 228
'Currie Doctrine,' 5, 228

D-Day, 226
Defence Scheme No. 1, 27
Defence Scheme No. 3, 27, 87

360 Index

de Guingand, Maj.-Gen. Francis, 177
Dempsey, Miles Christopher, 25, 36, 98–9, 101, 104, 158, 221, 226
Desire Trench, 16
Dewing, Maj.-Gen. Richard H., 41–2, 45–6
Dexter, Grant, 44, 208
Dieppe Raid, 3, 39, 63–4, 66, 74, 76, 79, 125, 152, 162, 171, 225–6, 236, 241
Dill, Sir John, 25–6, 45, 50, 55, 114, 152, 195–7, 226
Divisions: *British:* Mobile Division, 32, 35; 6th Airborne, 158; 1st Armoured, 97, 99, 106, 204; 7th Armoured, 49, 126, 158, 205–6; 8th Armoured, 108; 9th Armoured, 135, 138, 141; 10th Armoured, 126; 11th Armoured, 158; 42nd Armoured, 138; Guards Armoured, 129, 145, 158; 1st Infantry, 35; 2nd Infantry, 35, 115; 3rd ('Iron'), 119; 5th Infantry, 198; 8th Infantry, 35; 15th (Scottish), 106; 43rd, 147; 46th, 106; 47th (London), 18; 49th, 130; 50th (Northumbrian), 24, 206; 51st (Highland), 106, 126; 53rd, 126, 129; 61st, 130; 70th Infantry, 49; *Canadian* (First World War): 1st Division, 15; 2nd Division, 16; 3rd Division, 16; 4th Division, 42; (Second World War): 1st CID, 3, 35, 78, 87, 130, 173, 176, 183–4, 197, 200, 235–6, 239, 242, 284n41, 311n15; 2nd CID, 47, 63, 79, 129, 137–8, 162; 3rd CID, 123, 126, 129, 138, 205, 226; 4th CAD, 237–8; 5th CAD, 87, 127, 129, 137–8, 145, 147, 152, 164–5, 202, 205; *U.S.:* 2nd Armoured, 35

DOG, Exercise, 103, 108, 127, 235, 291n25
Dominion Arsenal, 44, 86
Douglas, Sholto, 64
Drocourt-Queant Line, 18
Dunkirk, 3, 45–7, 61, 91, 96, 112, 239
Dyde, H.A., 43, 192
DYNAMO, Operation, 45, 239

Eastern Command, 109
Eayrs, James, 30
Eden, Anthony, 50, 77
Eisenhower, Dwight David, 12, 36, 57, 59–62, 64, 72, 115, 120, 172–4, 176–9, 201, 204–5, 226, 228
Ellis, Maj. Lionel F., 222
English, John A., 19, 27, 84, 89, 92, 150–1, 166
Erskine, Maj.-Gen. Bobby, 126
European Theater of Operations (ETOUSA), 60
Evans, Maj.-Gen. Roger, 97, 106, 108

Fisher, Douglas, 223
Flers-Courcelette, battle of, 16
Fortune, Maj.-Gen. Victor, 106
Foulkes, Charles, 25, 90, 99, 164, 174, 180, 218, 224
FOX, Exercise, 100, 102–3, 108, 127, 235, 291n23
Fraser, David, 32, 62
Freyberg, Lt.-Gen. Bernard C., 45
Fuller, J.F.C., 32

Ganong, Brig. Hardy, 104, 123-4
Gardiner, James G., 50
Gatehouse, Maj.-Gen. Alex, 126
GAUNTLET, Operation, 63, 78-9, 220, 240, 276n47

German Army: First World War, 22; Second, 29, 61, 85
'Germany first' strategy, 54
Gerow, Maj.-Gen. Leonard T., 120
Gibson, Brig. Ralph Burgess, 209
Gibson, Lt.-Col. Thomas G., 98, 126, 129-30, 139, 156
GOLDFLAKE, Operation, 218
GOODWOOD, Operation, 158
Gort, Lord, 25, 36, 93, 174
Graham, Dominick, 6, 112
Graham, Lt.-Col. Howard, 87, 102, 183
Granatstein, J.L., 4, 26, 39, 192, 205, 208, 220, 224
Grant, G.M., 12
Great Depression, 27-8
Greece, 48, 220
Griffith, Paddy, 93
Grigg, Sir James, 69, 139, 212
GRIZZLY II, Exercise, 153
Guderian, Heinz, 29, 92, 143
GYMNAST, Operation, 61

Haig, Douglas, 20-2
Haining, Lieut.-Gen. Dick, 47
Haislip, Maj.-Gen. Wade H., 97
Haldenby, Brig. Eric W., 124
Hamilton, Nigel, 105, 118, 176-7
HAMMER, Operation, 41-3, 78-9, 91, 220, 236, 239
HARE, Exercise, 104-5
Harney, J.H., 14
Harris, Stephen J., 29, 33
Hayes, Maj.-Gen. Eric C., 125
Henderson, Lt.-Col. G.P., 120, 129-30, 139
Herdt, Dr. Louis, 14
Hertzberg, Maj.-Gen. Charles S.L., 90, 99, 101, 120, 130, 155

Hill 70, 17-18, 43
Hindenburg Line, 18
Hitler Line, 206
Hoffmeister, Maj.-Gen. Bert, 25, 87
Hollis, Sir Leslie, 56
Hong Kong, 39, 59
Horrocks, Gen. Sir Brian, 12, 24-5, 62, 98
Hughes, Sir Sam, 15
Huntington, Samuel P., 19
HUSKY, Operation, 73, 76-9, 171, 176, 182, 188, 197, 209, 212, 220, 242, 310fn4, 312fn28

IMPERATOR, Operation, 64
Imperial Defence College (IDC), 26, 34
Inglis, Brig.-Gen. Desmond, 175
Ironside, Gen. Sir William E., 25-6, 36, 41-2, 44-6, 55, 85, 96, 226
Ismay, Lt.-Gen. Hastings L., 80
Italy, 24, 29, 39, 46, 52, 87, 108, 153, 186-7, 190-4, 196-8, 200-7, 214-18, 227-8

Jacob, Sir Ian, 6
JUBILEE, Operation, 63-4, 66, 68, 71, 78-9, 220, 241, 266n40, 277n 51, 54, 57, 309n53
JUPITER, Operation, 61, 66-7, 72, 78-9, 220, 241, 278n67, 68, 279n70

Keller, Maj.-Gen. Rod F.L., 25, 90, 99-100, 102, 123-4, 127, 129, 138, 157, 159, 165
Kennedy, Maj.-Gen. John N., 48, 66, 115, 172-3, 197-8, 201, 226
King, William Lyon Mackenzie, 4, 34, 40, 43-4, 48, 50-1, 53-4, 59-60, 66-7, 73-7, 80, 94, 114, 184-6, 188,

192, 197–8, 200–2, 206–11, 214–15, 218, 225–7
Kitching, Maj.-Gen. George, 88, 110, 120

Laing, Brig. Horace V.D., 156
Lake, Maj.-Gen. P.H.N., 14
Latham, Brig. Harry, 158
Laurie, Lt.-Col. W.L., 98
Leach, Brig. R.J., 55
Lee, Gen. Raymond E., 91
Leese, Oliver, 15, 24–5, 36, 211, 221
Leigh-Mallory, Air Chief Marshal Sir Trafford, 139
Lett, Sherwood, 89
Lewin, Ronald, 120
Liddell Hart, Basil H., 32, 93, 150, 159, 163
LINK, Exercise, 163, 238, 309n58
London Controlling Station (LCS), 66
Louisiana Manoeuvres, 35, 98
Loyd, Maj.-Gen. H.C., 115, 154, 161
Ludendorff, Gen. Erich, 17
Lumsden, Maj.-Gen. Herbert, 126
Luton, Maj.-Gen. Robert R.M., 180
Lutz, James H., 32

MacArthur, Douglas, 15, 26, 28
MacBrien, Maj.-Gen. James H., 26–7
Macdonald, Sir John A., 12
Mackenzie, Dr C.J., 34, 180
Mallaby, Brig. Aubertin W.S., 48
Malone, Richard, 161, 176, 185, 209
Mann, Churchill C., 90, 99, 123, 159
MAPLE I, Exercise, 162
MAPLE II, Exercise, 162
Margesson, David, 48
Marshall, Gen. George C., 59–61, 63–4, 87, 98, 190, 198

Martel, Maj.-Gen. Giffard Le Q., 108, 152
Massie, Brig. R.H., 18
MAURICE, Operation, 41, 239
Massey, Vincent, 43–4, 69, 186, 194–5, 201–2, 205, 212, 219, 225
McCreery, Maj.-Gen. Richard L., 25, 36, 108–10
McNair, Lt.-Gen. Leslie J., 94, 98
McNaughton, Lt.-Gen. Andrew G.L.: affection for Currie, 20; affinity for British Empire, 13; artillery, views on, 112; as trainer/commander, 83–5, 88, 90–3, 96–7, 100–4, 108–11, 115–17, 119–20, 122, 125, 126–7, 221–3; Brooke, relations with, 55–8, 70, 78–9, 83, 106, 116–17, 167, 179–82, 200, 202, 212, 223–4; CGS, 26–30, 33; Crerar, relations with, 70–1, 73, 89, 117, 227; Dieppe, 63, 67; dream of joining Indian Army, 15; education, 14; focus on cross-channel operations, 52, 62, 74, 119, 201, 207; illness of, 114, 118; insufficient planning information, 54, 61, 68, 71, 77, 197–8, 202; Montgomery, relations with, 125, 174, 176–7, 183–6, 204, 224–5; Otter Committee, 23; Paget, relations with, 161, 163, 165, 213, 215–16, 224; problem of practical operations, 41, 46, 51, 69, 80; Ralston, relations with, 43–4, 77, 184–5, 187, 193–5, 203, 210, 212, 214, 219, 227–8; recognizes need for battle experience, 211; SPARTAN, 128–31, 133, 135, 137–41, 144–7, 149–61, 166, 221; Staff College Camberely, 24–5; Stuart, relations with, 173, 185, 188–90, 196, 208–9,

216, 219, 224; training difficulties, 86–8, 91–2, 94–5, 99; willingness to employ army, 39–41, 45, 47–8, 53, 67–9, 71–4, 76–7, 191, 193–4
McNaughton, Robert Duncan, 11
mechanized warfare, 30–2
Mediterranean, 46–7, 49, 57, 61, 70, 79–80, 171–4, 179–80, 187–91, 193–6, 198, 200, 203–6
Melville, Brig. James C., 155
Middle East, 46–7, 49–50, 58–9, 171, 175, 195
Middleton, Sir Frederick, 11
Middleton, Maj.-Gen. Troy H., 158
Miksche, Maj. Ferdinand O., 93
Military District No. 11, 26
Military District No. 13, 34
Militia Staff Course, 21
Milne, Sir George, 28
Montague, Brig. Price, 43, 111, 152
Montgomery, Bernard Law, 5, 6, 11–15, 18–19, 21, 23–5, 32, 34–6, 56, 64, 75, 84, 97–8, 103–5, 108–12, 114–15, 119–28, 139, 143, 147, 153–4, 158, 162, 164–5, 171–81, 183–6, 198, 204–6, 210–11, 219, 221–6
Morgan, Frederick E., 74, 152, 187, 189–90, 210, 219, 226
Morrison, Maj.-Gen. Edward, 17
Mountbatten, Lord Louis, 63, 66, 175, 183
Mount Sorrell, 16
Mowat, Farley, 223
Murchie, Maj.-Gen. John C., 209
Murison, Brig. Charles A.P., 98, 101
Murmansk convoys, 66

National Defence Headquarters (NDHQ), 5, 117

National Research Council (NRC), xiii, xviii, 3, 33
New Zealand Expeditionary Force, 45
Nicholson, Gerald W.L., 19, 172–3
Non-Permanent Active Militia (NPAM), 87
Normandy campaign, 31, 83, 98, 119, 126, 175, 205–6, 218, 222
North Africa, 46, 48–9, 52, 59, 61, 66, 73–6, 80, 111, 173, 179, 180–1, 191–2, 205, 220, 241–2
Norwegian campaign, 41, 42, 44, 57, 61, 66, 79, 220, 239, 241, 243
Nye, Archibald E., 40, 62, 86, 161, 171–3, 180, 189, 198, 200–1, 210, 216, 226

O'Connor, Lt.-Gen. Sir Richard, 25, 49, 111, 158
Odlum, Maj.-Gen. Victor W., 103, 105, 108–10, 115, 164–5
Orde, Brig. R.J., 33
Otter Committee, 23
OVERLORD, Operation, 172, 189, 198, 200, 202, 206, 220, 223–4, 228, 275n36, 316n7, 322n64
OVERTHROW, Operation, 67, 71, 241, 278n64, 279n72

Paget, Gen. Bernard Toliver, 5, 6, 15, 25, 36, 59, 61–2, 72–4, 108, 110–12, 114, 119, 128, 133, 135, 137, 139–40, 143–4, 150–7, 159, 161–6, 186, 189, 193, 197, 201, 203, 210–16, 221, 223–6
Passchendaele, 17, 20, 43, 182
Patton, Jr., George S., 15, 20, 23, 32, 35–6, 115, 143, 158, 183
Pearkes, Maj.-Gen. George R., 25,

34–5, 57, 90, 94, 97, 99–101, 103–5, 109–10, 115, 122–3, 125–7, 164–5
Pearl Harbor, 54
Pearson, Lester B., 42–4, 114
Penhale, Brig. Matthew H.S., 118, 120
Perley, Sir George, 21
Permanent Force, 3, 23, 29, 34, 87, 105, 249n14, 256n20, 260n49
Pershing, Gen. John J., 20, 33
Pile, Sir Frederick, 19
Pogue, Forrest C., 224
Pope, Col. Maurice A., 25, 46, 117–18, 225
Portal, Air Marshal Sir Charles, 139
Potts, Brig. Arthur E., 99, 104–5, 123–4
Power, Charles G., 50
Price, Maj.-Gen. Basil, 89–91, 94, 99, 122, 225
professional development, 24, 33
professionalism, 5, 6, 30, 106, 149

QUADRANT Conference, 196–8, 211

RAINBOW 5, 54
Ralston, James L., 5, 6, 43–4, 47–8, 50, 55, 59, 61, 68–71, 74–77, 79–80, 92, 115–17, 155, 157, 184–7, 190–7, 201–3, 205–21, 223–5, 227–8
RANKIN, Operation, 190
RATTLE, Exercise, 172
Reynolds, Quentin, 63
Richardson, Maj.-Gen. Charles, 205
Richmond, Sir Herbert, 26
Ridgway, Maj.-Gen. Matthew B., 97
Ritchie, Lt.-Gen. Neil, 115
Roberts, Maj.-Gen. James H., 64, 66, 99, 104, 115, 123–4, 129, 137–8, 159, 162–5

Rodger, Elliot, 34, 56, 90, 92, 223
Rogers, Norman, 34, 40, 43–4, 51, 55, 103, 228
Rommel, Erwin, 48, 119, 154
Roosevelt, Franklin D., 43, 49, 59–61, 63–4, 66–7
ROUNDUP, Operation, 60, 62, 68, 243, 275n35, 276n72, 280n73
Roy, Reginald H., 35, 91
Royal Military College (RMC), 14
Royal Navy, 13, 27, 48, 58
Rutherford, Ernest, 14
RUTTER, Operation, 63, 220, 225, 241, 276n48

Salisbury Plain, 32, 35
Salmon, Brig. Harry L., 25, 124–5, 127, 130, 165
Sandhurst, 14
Sansom, Maj.-Gen. Ernest W., 25, 90–1, 114, 127, 129, 133, 135, 137–41, 143–7, 149–53, 156–66
Schreiber, Shane B., 31
Second World War, 12, 14–15, 17, 19–20, 26, 29, 31, 35, 39, 55, 97, 105, 222
Seely, Brig. Gen. J.E.B., 20
SESAME, Operation, 64
Sicily, 3, 39, 73, 77–80, 87, 167, 171–9, 181, 184–6, 189–91, 193, 196–7, 209, 210, 224, 242, 282n22, 310n7, 311n18, 316n5, 318n38
Simonds, Maj.-Gen. Guy Granville, 14, 25, 32, 36, 67, 70, 83, 85–7, 88–92, 96, 99, 109, 119–20, 125, 129–30, 144, 146, 156, 159–60, 164–5, 176–7, 183–6
Sinclair, Sir Archibald, 139
Singapore, 58–9
SLEDGEHAMMER, Operation, 60,

Index

61, 63–4, 66, 68, 78–9, 220, 241, 274n30, 275n34, 277n52, 280n75
Smith, Brig. Armand A., 90, 94
Smith, Lt.-Col. James D.B., 157
Smith, Walter Bedell, 57, 173–4
Smuts, Field Marshal Jan Christian, 180
SNAFFLE, Exercise, 163
Somme, battle of, 15–16, 19, 43
Southam, Brig. William W., 123–4, 152
South-Eastern Command, 108, 114, 235–6
Southern Command, 109
Soviet Union, 47,
Spaatz, Gen. Carl A., 174
SPARTAN, Exercise, xiv–v, xxiii, 5, 73, 77, 113, 127–30, 144, 149, 150–9, 161–7, 212–13, 221, 224, 237, 301n4, 302n7, 308n45, 324n5, 325n12, 328n5
Spry, D.C., 44, 116, 206, 221
Stacey, C.P., 4, 21, 39, 41, 57, 61, 71, 74, 84–6, 88, 91, 98, 100, 102–3, 108, 111, 139, 144–5, 150–1, 153, 158, 160, 164, 173, 177–8, 180, 203, 206, 209–11, 224
Staff College (Camberley), 24–6, 88–9, 106, 110, 236, 254n5, 258n32
Stalingrad, 75
Stanley, Oliver, 42
Statute of Westminster, 42
Stein, Maj.-Gen. Charles R.S., 129, 152, 159, 164–5
Stewart, William A., 33
Stopford, Lt.-Gen. Montagu G.N., 129, 133, 137–40, 144, 146–7, 153, 210
St Eloi Craters, 16
Stewart, Brig. James C., 94

Stirling, Grote, 33
Stokes-Rees, Commander R.H., 92
Stuart, Kenneth, 3, 6, 14, 25, 32, 62, 68–70, 72–3, 76–7, 80, 160, 172–3, 175, 179, 181, 184–97, 201–3, 206–13, 215–21, 223–5, 227–8
SUPER-GYMNAST, Operation, 61
Swayne, Lt.-Gen. Jack, 74, 143
Swettenham, John, 12–13, 83, 141, 150, 173, 177, 180–1

Tank and Track Transport Experimental Establishment, 26
Tank Production Committee, 116
Templer, Lt.-Gen. Gerald W.R., 131
Tennessee manoeuvres, 35
'Ten Year Rule,' 28
TEWT, 24, 57, 88, 94, 122, 150
Thacker, Maj.-Gen. H.C., 27
TIGER, Exercise, 112, 125–7, 153, 162, 166, 236
TIGER, Operation, 49
TIGER No. 2, Operation, 49
TIMBERWOLF, Operation, 202–4, 206–7, 211, 220, 243
Tobruk, 61
TONIC, Operation, 69–72, 78–9, 182, 220, 242
TORCH, Operation, 61–4, 66–8, 71, 220, 241–2, 275n38, 39, 279n72, 280n4
Trench Warfare Training and Experimental Centre (TWTEC), 93
TRIDENT Conference, 189
Trondheim, 41
Turner, Maj.-Gen. Guy R., 90–1, 98–9, 101, 108, 114–15, 120, 129–30, 156, 180, 216

ULTRA, 47, 49

unemployment relief camps, 27, 33, 90
United States, 35, 43, 54, 61, 115, 184, 189
United Services Institute, 33

Vanier, Georges, 202
Verney, Maj.-Gen. Gerald L., 126
VICTOR, Exercise, 57, 83, 182, 226–7, 235
Victorian Canada, 11–13
Vimy Ridge, 17, 43, 55, 103
Vimy II, 80
Visiting Forces Act, 41
Vokes, Maj.-Gen. Chris, 25, 87, 100, 102, 105

Warren, Maj. Trumbull, 119, 159
WATERLOO, Exercise, 108–10, 127, 235, 295fn58
Watson, Maj.-Gen. David, 42

Wavell, Sir Archibald, 5, 18, 20, 25, 36, 240
WETBOB, Operation, 68, 242
Whitehead, Brig. G.V., 123–4
Whiteley, Maj.-Gen. John F.M., 179, 204
Winterton, Brig. Thomas J.W., 108
Woolich, Royal Military Academy, 14
Worthington, Frank F., 32, 153, 157, 164–5
Wyman, Brig. R.A., 129

'X' Mobile Composite Group RAF, 130

Young, Brig. Hugh A., 150, 152, 156, 159
Ypres, battle of, 15–16

'Z' Composite Group RAF, 130

www.ingramcontent.com/pod-product-compliance
Lightning Source LLC
Chambersburg PA
CBHW020349080526
44584CB00014B/947